# Fundamentals of Signals and Control Systems

*Series Editor*
*Maurice Charbit*

# Fundamentals of Signals and Control Systems

Smain Femmam

ISTE

WILEY

First published 2017 in Great Britain and the United States by ISTE Ltd and John Wiley & Sons, Inc.



ISTE Ltd
27-37 St George's Road
London SW19 4EU
UK

www.iste.co.uk

John Wiley & Sons, Inc.
111 River Street
Hoboken, NJ 07030
USA

www.wiley.com



Library of Congress Control Number: 2016957651

British Library Cataloguing-in-Publication Data
A CIP record for this book is available from the British Library
ISBN 978-1-78630-098-0

# Contents

# Contents

# Preface

This preface gives an overview of signals and systems, generalities and notions of process control and regulation.

## Introduction

The main objective of these two volumes is the analysis and the study of linear, time-invariant, causal signals and deterministic systems of finite dimensions. We will focus our efforts on defining a set of tools useful to automatic control and signal processing, after which we will discuss methods for the representation of linear dynamic systems for the analysis of their behavior. Finally, the identification and the synthesis of control laws will be addressed for the purpose of stabilization and regulation in systems control. Chapter 6 of the other volume [FEM 16] will be dedicated to the use of the Nao robot for a specific application; in this case, home care service.

## Signals and systems: generalities

Whether in the analog or digital field the study of the characteristic properties of signals and systems, the knowledge of mathematical tools, processing and analysis methods are constantly evolving and of late are increasingly significant. The reason is that the current state of technology, in particular of electronics and computer science, makes possible the implementation of very advanced processing systems, which are effective and increasingly less expensive in spite of their complexity. Aims and requirements generally depend on applications. Figure 1 presents the connections between the various disciplines, the scientific and technological resources for their

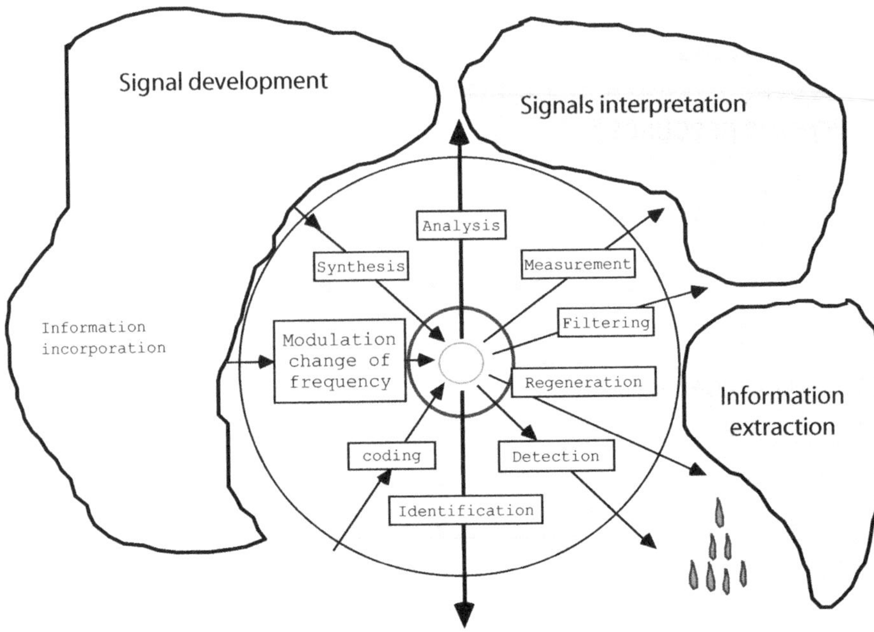


**Figure 2.** *Basic concepts of theoretical study and processing methods*

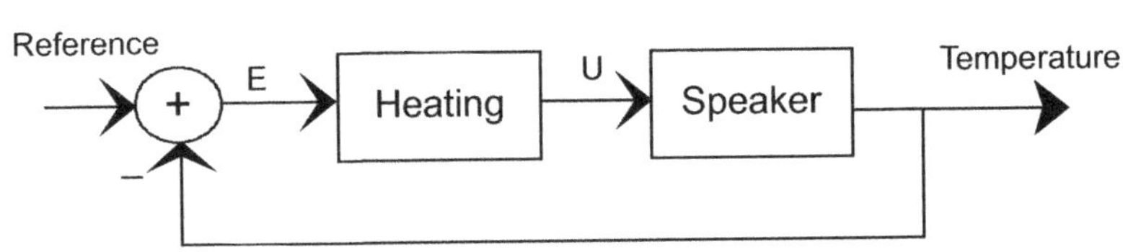


**Figure 3.** *Diagram of the model for vehicle operation*

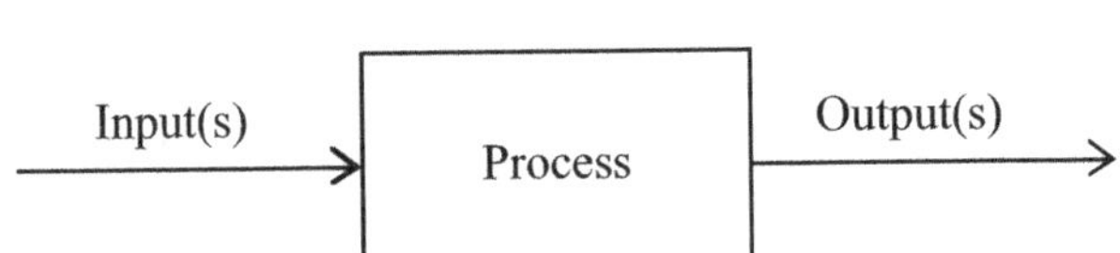


**Figure 4.** *Process with multiple input and output quantities*

# Preface

This preface gives an overview of signals and systems, generalities and notions of process control and regulation.

## Introduction

The main objective of these two volumes is the analysis and the study of linear, time-invariant, causal signals and deterministic systems of finite dimensions. We will focus our efforts on defining a set of tools useful to automatic control and signal processing, after which we will discuss methods for the representation of linear dynamic systems for the analysis of their behavior. Finally, the identification and the synthesis of control laws will be addressed for the purpose of stabilization and regulation in systems control. Chapter 6 of the other volume [FEM 16] will be dedicated to the use of the Nao robot for a specific application; in this case, home care service.

## Signals and systems: generalities

Whether in the analog or digital field the study of the characteristic properties of signals and systems, the knowledge of mathematical tools, processing and analysis methods are constantly evolving and of late are increasingly significant. The reason is that the current state of technology, in particular of electronics and computer science, makes possible the implementation of very advanced processing systems, which are effective and increasingly less expensive in spite of their complexity. Aims and requirements generally depend on applications. Figure 1 presents the connections between the various disciplines, the scientific and technological resources for their operation with the aim of processing signals or automatic control for the operation or the development of current applications[1].

In all areas of physics, for study, analysis and understanding of natural phenomena, a stage for modeling and for the study of the structure of the physical process is necessary. This has led to the development of techniques for modeling, representation and analysis of systems using a fairly general terminology. This terminology is difficult to introduce in a clear manner but the concepts which it relies upon will be defined in detail in the following chapters.

Signal processing concerns the various operations carried out on analog or digital physical quantities with the purpose of analyzing, interpreting and extracting information. These operations are illustrated in Figure 2.

The mastery and the implementation of signal processing techniques require the knowledge of a number of theoretical tools. The objective of this book is to establish the basic concepts of the theoretical study and clarify common processing methods.

A physical process is divided into several components or parts forming what is called a system. This is the case, for example, of an engine that consists of an amplifier, power supply, an electromagnetic part and a position and/or speed sensor. The input to the system is the voltage applied to the amplifier and the output is either the position or the speed of the rotation of the motor shaft.

Among the objectives of the control engineer is the modeling, behavior analysis and the regulation or control of a system, aiming for the dynamic optimization of its behavior. The operation of the system or control is designed to ensure that the variables or system outputs follow a desired trajectory (curve over time in general) or have dynamics defined by the specifications document. For temperature regulation of a speaker to a reference value, one of the following diagrams can be used. Details of vehicle operation is shown in Figure 3.

## Notions of process and operation control

The objective of automatic control is to design control and operation systems that are able to assign to a dynamic process (physical, chemical physical, biological, economical, etc.) a behavior defined in advance by the operator based on the requirements specifications. For example, we can consider speed regulation of a car that gives the process (the car) a previously determined speed, regardless of the disturbances that may occur (variation of the slope, etc.). Other examples include a

1 The author advises to avoid the book Traitement du signal Asservissement linéaires by I. Jelinsky in the series Vuibert Technologie in order to avoid giving readers any incorrect interpretation and confusion with the concepts presented in this book.

radar antenna alignment system for the monitoring of the trajectory of an airplane or a satellite, and an air conditioner designed to stabilize the temperature at a constant value fixed in advance.

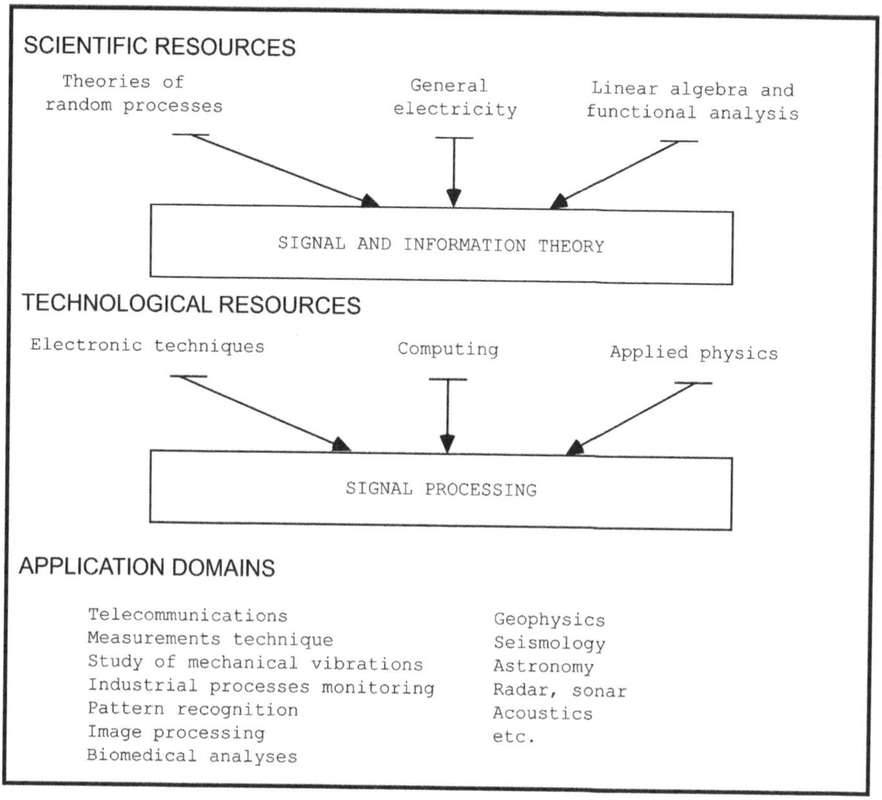


**Figure 1.** *Applied scientific and technological resources*

A process can be defined by establishing relationships between input and output quantities (this will be formally defined in different ways throughout this book. It is represented in Figure 4.

In the example of the car, the output is the speed and inputs may be the position of the accelerator pedal, the slope of the road and/or any other physical quantities that have an influence on the speed (the output of the system). Inputs consist of variables that can be manipulated (position of the pedal) and upon which no action is possible (the slope of the road). The latter are called disturbance inputs, they may be measurable or not accessible, random or deterministic. The variables that can be manipulated can be used as control inputs.

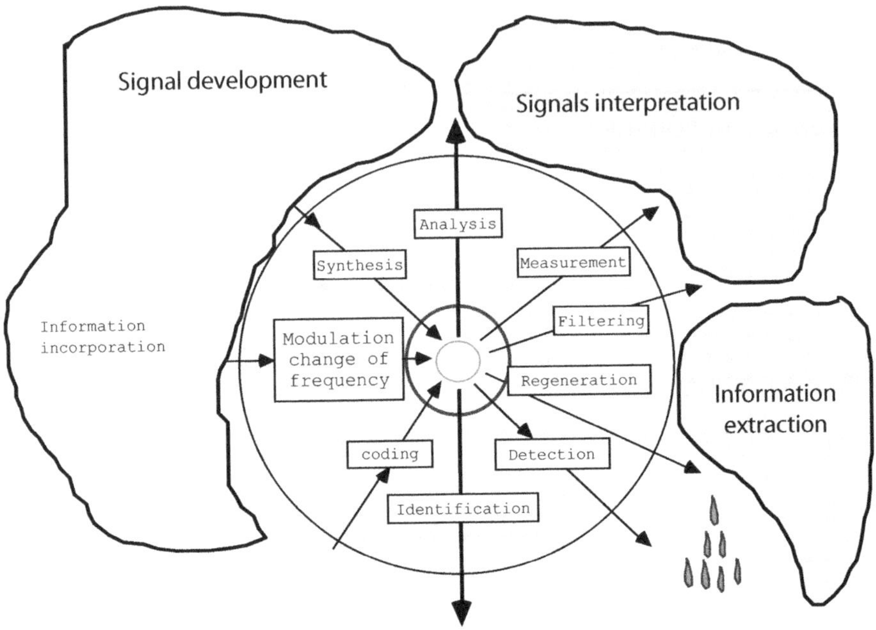


**Figure 2.** *Basic concepts of theoretical study and processing methods*

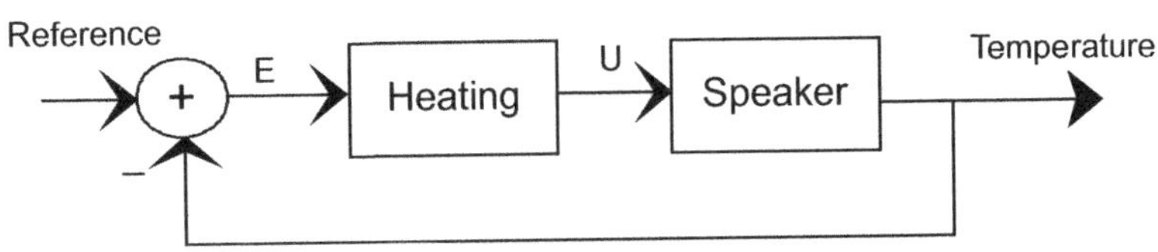


**Figure 3.** *Diagram of the model for vehicle operation*

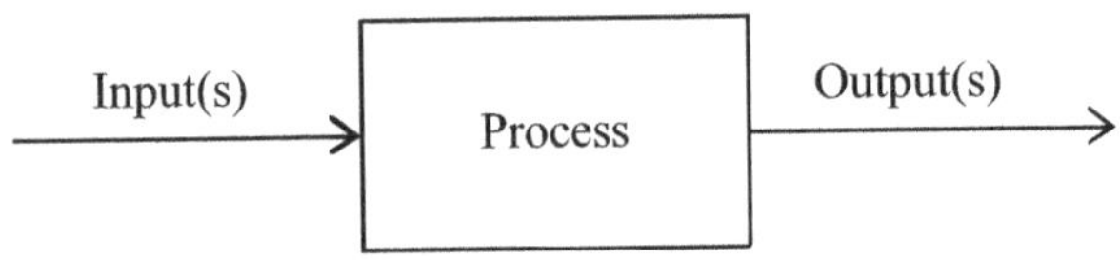


**Figure 4.** *Process with multiple input and output quantities*

In order to maintain the constant speed of a vehicle, a mathematical model of the process must be developed, in which the vehicle speed is linked to the position of the accelerator pedal, and then by inverting this model, the necessary input to obtain a specified speed can be derived. As a result, if the system output is not taken into consideration, an open-loop control is carried out (see Figure P.5).

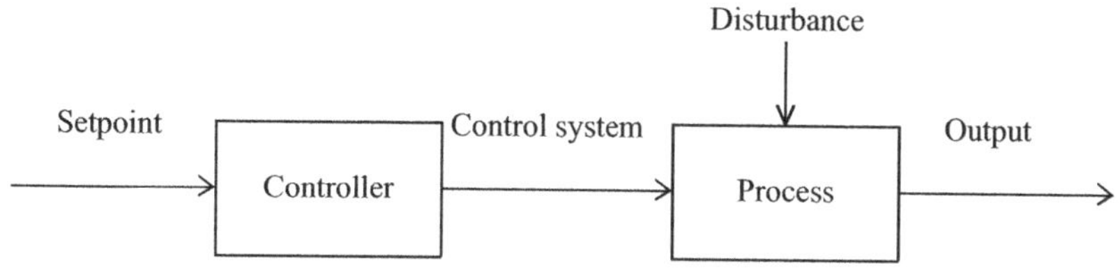


**Figure 5.** *Open-loop process*

This diagram shows that the control system does not account for disturbances; it cannot function properly. For example, if the vehicle is confronted with different slopes, the slope is considered a disturbance input for the process. It will be required that the model take the slope into account and thus a measurement system of the slope (which would result in a compensation of the measurable disturbances).

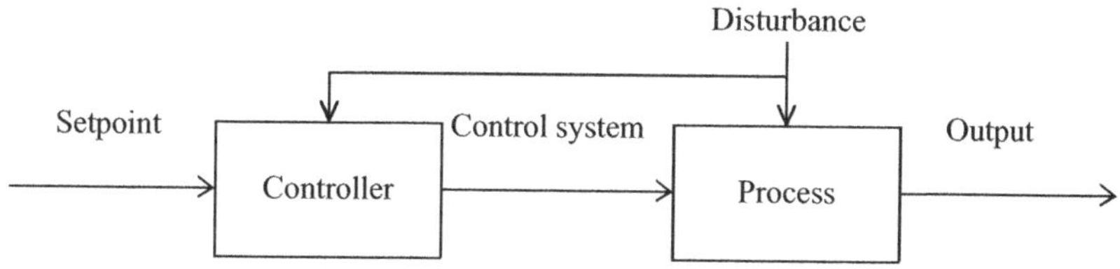


**Figure 6.** *Open-loop process with disturbances*

To improve the behavior, a control system can be defined that calculates, based on the desired speed obtained speed difference, the necessary action on the pedal to regulate (stabilize) the speed at a value specified by the operator. We will thus obtain an automatic control system of the speed of the vehicle. A sensor measuring the speed obtained is necessary. This system automatically performs what the driver does: it compares the target speed with the actual speed (displayed by the dashboard (sensor)) and acts upon the accelerator to reduce the speed difference to zero. The result will then be a *control system* or *loop system* or servo (servo system). The block diagram (functional) of the principle of a servo system is shown in Figure 7:

– $y^d$: setpoint is an electrical quantity that represents the desired value of the output of the system;

– $\varepsilon$: error signal between the setpoint and the actual output of the system;

– $u$: control signal generated by the control system;

– $y$: is a physical quantity that represents the system output.

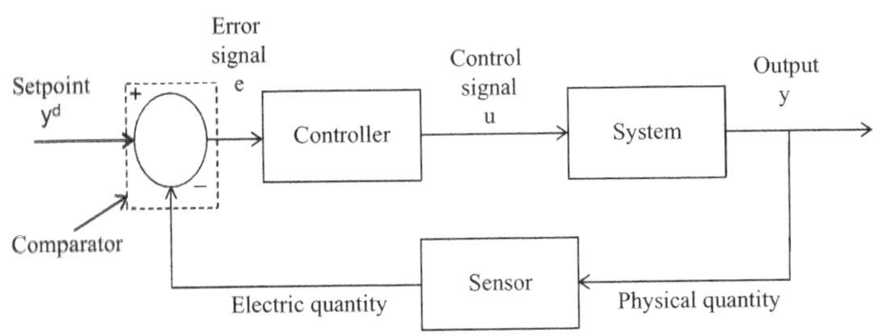


**Figure 7.** *Functional block diagram of a servo system*

The physical quantity $y$ is measured with a sensor that translates it into an electrical quantity. This electric quantity is compared to the setpoint using a comparator.

One of the great advantages of a looped system compared to an open-loop system lies in the fact that the loop system automatically rejects disturbances. In control systems, when the setpoint (reference) is constant, it is referred to *control* (for example oven temperature control and speed control of a motor); in the case where the reference is not constant, this is referred to as *tracking* (for example target tracking by an antenna).

An additional input (measure of the slope), with respect to the equipment existing in the previous example, would complete this diagram with an *anticipation* about the effect of the disturbance (due to the slope variation).

A system is said to be controlled when there is a loop between the output and the input or when the variable to be adjusted is the setpoint input of the system. For example, for the heating system of a house or of an enclosure, the input is the temperature setpoint and the output is the temperature in the enclosure. An open-loop heating system is a system that does not show any loopback, taking into account the effective temperature of the enclosure. Thus, it is sensitive to external shocks, a rise in the external temperature would cause an excess of heating.

*Examples*:

The examples are as follows: temperature system control of an oven, fluid flow servo control, speed system control of the trajectory of a vehicle. When the desired path is reduced to a point, this is referred to as regulation and not as system control because the aim here is to stabilize the output of the system at a point. A control system can be qualified by its degree of stability, accuracy, response speed, sensitivity to disturbances acting on the system, robustness with respect to disturbances on measures and errors or variations of the characteristic parameters of the system. The accuracy of a control system can be characterized by the maximum amplitude of the position error.

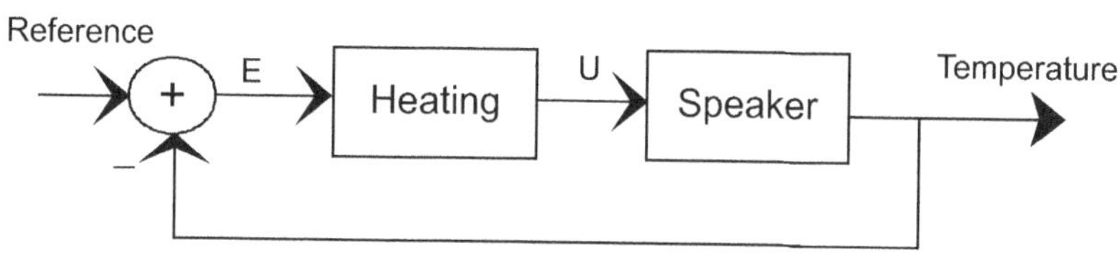


**Figure 8.** *Speed regulation of a motor*

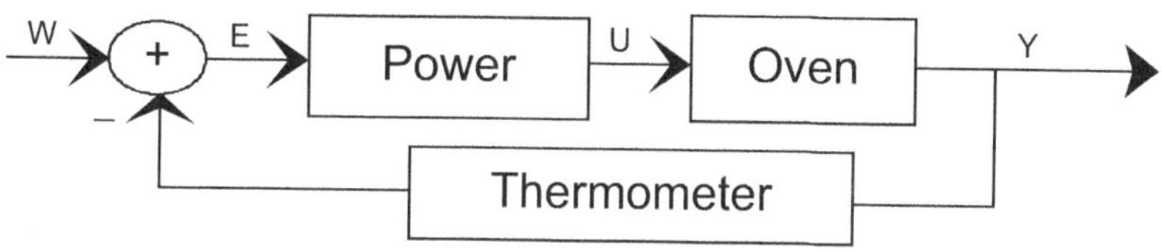


**Figure 9.** *Temperature regulation of an oven*

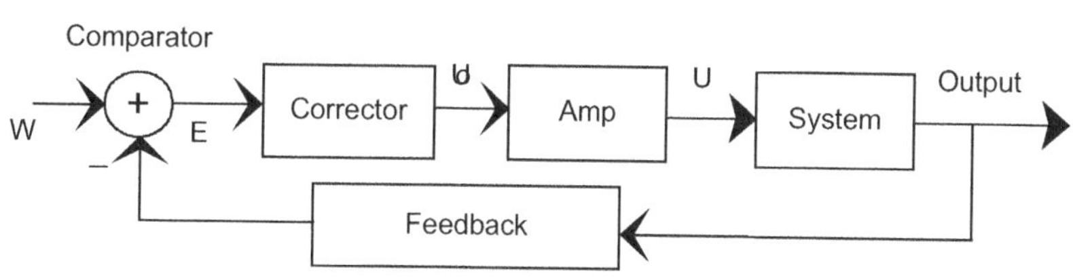


**Figure 10.** *Control system with correction*

In the definition of a control system, we will write transfer functions as follows: $H(p)$ transfer of the system to be controlled, $p$ is the Laplace operator; $R(p)$ transfer of the sensor or measurement unit, $C(p)$ transfer of the corrector or controller. The setpoint is $\omega(t)$ and the output to be controlled $y(t)$. The direct chain consists of $C(p)$ and $H(p)$. Block $R(p)$ constitutes the feedback chain. $e(t)$ is the difference between output and setpoint also called control error or trajectory tracking error. In order to simplify study, we consider a unity feedback scheme in which $R(p) = 1$.

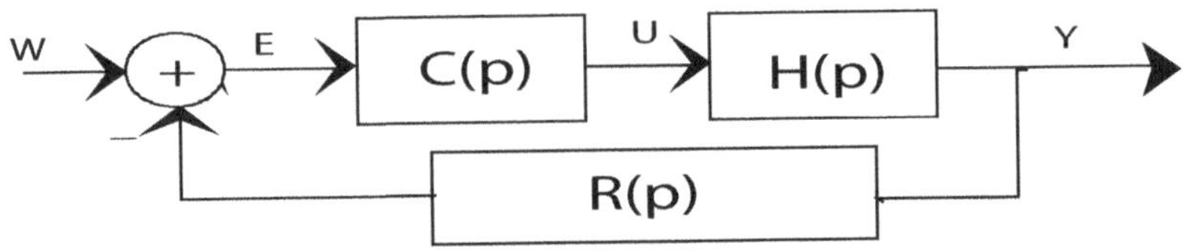


**Figure 11.** *Controlled transfer function with feedback*

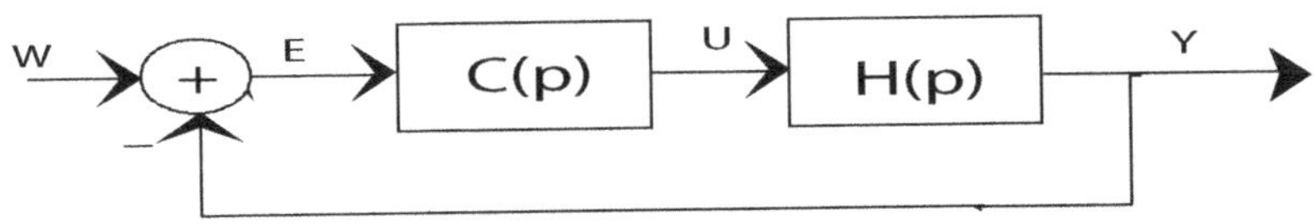


**Figure 12.** *Controlled transfer function without feedback*

In general, transfers $H(p)$ and $R(p)$ are known, estimated or can be obtained and the goal is the determination of a corrector $C(p)$ that can satisfy the required performances for the closed-loop system (transfer from $w$ to $y$).

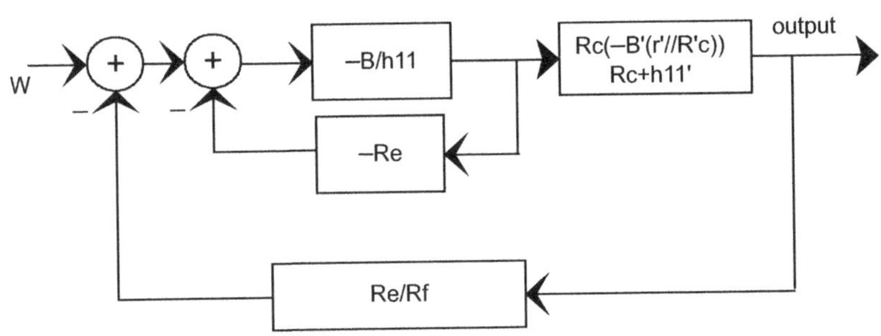


**Figure 13.** *Control system with closed-loop system controller*

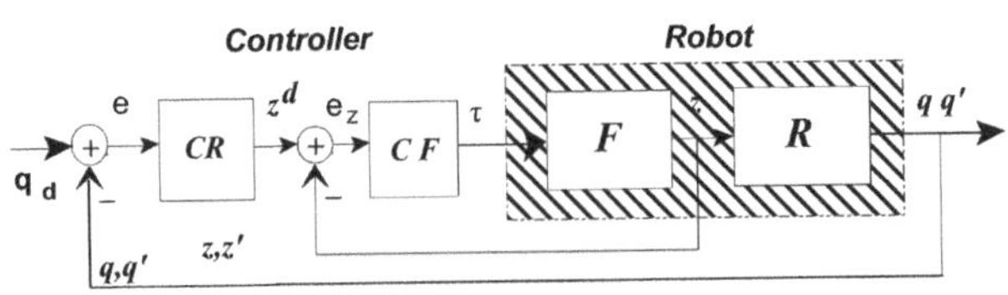


**Figure 14.** *System control with correction of a robotized system*

Several types of systems can be distinguished:

– continuous systems for which all measured quantities are continuous;

– discrete systems for which all measured quantities are only measured at very specific times (discontinuous or discrete); these are referred to as sampled data or digital systems;

– linear systems (they can be described by linear differential equations);

– nonlinear systems (described by nonlinear differential equation). It is possible, often in the first approximation, to linearize nonlinear systems based on an operation point (equilibrium), considering small variations around this point;

– time-invariant systems (described by differential equations with constant coefficients) and time-variable systems (described by differential equations with time-variable coefficients).

In this book, we consider time-invariant linear, continuous and sampled-data systems.

NOTATION 1.– *Consider a continuous $r$-input system denoted $u$ and $m$ outputs $y$,*

$$u \in U \subset R, y \in Y \subset R \qquad [P.1]$$

$$u \in U \to . \to y \in Y \qquad [P.2]$$

*A minimal state representation of this system will be written as:*

$$\dot{X} = A.X + Bu \qquad [P.3]$$

$$y = C.X + D.u \qquad [P.4]$$

*Its transfer function*

$$G(p) = C(pI - A)^{-1}B + D \qquad [P.5]$$

*is denoted as:*

$$G(p) = \begin{bmatrix} A & B \\ C & D \end{bmatrix} = [A, B, C, D] \qquad [P.6]$$

*A is stable: the eigenvalues of A have a real part* $< 0$.

$G(p)$ *is stable: the poles of* $G(p)$ *are in* $Re(p) < 0$.

system for home care. Without claiming that it is capable of recreating or replacing real human contact, the proposed system, based on the integration of several technologies, will create a permanent, friendly and reassuring, presence at home. Due to the use of modern means of audio/video communication (speech recognition, speech synthesis, etc.), by the mobile humanoid Nao robot from Aldebaran Robotics and that of fixed cameras allowing the environment to be perceived, the objective is to provide a set of services that makes it possible to assist senior citizens in their daily tasks (engaging with relatives, remote medical consultation, looking for objects, medication reminders, etc.), entertain them (games, physical exercises, friendly presence, etc.) and detect dangerous or abnormal situations (falls, prolonged inactivity, unusual behavior, etc.) to, in the end, alert the doctor and relatives.

system for home care. Without claiming that it is capable of recreating or replacing real human contact, the proposed system, based on the integration of several technologies, will create a permanent, friendly and reassuring, presence at home. Due to the use of modern means of audio/video communication (speech recognition, speech synthesis, etc.), by the mobile humanoid Nao robot from Aldebaran Robotics and that of fixed cameras allowing the environment to be perceived, the objective is to provide a set of services that makes it possible to assist senior citizens in their daily tasks (engaging with relatives, remote medical consultation, looking for objects, medication reminders, etc.), entertain them (games, physical exercises, friendly presence, etc.) and detect dangerous or abnormal situations (falls, prolonged inactivity, unusual behavior, etc.) to, in the end, alert the doctor and relatives.

$\Lambda_{Hg} = H(p).g(p)$ *is the Laurent operator or multiplication in the frequency domain.*

$\oplus$ *Direct sum of two spaces.*

$H_2^{\perp}$ *Orthogonal space at* $H_2$*;*

with the set $H_2$={$H(p)$ analytic matrix function in $Re(p) > 0$}

$\Pi_1$ is the orthogonal projection on $H_2$ and $\Pi_2$ is the orthogonal projection on $H_2^{\perp}$.

NOTATION 2.– $\mathcal{B}(x, r) \equiv$ *Globe of radius* $r$*, centered in* $x$ *of the space; example* $\mathcal{B}(0, 1)L_2 = \mathcal{B}L_2$ *refers to the Globe of unit radius of the space* $L_2$*.* $L_2$ *is the set of square-integrable functions.*

In the literature, a real rational transfer function refers to a rational transfer function with real coefficients.

## Nao robot: application for home care

In view of these considerations and following previous and present works, the purpose of this section is the perception for the joint understanding of space and surroundings of an autonomous, cognitive and sociable personal robot. This Nao robot will typically act according to a planned action scheme, ensuring their viability, the consistency of current models and the presence of entities opposed to its initial plan, while interacting with the physical world through perception. The selected approach, which is clearly aimed at integrating perceptual functions on robotic platforms, relies on probabilistic modeling to consider multiple and uncertain percepts. At the sensory level, these percepts will be mainly originating from the vision embedded in the robot. These perceptual functions are to be derived from autonomy, cognition and sociability capabilities outlined for our Nao robot.

An increasingly more worrying situation due to the aging of the population is the increase in the number of elderly people living alone, far away from their loved ones, sometimes physically or mentally vulnerable. Studies recommend that, at the social level as well as at the economic level, people stay at home, eventually resorting to help (cleaning, care, gardening, etc.), to preserve as much as possible the references and the social network of the person. However, with the cost of home care being relatively burdensome, assistance times are limited and are not always enough to fight solitude. In addition, when the elderly person presents risks of accidents (fall, stroke, etc.), or when undergoing medical treatment, monitoring solely based on domestic help has become risky, or even insufficient (admission to hospital then being preferable).

To fight more effectively against isolation and the dependence of elderly people, the presented system aims to implement a prototype for an automated and robotic

$G(p)$ *is instable: the poles of* $G(p)$ *are in* $Re(p) > 0$.

$\tilde{G}(p) = G(-p)$

$G(p)$ *is an eigen transfer function if* $G(\infty)$ *is finite.*

$G(p)$ *is a strictly eigen transfer function if* $G(\infty) = 0$.

$A^T$ *is the transpose of matrix*

$$A = [a_{ij}] = \begin{bmatrix} a_{11} & a_{12} \\ a_{21} & a_{22} \end{bmatrix} \quad [P.7]$$

$$A^T = \begin{bmatrix} a_{11} & a_{21} \\ a_{12} & a_{22} \end{bmatrix} \quad [P.8]$$

$A^{-T}$ *is the inverse transpose matrix of* $A$.

$A^*$ *is the conjugate transpose matrix of* $A = [a_{ij}]$ *(or Hermitian transpose matrix of* $A$*).* $A^* = [a_{ij}^*]$

$\lambda_i(A)$ *is the ith eigenvalue of* $A$.

$\sigma_i(A)$ *is the ith singular value of* $A$.

$\underline{\sigma}(A)$ *and* $\bar{\sigma}(A)$ *are the minimal and maximal singular values of* $A$.

$Diag(a_i)$ *is the diagonal matrix whose diagonal elements are the* $a_i$.

$C_-$ *is the set of complex numbers with a negative real part.*

$C_+$ *is the set of complex numbers with a positive real part.*

$C^n$ *is the set of complex vectors with elements in* $C$.

$C^{nxm}$ *is the set of complex matrices of dimensions* $(nxm)$ *with elements in* $C$.

$< x, y >$ *is the scalar product of* $x$ *and* $y$.

$h \circledast u$ *is the convolution product of* $h(t)$ *and* $u(t)$.

$\mathcal{F}$ *is the Fourier transform operator.*

$\theta_h$ *convolution operator by* $h(t)$, $H(p)$ *should denote the Laplace transform of* $h(t)$.

Several types of systems can be distinguished:

– continuous systems for which all measured quantities are continuous;

– discrete systems for which all measured quantities are only measured at very specific times (discontinuous or discrete); these are referred to as sampled data or digital systems;

– linear systems (they can be described by linear differential equations);

– nonlinear systems (described by nonlinear differential equation). It is possible, often in the first approximation, to linearize nonlinear systems based on an operation point (equilibrium), considering small variations around this point;

– time-invariant systems (described by differential equations with constant coefficients) and time-variable systems (described by differential equations with time-variable coefficients).

In this book, we consider time-invariant linear, continuous and sampled-data systems.

NOTATION 1.– *Consider a continuous $r$-input system denoted $u$ and $m$ outputs $y$,*

$$u \in U \subset R, y \in Y \subset R \tag{P.1}$$

$$u \in U \rightarrow . \rightarrow y \in Y \tag{P.2}$$

*A minimal state representation of this system will be written as:*

$$\dot{X} = A.X + Bu \tag{P.3}$$

$$y = C.X + D.u \tag{P.4}$$

*Its transfer function*

$$G(p) = C(pI - A)^{-1}B + D \tag{P.5}$$

*is denoted as:*

$$G(p) = \begin{bmatrix} A & B \\ C & D \end{bmatrix} = [A, B, C, D] \tag{P.6}$$

*A is stable: the eigenvalues of A have a real part* $< 0$.

*$G(p)$ is stable: the poles of $G(p)$ are in $Re(p) < 0$.*

# 1

# Introduction, Generalities, Definitions of Systems

This chapter explores the representation, modeling and identification of signals, transmission systems and filtering concepts.

A number of mathematical concepts introduced in this chapter (distributions) are an extension of the conventional notion of functions. The comprehensive study of distributions is not absolutely essential if some unproven results are accepted. The objective of this chapter is to introduce basic notions, temporal relations and transformations that enable this correspondence to be established.

## 1.1. Introduction

Modeling is a very important step in linear systems control. To properly control a system, the knowledge of a good model is necessary. For example, to drive a car, the more accurate the knowledge of its dynamic behavior or its model is (by learning or training), the better it will be controlled at high speed and therefore the better it will be driven and will show the best performance. The dynamic model is acquired by learning or by identifying the system after knowing the structure of this model.

During the development of an application for automation purposes, we follow the following steps:

1) modeling;

2) identification;

3) behavior analysis;

4) controller synthesis;

A communication system is usually a means of communication between an information source and a recipient (electronic, optical or mechanical system)

Communication sequence:

Emission: Message $\longrightarrow$ Coder $\longrightarrow$ Electronic system $\longrightarrow$ Signal $\longrightarrow$ Transmitter $\longrightarrow$ Channel

Reception:

Channel $\longrightarrow$ Receiver $\longrightarrow$ Electronic system $\longrightarrow$ Decoder $\longrightarrow$ Signal $\longrightarrow$ Processing

Processing:

Signal processing $\longrightarrow$ Information processing $\longrightarrow$ Decision

– The variable is usually time $t$ (it can be space $\xi$ or any other physical parameters).

– The signal is denoted as $s(t)$, $y(t)$ (or $z(\xi)$).

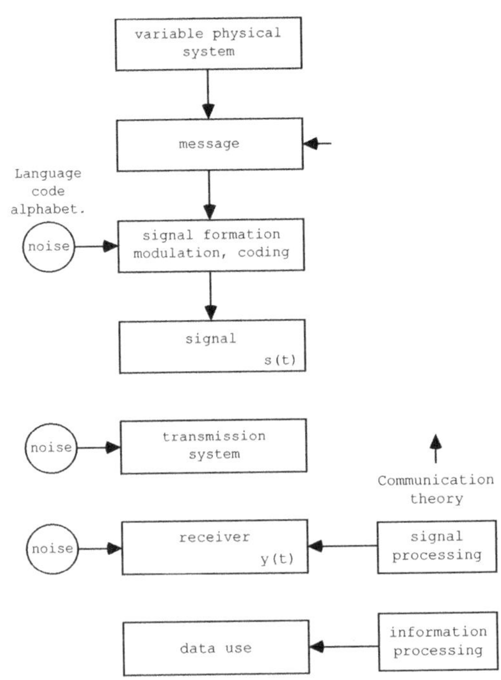


**Figure 1.4.** *Communication sequence*

1

# Introduction, Generalities, Definitions of Systems

This chapter explores the representation, modeling and identification of signals, transmission systems and filtering concepts.

A number of mathematical concepts introduced in this chapter (distributions) are an extension of the conventional notion of functions. The comprehensive study of distributions is not absolutely essential if some unproven results are accepted. The objective of this chapter is to introduce basic notions, temporal relations and transformations that enable this correspondence to be established.

## 1.1. Introduction

Modeling is a very important step in linear systems control. To properly control a system, the knowledge of a good model is necessary. For example, to drive a car, the more accurate the knowledge of its dynamic behavior or its model is (by learning or training), the better it will be controlled at high speed and therefore the better it will be driven and will show the best performance. The dynamic model is acquired by learning or by identifying the system after knowing the structure of this model.

During the development of an application for automation purposes, we follow the following steps:

1) modeling;

2) identification;

3) behavior analysis;

4) controller synthesis;

5) control implementation;

6) analysis and study of the closed-loop system;

7) verification of the performance and eventually repeat steps 2, 3 or 4.

The modeling stage becomes crucial when the requirements are strict regarding performances and when the control implemented proves to be complex.

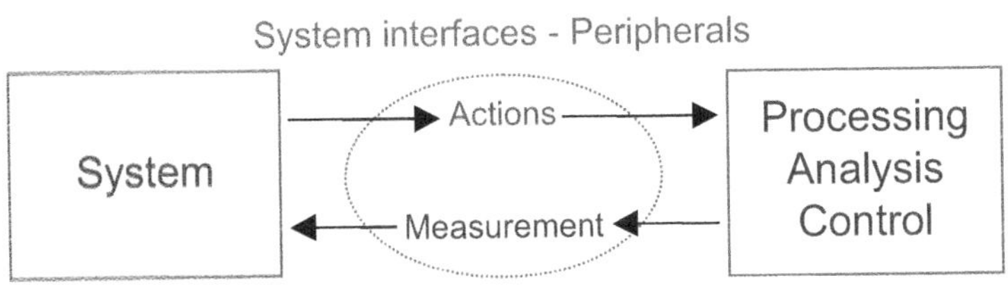


**Figure 1.1.** *Peripherals of a system*

## 1.2. Signals and communication systems

In electronics, as for most other areas, a signal designates any electromagnetic or physics phenomenon used as medium for information to be transmitted. This signal is used to characterize a physical quantity captured by reflecting its evolution over time and or in space, amplitude, energy or power. For example, this is the case in a thermometer with the indicated level or the electrical signal supplied by a thermocouple.

Electrical signals are typically provided by a system called sensor, consisting of an element sensitive to a physical effect that it converts into an electrically measurable quantity and an adapter amplifier that thereof provides the equivalence in the form of a signal. This is the case for pressure, temperature, radiation, speed, position and acceleration sensors. This signal is then manipulated by an analog or a digital system.

Sensor: Physical effect$\longrightarrow$Converter$\longrightarrow$Amplifier$\longrightarrow$Adaptor$\longrightarrow$Signal

In general, in the case of an observation of a physical phenomenon, a signal can be defined as being the variation of a physical quantity (measured by a sensor) in time ($t$) or in space ($x$) - (observation). In this case, the processing may involve the separation of signal and noise, extraction of information, extraction of frequency and temporal characteristics, etc.

Observation: Sensor $\longrightarrow$ Electronic system $\longrightarrow$ Signal $\longrightarrow$ Information

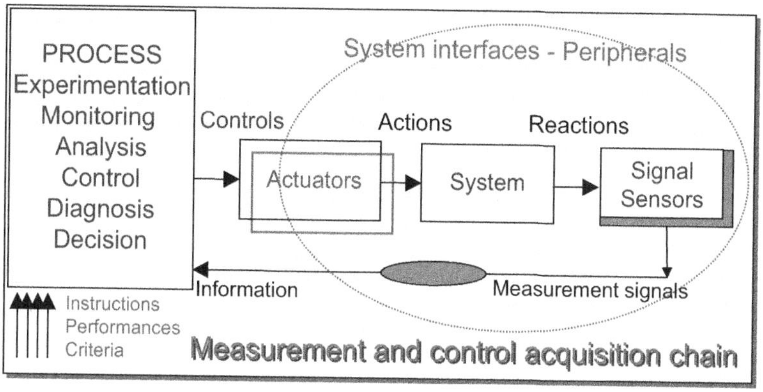


**Figure 1.2.** *Acquisition, sensors and triggers*

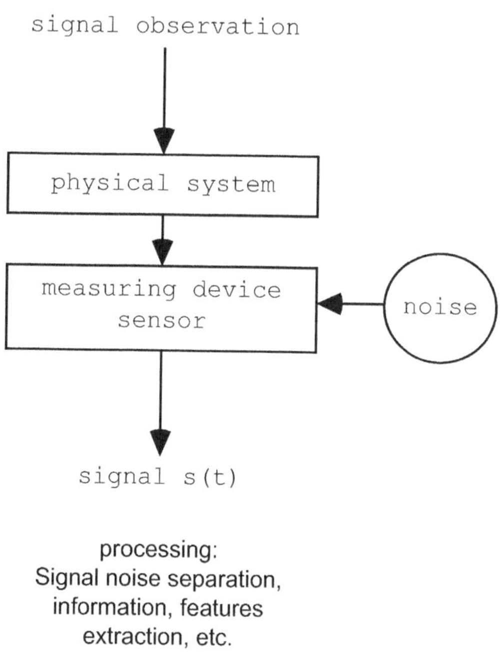


**Figure 1.3.** *Observation sequence*

A communication system is usually a means of communication between an information source and a recipient (electronic, optical or mechanical system)

Communication sequence:

Emission: Message $\longrightarrow$ Coder $\longrightarrow$ Electronic system $\longrightarrow$ Signal $\longrightarrow$ Transmitter $\longrightarrow$ Channel

Reception:

Channel $\longrightarrow$ Receiver $\longrightarrow$ Electronic system $\longrightarrow$ Decoder $\longrightarrow$ Signal $\longrightarrow$ Processing

Processing:

Signal processing $\longrightarrow$ Information processing $\longrightarrow$ Decision

– The variable is usually time $t$ (it can be space $\xi$ or any other physical parameters).

– The signal is denoted as $s(t)$, $y(t)$ (or $z(\xi)$).

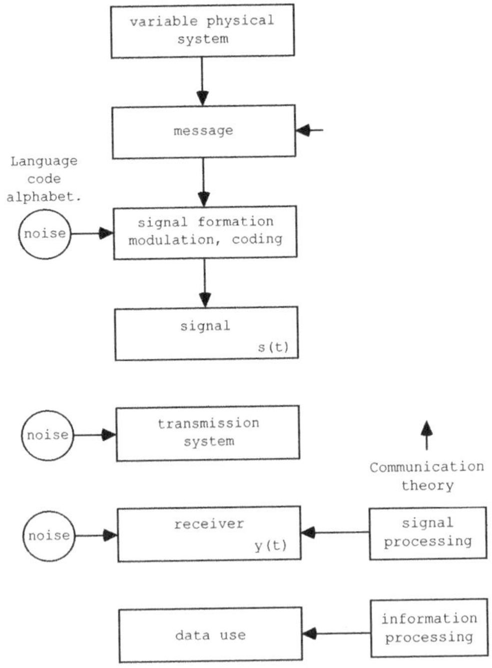


**Figure 1.4.** *Communication sequence*

## 1.3. Signals and systems representation

### 1.3.1. *Signal*

A signal is a physical representation of a phenomenon that evolves in time or in space. It can be represented in time and frequency domains. The frequency representation of a signal is interesting because it provides more information about the signal. The tool used to shift from the time domain to the frequency domain is the Fourier transform (FT).

Analog macroscopic measurements provide curves (in time) of the relevant signal. This signal is represented by a mathematical function $x(t)$, most often, with real values of one real variable (time $t$).

These functions, since they reflect physical quantities with finite energy or finite average power, are:

– real;

– bounded functions and with bounded support (integration without difficulties);

– continuous and differentiable in any point (derivation with no difficulties);

– they can be periodical $x(t) = x(t + T)$ if $T$ is the period;

– in addition, for causal physical signals, we get: $x(t) = 0$ for $t < 0$.

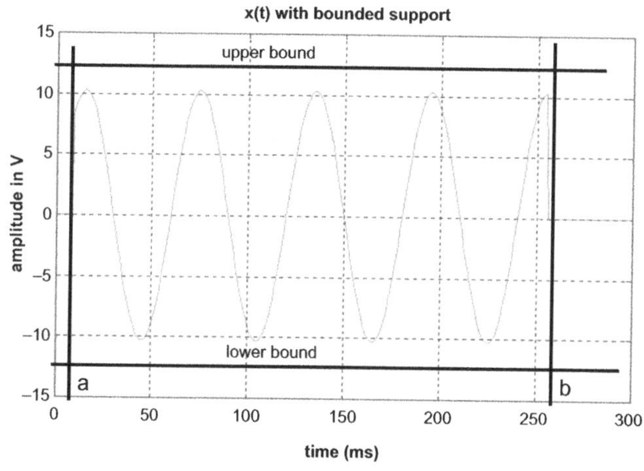


**Figure 1.5.** *Time signal with bounded support*

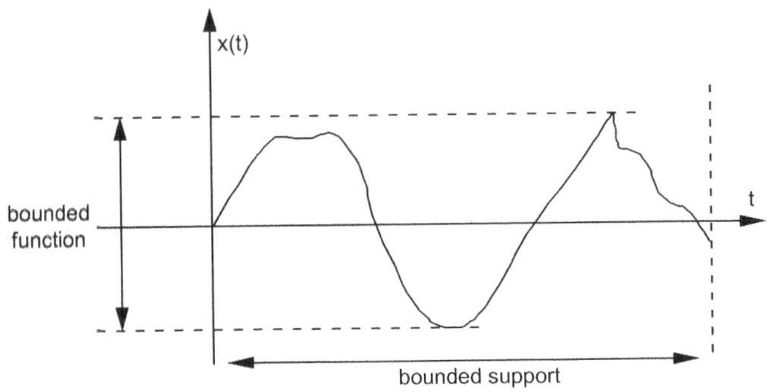


**Figure 1.6.** *Signal defined over a bounded medium*

Subsequently, in order to simplify calculations and mathematical manipulations, we will consider that signals can be represented (or modeled) by functions having the following properties:

– bounded functions;

– functions defined in $t \in R = ]-\infty, +\infty[$;

– functions with discontinuities or piecewise continuous (rectangle or gate, sawtooth signals, etc.);

– real- or complex-valued functions.

Often, the set of all functions having the above properties is restricted either to that of absolutely summable functions, $x(t)$ such that $\int_{-\infty}^{+\infty} |x(t)|\, dt$ does exist, or that of square-integrable functions such that quantity $\int_{-\infty}^{+\infty} |x(t)|\, dt$ exists and is finite. The advantage of square-integrable functions is that they make it possible to represent finite-energy signals (case of most frequently encountered real signals). It can be shown that this set constitutes a vector space.

### 1.3.2. *Functional space* $L_2$

Consider the vector space of square-integrable functions:

$$L_2 = \{\,\text{function}\, f(t) : \mathbb{R} \longrightarrow \mathbb{C}\,\text{such that} : \int_{-\infty}^{+\infty} |f(t)|^2\, dt < \infty\} \qquad [1.1]$$

If $f^*(t)$ is the conjugate of $f(t)$, we get

$$< f(t), g(t) >= \int_{-\infty}^{+\infty} f(t)g^*(t)dt \qquad [1.2]$$

the scalar product of the two functions. If $f(t)$ and $g(t)$ are real, it is a scalar product, otherwise a Hermitian product.

Schwartz inequality:

$$|< f, g >|^2 \leq < f, f >< g, g > \qquad [1.3]$$

The equality is obtained if and only if $f = \lambda.g$ with $\lambda$ scalar.

PROOF.– $f(t)$ and $g(t)$ are two functions of the space of finite-energy functions $L_2$, and $\lambda$ is a constant parameter arbitrarily chosen. Let $\lambda \in \mathbb{C}$ such that $q =< f + \lambda g, f + \lambda g >^2 \geq 0$.

Let: $\alpha = \int_{-\infty}^{+\infty} |f(t)|^2 \, dt$ , $\gamma = \int_{-\infty}^{+\infty} |g(t)|^2 \, dt$ and $\beta = \int_{-\infty}^{+\infty} f(t)g^*(t)dt$.

$\alpha =< f, f > \ \gamma =< g, g >$

$\beta =< f, g > \ \beta^* =< g, f >$

$$\forall \lambda \in \mathbb{C}, \ q = \alpha + \lambda^* \beta + \lambda \beta^* + \lambda \lambda^* \gamma \geq 0. \qquad [1.4]$$

1) if $\gamma = 0 \Longrightarrow \forall \lambda \in \mathbb{C}, \ q = \alpha + 2Re(\lambda^* \beta) \geq 0 \Longrightarrow \beta = 0$

the inequality is then verified: $q =< f, f >= \alpha$

2) if $\gamma \neq 0$, multiply by $g$ the two members of the equation [1.4]: $\gamma\alpha + \lambda^*\beta\gamma + \lambda\beta^*\gamma + \lambda\lambda^*\gamma\gamma > 0$

$$(\lambda\gamma + \beta)\lambda^*\gamma + \beta^*\lambda\gamma + \alpha\gamma + \beta\beta^* - \beta^*\beta \geq 0 \qquad [1.5]$$

$$(\lambda\gamma + \beta)(\lambda^*\gamma + \beta^*) + \alpha\gamma - \beta\beta^* \geq 0 \qquad [1.6]$$

$$|\lambda\gamma + \beta|^2 + \alpha\gamma - |\beta|^2 \geq 0 \qquad [1.7]$$

or even

$$\lambda\gamma + \beta 2 + \alpha\gamma \geq |\beta|^2 \qquad [1.8]$$

taking $\lambda = -\frac{\beta}{\gamma}$, we get the Schwartz inequality: $\alpha\gamma \geq |\beta|^2$

The equality corresponds to $q = 0$ where

$$\int_{-\infty}^{+\infty} |f(t) + \lambda g(t)|^2 \, dt = 0 \Longrightarrow f(t) + \lambda g(t) = 0 \text{ where } f(t) = -\lambda g(t) \quad [1.9]$$

EXAMPLE 1.1.– *The output signal of a harmonic oscillator can be represented by the sinusoidal function represented by the curve below. This signal (assumed as deterministic) can be modeled by the following equation:* $y(t) = \sin(2\pi f_o t + \varphi)$.

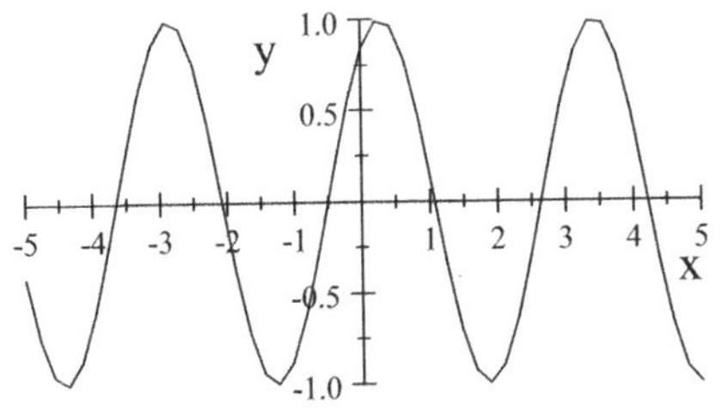


**Figure 1.7.** *Harmonic oscillator output signal*

This function is continuous, differentiable, bounded, and periodic of period $T = 1/f_o$ but with unbounded support and square non-integrable (infinite energy and finite mean power). In this case, for the study, we will rather consider *the average power* in a period and for the spectral representation either the FT in the sense of distributions or Fourier series will be used.

Therefore, for the study of signals, abstract mathematical modeling will be used in order to take advantage of the power of the theoretical tools available. Often, this abstract representation expands the properties of the signal being considered and its definition (finite average power signals, distributions). The interpretation of the results obtained with such mathematical models must be made with care, taking into account physical considerations of the problems in order to face realistic situations after a theoretical study.

### 1.3.3. *Dirac distribution*

It is possible to summarily define $\delta(t)$ by its properties because they are most often sufficient for the processing technique under consideration.

DEFINITION 1.1.– *Consider the function defined by* $f_l(t) = 0$ *if* $t < -l/2$ *or* $t > l/2$ *and* $f_l(t) = 1/l$ *if* $t \geq -l/2$ *and* $t \leq l/2$. *The appearance of this function is*

*represented here for different values of l. The limit of this function as l tends to zero gives us the Dirac distribution. When the width of the curve l is made to tend to zero, an infinitely high and narrow rectangle is obtained whose area is always equal to* 1. *When the limit is reached, we get a mathematical object that is not a function of R, because it is undefined for t* $= 0$. This is the Dirac delta function or impulse symbol *denoted as* $\delta(t)$.

*Conventionally, it is represented by an arrow of height* 1 *at* $t = 0$.

$\delta(t-t_1)$ represents the impulse translated of $t_1$ on the time axis. $\delta(t).x(t)$ is also a distribution of the same kind that represents a mass point $x(0)$ concentrated in $t = 0$. It shows the following features.

For its application in physics, the interesting particularity of Dirac's delta function is the finite area (equal to 1), representing, for example, a point mass concentrated in $t = 0$. This distribution is very useful for the mathematical modeling of physical phenomena: point sources in optics, point mass in mechanics, percussion in acoustics, point charge in electricity, acceleration during shock. Before stating its properties, we define the convolution product $*$ by the composition product.

The delta function is graphically represented by convention as an arrow with a unit height. It should be noted that the height is here connected with the mass of the distribution (surface under the curve) and not with amplitude, as is the case for functions.

DEFINITION 1.2.– *The convolution product of two functions* $f(t)$ *and* $g(t)$ *denoted as* $f(t) * g(t)$ *is defined by the integral (when it makes sense):*

$$h(t) = x(t) * y(t) = \int_{-\infty}^{+\infty} x(t-\tau)y(\tau)d\tau \quad [1.10]$$

PROPOSITION 1.1.– The delta function shows the following particularities (Table 1.1).

| The area under the curve is equal to 1: $\int_{-\infty}^{+\infty} f_l(t)dt = 1$ for any width $l$. | |
|---|---|
| $x(t) * \delta(t) = x(t)$ | $\delta(t)$neutral element of the convolution |
| $x(t) * \delta(t-t_o) = x(t-t_o)$ | (offset) |
| $x(t-t_1) * \delta(t-t_2) = x(t-t_1-t_2)$ | |
| $\delta(t-t_2) * \delta(t-t_1) = \delta(t-t_2-t_1)$ | |
| $\delta(at) = \lvert a\rvert^{-1}\,\delta(t)$ | |

**Table 1.1.** *The delta function*

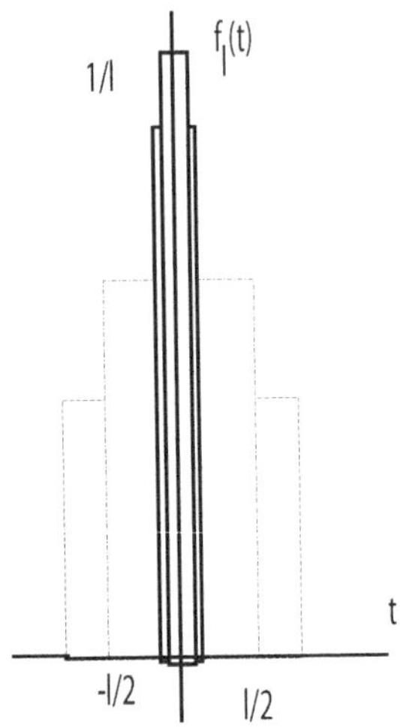


**Figure 1.8.** *Unit impulse*

## 1.4. Convolution and composition products – notions of filtering

In Table 1.1, let us recall that $*$ denotes the convolution product and $\delta$(t) is the delta function (impulse symbol), which is equal to zero everywhere except at zero where it is infinite.

### 1.4.1. *Convolution or composition product*

PROPOSITION 1.2.– Convolution or product composition has the following proprieties:

– commutativity: $x(t) * y(t) = y(t) * x(t)$ (change in variable $u = t - \tau$);

– distributivity: $x(t) * (y(t) + z(t)) = x(t) * y(t) + x(t) * z(t)$;

– associativity: $(x(t) * y(t)) * z(t)) = x(t) * (y(t) * z(t))$.

*Dirac function properties*:

The graphic representation of a pulse is by convention as shown in Figure 1.9 for $\delta(t - t_o)$.

a) $\delta(t) = 0 \; \forall t \neq 0$ and $\delta(t) = \infty$ for $t = 0$;

b) $\int_{-\infty}^{+\infty} \delta(t)dt = 1$ $\int_{-\infty}^{+\infty} \delta(t)x(t)dt = x(0)$; $\int \delta(t - t_o)x(t)dt = x(t_o)$

$\delta$(t-t$_o$) represents the impulse offset by $to$ on the axis of time $t$. The quantity $\delta$($t$ $t_o$)x(t) is also a distribution of the same kind representing a point mass $x(to)$ concentrated in $t = t_o$.

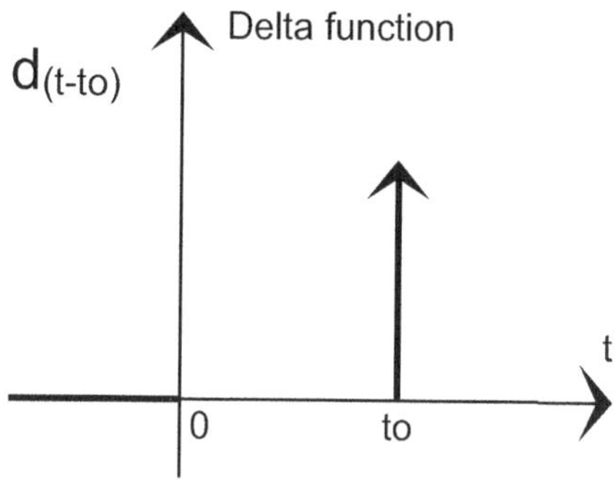


**Figure 1.9.** *Dirac's delta function*

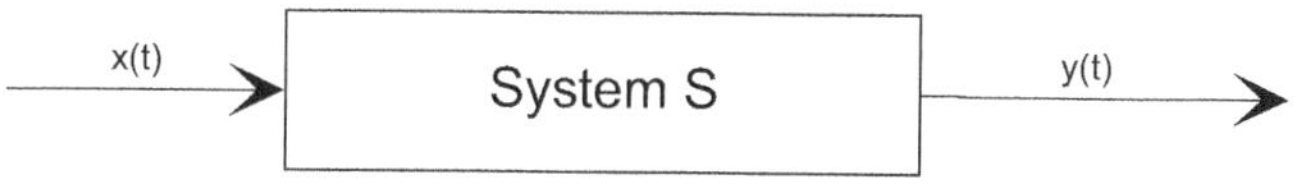


**Figure 1.10.** *Input/output of a system*

| | |
|---|---|
| (c) $x(t) * \delta(t) = x(t)$ | Neutral element of the convolution |
| (d) $x(t) * \delta(t - t_o) = x(t - t_o)$ | Offset of a time function |
| (e) $x(t - t_1) * \delta(t - t_o) = x(t - t_1 - t_o)$ | Offsets cumulation |
| $\delta(t - t_1) * \delta(t - t_o) = \delta(t - t_1 - t_o)$ | |
| (f) $\delta(at) = \lvert a \rvert^{-1}\, \delta(t)$ | Change of scale |
| (g) $\delta(t - t_o) = \delta(t_o - t)$ | Symmetry of the pulse |
| (h) $\delta(t - t_o) = \frac{d}{dt}(u(t - t_o))$ | Where $u(t - t_o)$ is the unit level offset in $t_o$ |

**Table 1.2.** *Properties of the delta function*

The delta function can also be physically approximated by a triangular or exponential function whose area under the curve is equal to the unit.

### 1.4.2. *System*

A system can be represented by a filter whose response is the convolution product of its transfer function and the input signal. A system can be described by time-differential equations and algebraic equations or by a transfer function in the complex plane (frequency domain). The transfer function represents the Laplace transform of the impulse response of the system.

## 1.5. Transmission systems and filters

Consider a system $S$, with an input a signal $x(t)$ assumed as real (for the moment) and an output $y(t)$ (the response of the system to input $x(t)$). The system $S$ can be represented by an operator that we will also define as $S : X \longrightarrow Y$. We will then write $y(t) = S(x(t))$ to mean that $S$ transforms $x(t)$ of the input signals vector space $X$ into $y(t)$ of the output vector space $Y$ (under certain assumptions of reliability, stability, signals boundedness, etc.):

– the system is linear if the operator associated with it is linear:

$x_1(t) \longrightarrow y_1(t)$ and $x_2(t) \longrightarrow y_2(t)$, then at $x_1(t)+b.x_2(t) \longrightarrow a.y_1(t)+b.y_2(t)$ $\forall a, b \in R$

– the system is time invariant (stationary) if its behavior is independent of the time origin:

$x_1(t) \longrightarrow y_1(t)$, then $x_1(t-\tau) \longrightarrow y_1(t-\tau)$ $\forall \tau \in R$

– a system is known as causal if its impulse response (response to an impulse) is zero for negative times ($h(t) = 0$ if $t < 0$).

REMARK 1.1.– *As a first approximation, almost all systems are (very often) considered as linear for weak signals. A linear system realizes an application from a vector space $X$ in an another $Y$. Generalizing to the spaces of complex signals (spaces defined in $C$). In general, $X$ and $Y$ define the same vector space (or two subspaces), the application $S$ is then a linear operator and $Y$ is the image of $X$ by $S$.*

Consequently, any signal, $x(t) \in X$, can be written according to the base elements of the vector space $x_i(t)$:

$x(t) = \sum_i a_i x_i(t) \longrightarrow y(t) = \sum_i a_i y_i(t)$; with $S : x_j(t) \longrightarrow y_j(t)$ for $j = 1, 2, 3...$

The response of the system therefore will be written based on the $y_i(t)$ images of the elements of the basis of the space $X$ by the operator $S$ (linear combination with the same coefficients as the decomposition of $x(t)$). To know the answer to any random action, is suffices that the image by operator $S$ is known for an enumerable collection of function $x_i(t)$ (basis of $X$).

### 1.5.1. *Convolution and filtering*

The physical justification of the convolution or composition product for filtering uses the concept of impulse response of a system or filter:

– $h(t)$ is the *impulse response* of the filter or system $(H)$ that is the response of this system when an impulse is applied on its input, such as Dirac's delta function $\delta(t)$, $S : \delta(t) \longrightarrow h(t)$.

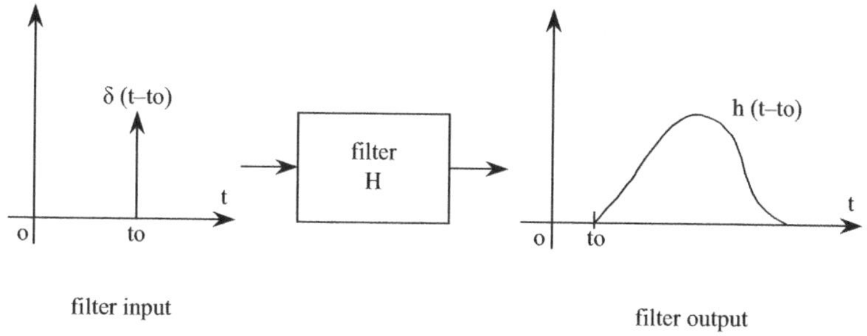


**Figure 1.11.** *Impulse response of the filter or system*

– When an impulse is applied to the system offset by $t_o$, its response will be shifted as much: $S : \delta(t - t_o) \longrightarrow h(t - t_o)$.

– When applying a signal $x(t)$ on input of this system, the input signal is subdivided with a step $\Delta\tau$, into a set of elementary impulse, as shown in Figure 1.12.

– Considering the $k$th shifted impulse of $k\Delta\tau = \tau$, it has mass (weight or area under the curve) $x(k\Delta\tau).\Delta\tau$; the response to this impulse will also be shifted in time and amplitude $(x(k\Delta_\tau).\Delta\tau).h(t - k\Delta\tau)$.

– Then, the response to any random signal $x(t)$ will be the sum of the terms consisting of the responses to all the impulses that constitute $x(t)$; that is the sum of $x(k\Delta\tau).\Delta\tau.h(t - k\Delta\tau)$ for all values of $k$ obtained during the subdivision, that is $\sum_k (x(k\Delta\tau).\Delta\tau.h(t - k\Delta\tau))$ :

- whence by passing to the limit when $\Delta\tau$ tends to zero, and by replacing $k\Delta\tau$ by $\tau$ and $\Delta\tau$ by $d\tau$, we get the response of the filter on input x(t): $y(t) = \int_0^{+\infty}(x(\tau).h(t - \tau))d\tau$, that is, because signals are causal:

$$y(t) = \int_{-\infty}^{+\infty} (x(\tau).h(t - \tau))d\tau = x(t) * h(t) = h(t) * x(t) \qquad [1.11]$$

In conclusion, for a system the response $y(t)$ to an input $x(t)$ is expressed as the convolution product of the input and the impulse response of the system. Therefore, it is possible to represent a system by its impulse response inasmuch as it is sufficient to determine its response to any input $x(t)$. It is the *representation of a system by the impulse response*.

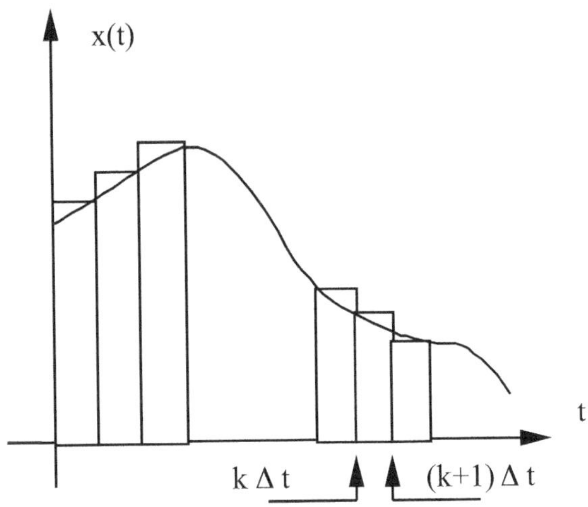


**Figure 1.12.** *Input signal with a step $\Delta\tau$*

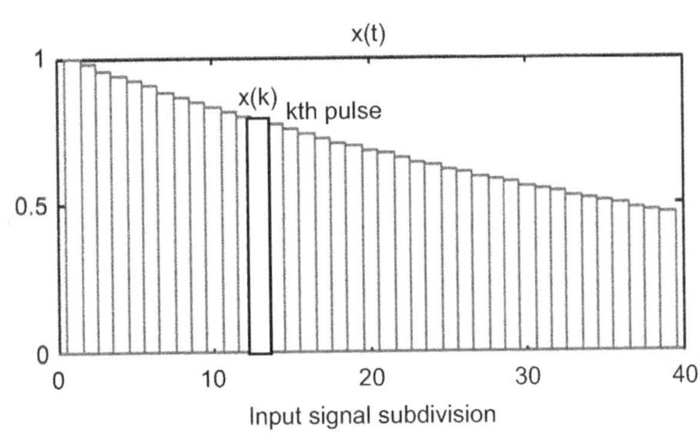


**Figure 1.13.** *Signal subdivision*

In this section, we have shown the interest of functions, distributions and operators and mathematical tools for modeling signals and systems. In the following section, we will present some types of signals and systems as well as their modeling and representations.

## 1.6. Deterministic signals – random signals – analog signals

### 1.6.1. *Definitions*

#### 1.6.1.1. *Deterministic signals*

Signals originate from phenomena for which the knowledge of initial conditions and physical laws allows the anticipation of the result of the measurement and represents a set of results in the form of a function $x(t)$ (for example falling body, filter response with a known $x$). Such a signal is considered as finite when it is possible to determine its value at any time $t$. The description of a signal may be non-parametric (recording, graphics, etc.) or parametric.

We refer to parametric representation or signal model when one is able to define a set of parameters that make it possible to trace the evolution of the signal in time and determine its values at any moment.

| | | |
|---|---|---|
| Sinusoid | $x(t) = A.sin(\omega t) + B$ | Parameters: A, $\omega$, B |
| Damped oscillating signal | $s(t) = A(1 - e^{-b.t}).cos(\omega t + \varphi)$ | Parameters: A, b, $\omega$,$\varphi$ |
| Square signal | $y(t) = A.signe(sin(\omega t)) + b$ | Parameters: A, b, $\omega$ |

**Table 1.3.** *Deterministic signals*

EXAMPLE 1.2.–

A deterministic signal may, in principle, be rigorously reproduced identical to itself.

#### 1.6.1.2. *Random signals (or probabilistic)*

Signals for which the result of a measurement (test) is not predictable and that can only be characterized by using statistical quantities (random distribution), cannot be determined by instantaneous values. For example, the value of the temperature in a geographical point is impossible to determine in advance (before measurement), it constitutes a random signal. A prediction can only be made in the statistical sense with a probability not equal to the unity. A range of values is associated with a probability of occurrence (for example probability $(10 < \theta \leq 20) = 0.3$). *A random signal is not strictly reproducible.*

#### 1.6.1.3. *Signal and noise*

Signal = a quantity carrying information.

Noise = a quantity carrying no information or unnecessary to the user.

It is the recipient who considers if there is or isn't information. For instance, thermal noise from the sky is a signal for the radio astronomer; on the other hand, it is a disruptive noise for telecommunication engineers.

### 1.6.2. *Some deterministic analog signals*

Heaviside signal (unit step), denoted as $H(t)$, $G(t)$ or $u(t)$, will subsequently be written as $u(t)$. It is defined by:

$$u(t) = \begin{cases} 1, \text{ if } t \geq 0 \\ 0, \text{ if } t < 0 \end{cases} \qquad [1.12]$$

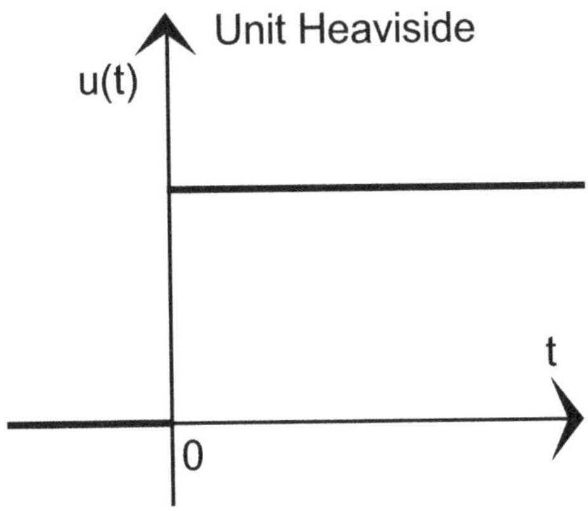


**Figure 1.14.** *Heaviside*

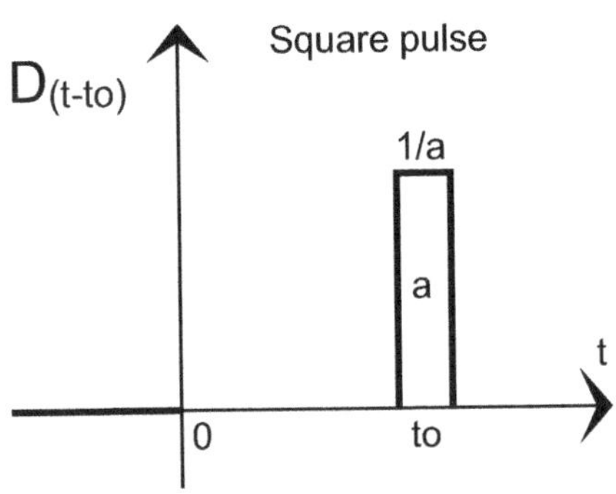


**Figure 1.15.** *Rectangular pulse*

Its derivative is the Dirac delta function $\delta(t)$

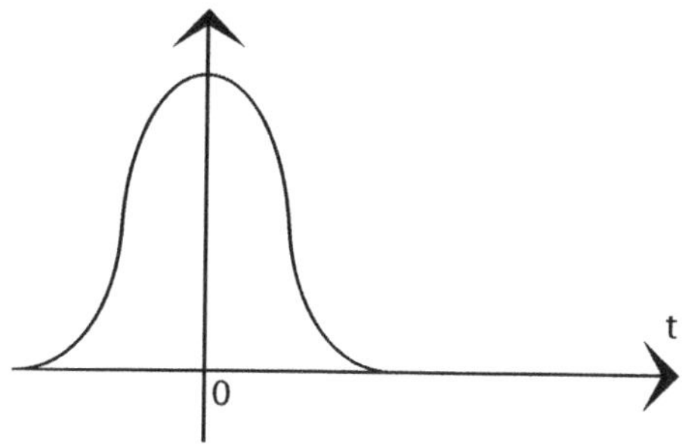


**Figure 1.16.** *Gaussian*

#### 1.6.2.1. *Exponential signals*

This class includes polynomial and sinusoid signals. There are used in the solution of differential equations with constant coefficients; if $\rho=\frac{d}{dt}$ denotes the derivation operator:

$$\rho^n x(t) + a_{n-1}.\rho^{n-1}x(t) + ... + a_1.\rho x(t) + a_0.x(t) = 0 \quad \rho x(t) = \frac{d}{dt}x(t) \qquad [1.13]$$

The general solution of this type of equation is of the form (a combination of particular solutions $e^{\alpha_n.t}$):

$$x(t) = \sum_{i=1}^{i=q} e^{\alpha_i.t} \sum_{j=1}^{n_i-1} c_{ij}.t^j \qquad [1.14]$$

where $c_{ij}$ are constants depending on initial conditions and are thereof the complex or real roots of the characteristic equation:

$$\rho^n + a_{n-1}.\rho^{n-1} + ... + a_1.\rho + a_0 = 0 = \prod_{i=1}^{q}(\alpha - \alpha_i)^{n_i} \quad , \sum_{i=1}^{q} n_i = n \qquad [1.15]$$

#### 1.6.2.2. *Impulse signals*

Different forms of impulse functions can be considered. Figures 1.17 and 1.23 give some examples where pulse height and width are chosen such that the area under the curve is unitary. All these functions have in common the properties of Dirac's delta function previously defined.

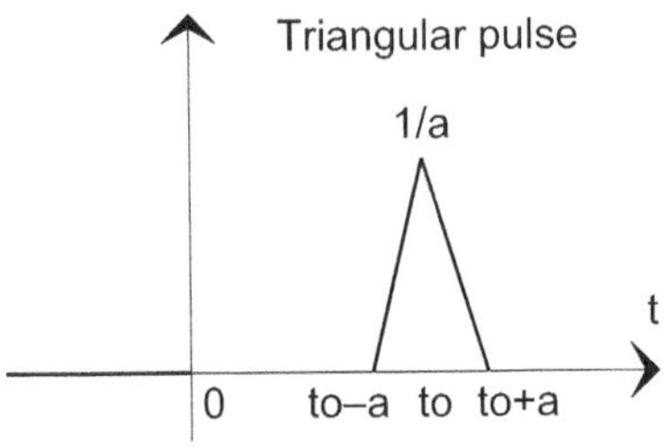


**Figure 1.17.** *Triangular pulse*

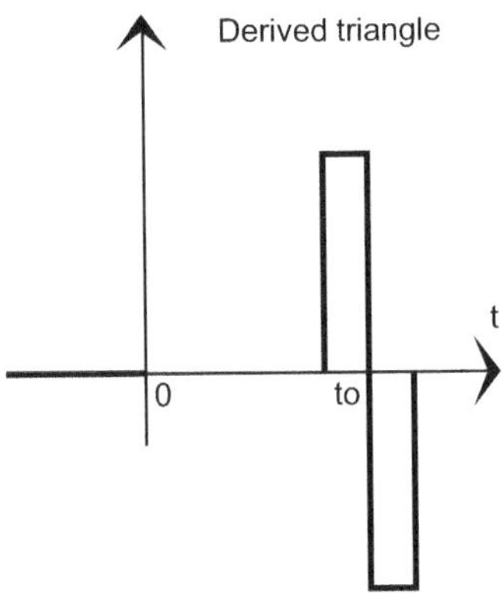


**Figure 1.18.** *Derivative of the triangular signal*

#### 1.6.2.2.1. Examples

1) Rectangular function centered in $t_o$ $f_l(t - t_o)$

2) Triangular function of height $1/a$ and width at the base $2a$, also centered in $t_o$.

3) Note that the derivative of this impulse yields two square impulses of opposite signs.

4) Exponential function, for example, if $u(t)$ is the step function: $\Delta(t) = (1/a)e^{-t/a}.u(t)$. All these functions can be used to decompose a signal into a series of pulses because at the limit they give $\delta(t)$, that is: $\int_{-\infty}^{+\infty} \delta(t - t_o)x(t)dt = x(t_o)$.

5) Polynomial function: $\Delta(t) = (1/2\pi)a/(a^2 + t^2)$.

6) Gaussian-shaped function: $\Delta(t) = (\frac{1}{a\sqrt{2\pi}})e^{-t^2/a^2}$ Gaussian form

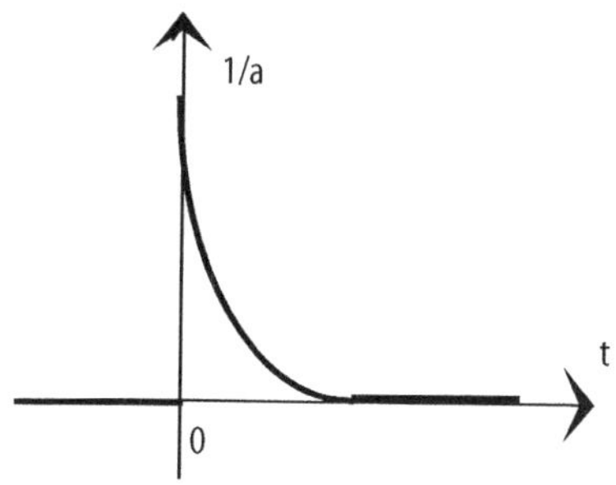


**Figure 1.19.** *Exponential*

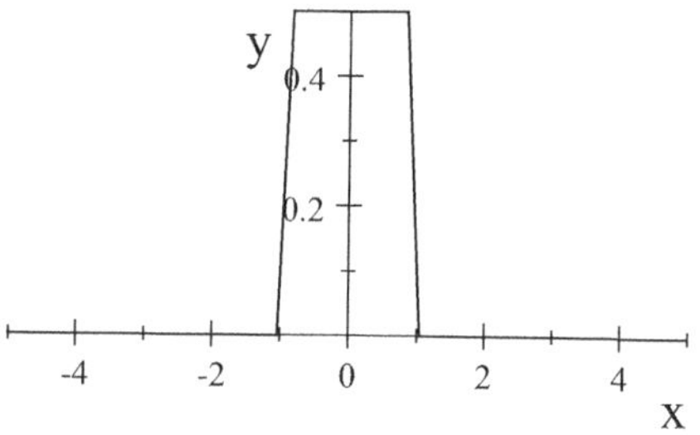


**Figure 1.20.** *Rectangular pulse*

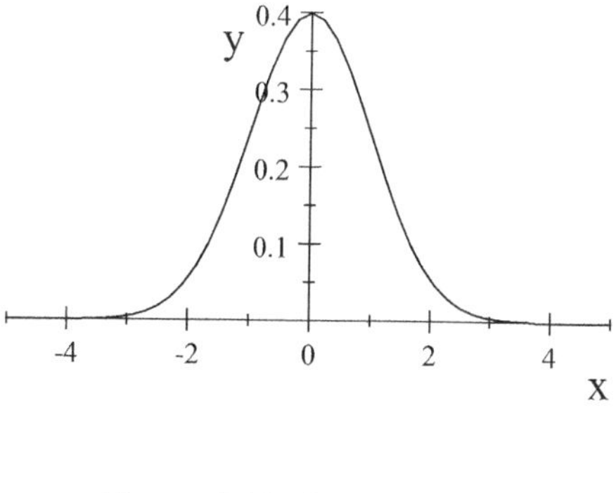


**Figure 1.21.** *Gaussian form*

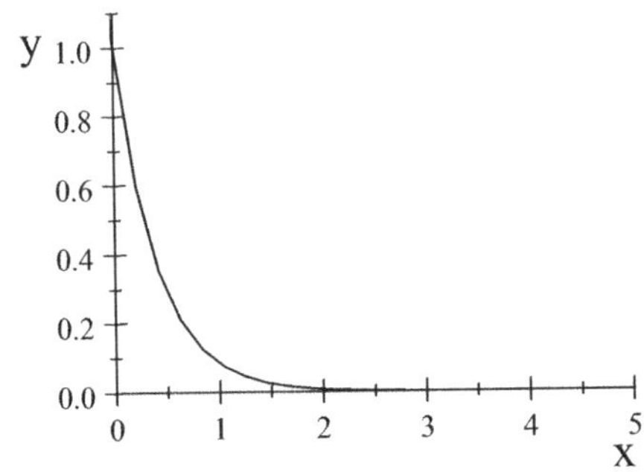


**Figure 1.22.** *Exponential form*

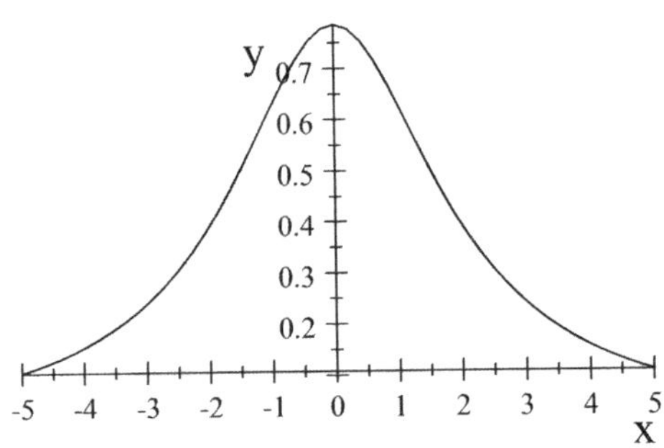


**Figure 1.23.** *Polynomial form*

### 1.6.3. *Representation and modeling of signals and systems*

In order to introduce the different types of modeling, we are going to study a few examples.

#### 1.6.3.1. *Representation by polynomial equations*

The signal is defined by a polynomial in $t$ or a function of $t$ such as exponential functions. These signals are generally of the same type as sinusoid or exponential signals.

EXAMPLE 1.3.–

*a)* $y(t) = \frac{K}{(a^2+t^2)^n}$

*b)* $y(t) = \cos(\omega t) = (e^{j\omega t} + e^{-j\omega t})/2$

#### 1.6.3.2. *Representation by differential equations*

They are defined by differential equations and they are their solutions; for example, for a signal:

$$\rho^n x(t) + a_{n-1}.\rho^{n-1} x(t) + ... + a_1.\rho x(t) + a_0 x(t) = 0, \; \rho x(t) = \frac{dx(t)}{dt} \quad [1.16]$$

and for an input $u(t)$ and output system $x(t)$, we get:

$$\begin{aligned} \rho^n x(t) + a_{n-1}.\rho^{n-1} x(t) + ... + a_0 x(t) &= b_m.\rho^m u(t) \\ &+ b_{m-1}.\rho^{m-1} u(t) + ... + b_1.\rho u(t) + b_0 u(t) \end{aligned} \quad [1.17]$$

#### 1.6.3.3. *Representation by state equations*

Let a vector of dimension $n$ be defined (minimal representation for a system of order $n$), the knowledge of which together with the initial state makes it possible to determine the state of the system at any time using the equation of the system. The state equation of a system for the linear case is defined by a vector differential equation of order one of the original differential equation linking the input to the output. It is presented in the following form for a system whose input is $u(t)$ and output is $y(t)$.

DEFINITION 1.3.– *The minimal dimension of the state of a system corresponds to the number of initial conditions necessary to integrate its differential equation (evolution equation). For a system of order $n$ (degree of its differential equation), this dimension is equal to $n$. The first-order vector differential equation describing the dynamics of a state vector is a state representation of the system.*

State equation of the system with input $u$ and output $y$ is written as

$$\dot{X} = A.X + Bu(t) \quad [1.18]$$

Observation equation

$$y(t) = C.X \quad [1.19]$$

In the case of a signal, it suffices to cancel input $u$ in the above-mentioned equation.

#### 1.6.3.4. *Graphic representations*

Among graphic representations, the most important and the most commonly used are as follows:

– the time representation of the evolution of the signal (plot of the impulse or step response, or the evolution of the signal in time);

– representation in the phase plane (the plot in a coordinate system defined by the components of the system state vector, for example, the derivative or velocity as a function of the position for a second-order system), which is rather interesting in automatic control as in signal processing;

– frequency representations (Bode, Black, Nyquist) that are studied in automation after the time-frequency transformations;

– the time-frequency representation (three-dimensional) that is rather useful for non-stationary signals.

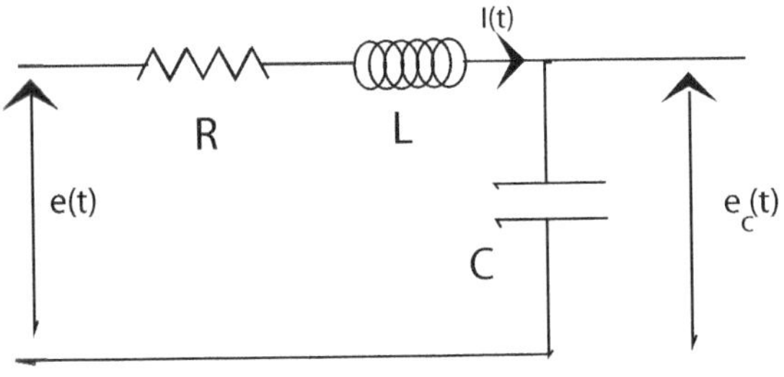


**Figure 1.24.** *A R L C circuit*

##### 1.6.3.4.1. Examples

EXAMPLE 1.4.– Electric RLC circuit

The differential equation giving the behavior model of this circuit is written as:

$$C\frac{de_c(t)}{dt} = i(t) \quad \text{and} \qquad [1.20]$$

$$e(t) = Ri(t) + L\frac{di(t)}{dt} + e_c(t) \qquad [1.21]$$

The state vector of the system is given by $X(t) = \binom{e_c(t)}{i(t)}$. This vector defines the internal state of the system at the moment $t$. This allows us to obtain the system state representation, which is a temporal representation. This representation is not unique. In this case, matrix $A$ and the state representation are defined by:

$$\dot{X} = A\begin{pmatrix} e_c(t) \\ i(t) \end{pmatrix} + Bu \qquad [1.22]$$

$$e_c = (1;0)X = (1;0)\begin{pmatrix} e_c(t) \\ i(t) \end{pmatrix}; \qquad [1.23]$$

with $A = \begin{pmatrix} 0 & \frac{1}{C} \\ -\frac{1}{L} & -\frac{R}{L} \end{pmatrix} \quad B = \begin{pmatrix} 0 \\ \frac{1}{L} \end{pmatrix}$

EXAMPLE 1.5.– Butterworth filter case

A normalized low-pass, fourth-order Butterworth filter, with a cut-off pulse 1 rad/s, whose input is $u(t)$ and output $y(t)$, has a differential equation as:

$$(\rho^4 + 2.6131\rho^3 + 3.4142\rho^2 + 2.6131\rho + 1)y(t) = u(t)\rho \text{ and } \quad x(t) = \frac{dx(t)}{dt} \qquad [1.24]$$

It can be written in the state form:

$$\dot{X} = AX + Bu; \quad y = CX \qquad [1.25]$$

with $A = \begin{pmatrix} -2.6131 & -3.4142 & -2.6131 & -1 \\ 1 & 0 & 0 & 0 \\ 0 & 1 & 0 & 0 \\ 0 & 0 & 1 & 0 \end{pmatrix} B = \begin{pmatrix} 1 \\ 0 \\ 0 \\ 0 \end{pmatrix} C = (0\,0\,0\,1)$

### 1.6.4. *Phase–plane representation*

Consider the case of the second-order signal or system in free regime, defined by its differential equation:

$$\ddot{y}(t) = f(\dot{y}(t), y(t)) \qquad [1.26]$$

By considering $x_1 = y(t)$ and $x_2 = \dot{y}(t)$ as components of the state vector, it is possible to associate thereto the following state representation:

$$\dot{X}(t) = \begin{pmatrix} x_2 \\ -f(x_1, x_2) \end{pmatrix} = AX; \quad y(t) = [1;0].X \qquad [1.27]$$

State variables $x_1$ and $x_2$ are also called phase variables; they determine at every moment the state of the system that can be represented by a point $M$ in the plane $(x_1, x_2)$. This plane is called the phase plane. Point $M$ is characterized by state vector $X(t)$ and its evolution in time from an initial point $M_o$ describes a phase curve. Phase curves can be graded in time and depend on the initial point $M_o$ (the initial state). By eliminating the time variable, an equation is obtained that is parameterized according to the coordinates $(x_1, x_2)$, defining a network of curves. The plot of these curves provides a visualization of the evolution of the system.

A large number of linear or nonlinear systems can be approximated by a second order. This method of representation allows us to easily conclude about the evolution and stability of a system. These disadvantages lie in the fact that graphical representation is impossible for systems with an order higher than 2 and is only applicable for autonomous systems (without input control and whose differential equation does not explicitly depend on time) having a unique solution (for the state equation). It should be noted that the principle remains valid for systems of order greater than 2, although the graphical representation is impossible. In the following section, we illustrate this method for the case of a second-order linear system.

#### 1.6.4.1. *Case of a second-order linear system*

For the representation of a second-order system in the phase plane, we consider the equation:

$$\ddot{y}(t) + 2z\omega_o\dot{y}(t) + \omega_o^2 y(t) = 0 \qquad [1.28]$$

$z$ is the damping and $\omega_o$ is the angular frequency of the system. The poles $p_1$, $p_2$ are the roots of the characteristic equation:

$$r + 2z\omega_o r + \omega_o = 0 \qquad [1.29]$$

According to damping values $z$, several possible cases can be distinguished.

*Case 1*: If $p_1$and $p_2$ are two real negative roots of the characteristic equation:

$$y(t) = x_1 = C_1 e^{p_1 t} + C_2 e^{p_2 t} \qquad [1.30]$$

$$\dot{y}(t) = x_2 = p_1 C_1 e^{p_1 t} + p_2 C_2 e^{p_2 t} \qquad [1.31]$$

hence

$$e^{p_1 t} = \frac{[p_2 x_1 - x_2]}{[p_2 - p_1]C_1} \quad \text{and} \quad e^{p_2 t} = \frac{[p_1 x_1 - x_2]}{[p_1 - p_2].C_2} \qquad [1.32]$$

wherefrom by raising these two equations to the power $p_1$ and $p_2$, we get:

$$(p_2x_1 - x_2)^{\frac{p_2}{p_1}} = C(p_1x_1 - x_2) \qquad [1.33]$$

Constant $C$ depends on constants $C_1$ and $C_2$. This equation represents a family of parabolas tending to zero in $t$, as shown in Figure 1.25.

*Case 2*: If $p_1$ and $p_2$ are two real positive roots, then it yields, by proceeding as previously shown, the result in Figure 1.26.

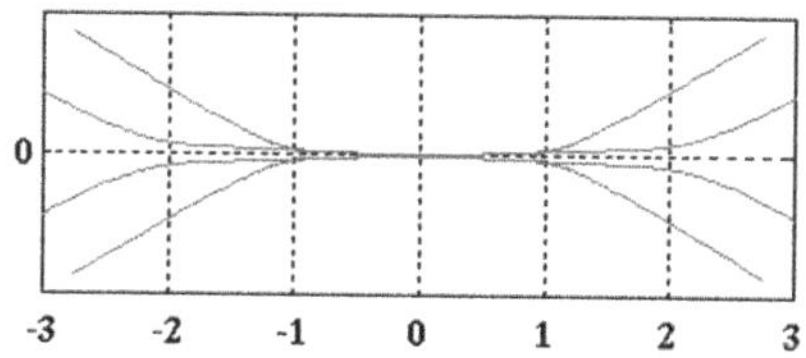


**Figure 1.25.** *$p_1$ and $p_2$ are two real negative roots of the characteristic equation*

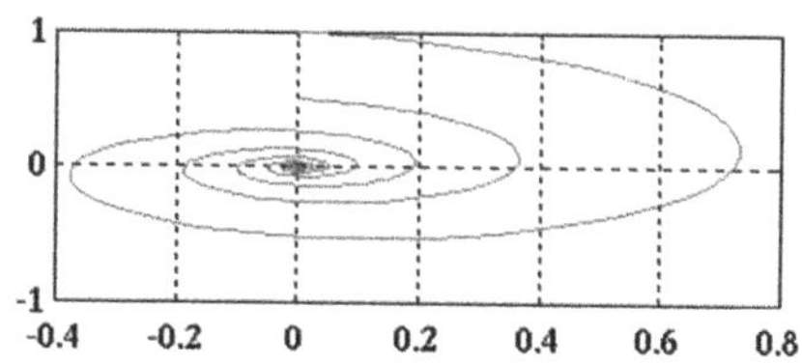


**Figure 1.26.** *$p_1$ and $p_2$ are two real positive roots*

*Case 3*: In the case where $p_1$ and $p_2$ are two real roots of opposite signs, equation [1.33] this time belongs to a family of hyperbolas, because $p_2/p_1 < 0$, whose asymptotes have equations:

$$x_2 = p_1.x_1 \quad \text{and} \qquad x_2 = p_2.x_1 \qquad [1.34]$$

One of the asymptotes (the one corresponding to $p_i$ <0) has its branches directed toward the origin ($x_1$=0, $x_2$=0). The representation is shown in Figure 1.25.

*Case 4*: If $p_1$and $p_2$ are two complex roots with real negative parts, the system is stable and damped oscillating. The solutions are of the form:

$$x_1 = Ae^{p_1}.\sin(\omega t + \varphi) \text{ and } x_2 = Be^{p_2}.\sin(\omega t + \psi) \qquad [1.35]$$

$A$, $B$, $\varphi$ and $\psi$ depend on initial conditions. Trajectories have the shape of spirals going toward zero as shown in Figure 1.26.

*Case 5*: If $p_1$ and $p_2$ are two complex roots with real positive parts, this yields spirals ranging from zero to infinity (see Figure 1.27).

*Case 6*: For two pure imaginary roots, we then obtain ellipses as presented in Figure 1.28:

$$x_1 = A.\sin(\omega t + \varphi) \text{ and } x_2 = B.\sin(\omega t + \psi) \qquad [1.36]$$

We have previously presented the phase–plane representation. This representation is useful for the stability analysis of a system but it is limited to the second-order case.

### 1.6.5. *Dynamic system*

#### 1.6.5.1. *Definitions relating to the equilibrium and stability of a dynamic system*

Consider a free dynamic system described by the differential vector equation:

$$\dot{x} = f(x,t)\, \forall t \in R \qquad [1.37]$$

where $x(t)$ is the state vector of dimension $n$ of the system, defining a point in the phase plane, and $f(x,t)$ is a vector function in $R$ that may be nonlinear. We consider here the case of free signals or systems. Note that this system is free but it is not generally autonomous since its equation depends explicitly on time. The equation of an autonomous system can be written in the form $\dot{x} = f(x)$ without explicit dependence of time. Thus, a free time variant system is not autonomous.

Let $\Phi(t_o, x_o, t)$, the vector function, non-perturbated solution of equation [1.37], differentiable with respect to time and such that for any $x \in R^n$ , $t \in R$ ($x(t_o) = x_o$ initial state and $t_o$ initial moment), we have:

$$\Phi(t_o, x_o, t) = x = x(t) \qquad [1.38]$$

$$\frac{d(\Phi(t_o, x_o, t))}{dt} = f[\Phi(t_o, x_o, t), t] \qquad [1.39]$$

$\Phi(t_o, x_o, t)$ defines a curve or trajectory of the system in the phase space for an initial state $x_o \in \mathbb{R}^n$ and $t \in \mathbb{R}$.

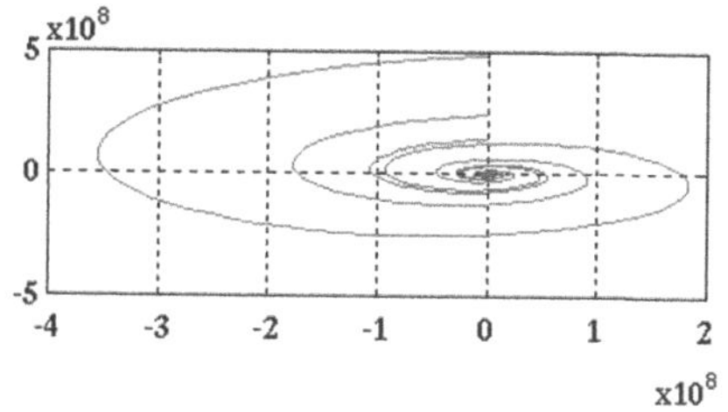


**Figure 1.27.** *$p_1$ and $p_2$ are two complex roots with real positive parts*

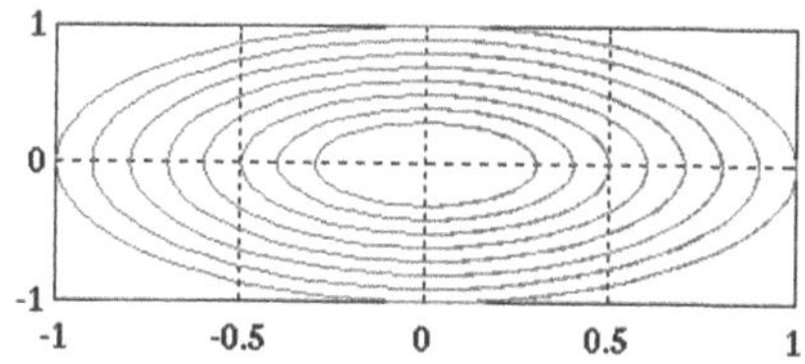


**Figure 1.28.** *Two pure imaginary roots*

#### 1.6.5.2. *Equilibrium state of a system*

DEFINITION 1.4.– *A state $x_e$ of the free system is called equilibrium point if:*

$$f(x_e, t) = 0 \, \forall t \in \mathbb{R}, t > to \quad [1.40]$$

This expresses that if $x_e$ is an equilibrium point, then, in a case without disturbance, the system tends to remain in a nearby neighborhood because at this point velocity $\dot{x}_e = f(x_e, t)$ is zero. When the equilibrium point is $x_e \neq 0$, it can be brought by transformation to the origin of the phase space. For a linear time-invariant system, $\dot{x} = f(x, t) = A.x$, if $A$ is not singular, then $x_e = 0$ is the point of equilibrium. If $A$ is singular, there is an infinite number of equilibrium points. The search for equilibrium points does not require solving differential equations $(\dot{x} = f(x, t) = 0)$.

##### 1.6.5.2.1. Stable equilibrium state

DEFINITION 1.5.– *An equilibrium state is stable if, after deviating from this equilibrium, the system remains in the neighborhood or tends to return to this equilibrium point. An equilibrium point will be known as unstable equilibrium if after deviating from this point, the system tends to move away from it.*

For example, a pendulum has a stable equilibrium (low vertical position) and an unstable equilibrium (high vertical position, modulo $2k\pi$).

#### 1.6.5.3. *Stable system: bounded input bounded output*

DEFINITION 1.6.– *A system is said to be stable if for any bounded input $e(t)$ it establishes a correspondence to a bounded output $s(t)$. This notion of stability is called "Bounded Input Bounded Output stability".*

If $h(t)$ is the impulse response of the system, the following can be stated:

$$s(t) = \int_{-\infty}^{+\infty} e(\tau).h(t-\tau)d\tau = e(t) * h(t) \qquad [1.41]$$

Knowing that $e(t)$ is bounded, $M \in R$ we then obtain:

$$|s(t)| \leq \int_{-\infty}^{+\infty} |e(\tau)| . |h(t-\tau)|\, d\tau \leq M. \int_{-\infty}^{+\infty} |h(\tau)|\, d\tau \qquad [1.42]$$

hence the system is stable if $h(t)$, its impulse response, is a summable function. A sufficient condition for stability is that $h(t)$ is summable ($\int_{-\infty}^{+\infty} |h(\tau)| \mathrm{d}\tau$ is finite).

## 1.7. Comprehension and application exercises

EXERCISE 1.–

1) Write the differential equation of a system composed of mass $M$ suspended by a spring of stiffness $k$. We consider that the mass moves along a vertical axis without friction.

2) Represent this system in the form of state taking a state vector composed of the position and the velocity of mass $M$.

EXERCISE 2.–

An oscillator delivers a signal composed of two frequencies $f_1$ and $f_2$:

1) give all possible representations to define this signal and the parameters corresponding to them;

2) the amplitude of the oscillations of both frequencies decreases by 10% after 5 h; what happens to the representations of question 1?;

EXERCISE 3.–

1) a low-pass filter has impulse response $h(t) = Ae^{\frac{-t}{t_1}} + Be^{\frac{-t}{t_2}}$ to determine other representations capable of describing this system;

2) give the conditions about the parameters of the system providing the stability of the filter;

3) give different graphical representation of this system.

# 2

# Transforms: Time – Frequency – Scale

This chapter explores the Fourier series and its transforms, and Laplace transforms (LTs).

## 2.1. Fourier series applied to periodic functions

### 2.1.1. *Fourier series*

Given the signal represented by the periodic function $x(t)$ of period $T$, $(x(t+T)=x(t))$. If this function satisfies the Dirichlet conditions, then it can be formally decomposed in the form of a Fourier series (under certain regularity conditions, the function must be summable over a period).

$$x(t)=\frac{a_o}{2}\sum_{n=1}^{n=+\infty}\left(a_n.\cos(\frac{2\pi nt}{T})+b_n.sin(\frac{2\pi nt}{T})\right) \qquad [2.1]$$

$\frac{a_o}{2}$ is the average value of the signal $x(t)$, and coefficients $a_n$ and $b_n$ are given by the following formulas ($\upsilon_o=\frac{1}{T}$):

$$a_n=\frac{2}{T}\int_0^T x(t).\cos(\frac{2\pi nt}{T}).dt \quad for,\ n=1,2,3,... \qquad [2.2]$$

$$b_n=\frac{2}{T}\int_0^T x(t).\sin(\frac{2\pi nt}{T}).dt \quad for,\ n=1,2,3,... \qquad [2.3]$$

$$\text{and} \quad a_o=\frac{2}{T}\int_0^T x(t).dt \qquad [2.4]$$

REMARK 2.1.– *If x(t) introduces first-kind discontinuities, the sum of the Fourier series (denoted S(t)) gives at the times of these discontinuities $t_i$: S($t_i$) = $\frac{x(t_i+0)+x(t_i-0)}{2}$; S(t) in $t_i$ is the half sum of the right limit and of the left limit at discontinuities times.*

In general, the Fourier series is formulated in exponential form.

Let

$$g_n(t) = e^{\frac{2\pi nt}{T}} \qquad [2.5]$$

These complex functions are periodic of period $T$ and are orthogonal two-by-two:

$$< g_n(t), g_m(t) >= \frac{1}{T}\int_0^T g_n(t).g_m^*(t)dt = 0 \text{ if } m \neq n \text{ or } 1 \text{ if } m = n. \qquad [2.6]$$

$< g_n(t), g_m(t) >$ denotes the scalar product of $g_n(t)$ and $g_m(t)$ . The Fourier series is written as:

$$S(t) = \sum_{n=-\infty}^{n=+\infty} C_n.g_n(t) = x(t) \qquad [2.7]$$

if $x(t)$ is a periodic function (or defined in $[0, T]$) having only a finite number of jumps per period, there is a suite $C_n$ and the Fourier series converges to $x(t)$ (or $\frac{x(t+0)+x(t-0)}{2}$ if $x(t)$ shows a jump in $t$). The coefficients $C_n$ are given by:

$$C_n =< x(t), g_n(t) >= \frac{1}{T}\int_0^T x(t).g_n^*(t).dt \qquad [2.8]$$

The relations between coefficients $a_n$ and $b_n$, on the one hand, and $C_n$, on the other hand, for $x(t)$ a real signal are:

$$C_n = \frac{a_n - i.b_n}{2} \quad \text{and} \quad C_n^* = \frac{a_n + i.b_n}{2} = C_{-n} \qquad [2.9]$$

The coefficients $C_n$ are complex and their modulo can be represented as

$$|C_n| = \frac{\left(|a_n|^2 + |b_n|^2\right)^{1/2}}{2} \text{ where } |C_n|^2 = \frac{\left(|a_n|^2 + |b_n|^2\right)}{2} \qquad [2.10]$$

and their argument

$$Arg(C_n) = Arctg(-b_n/a_n) \quad [2.11]$$

with respect to $n$ (or to frequency $f_n = n\upsilon_o = \frac{n}{T}$). By plotting them we are able to obtain the amplitude (modulo), the power (squared modulo) or the phase spectrum (the argument according to the frequency).

### 2.1.2. *Spectral representation (frequency domain)*

– The amplitude spectrum is obtained by plotting $|C_n|$ with respect to $n$. The amplitude spectrum is symmetrical because $C_n = C_n^* = C_{-n}$ for a real signal;

– the phase spectrum is obtained by plotting the argument of $C_n = f(n) = Arg(C_n) = Arctg(-b_n/a_n)$;

– the power spectrum is obtained by plotting $|C_n|^2$with respect to $n$;

– amplitude and power spectra are symmetrical for a real signal;

– for a periodic signal, there are a line spectrum, $\upsilon_o$=$\frac{1}{T}$ is the frequency of the fundamental, the distance between lines is $\upsilon_o$=$\frac{1}{T}$.

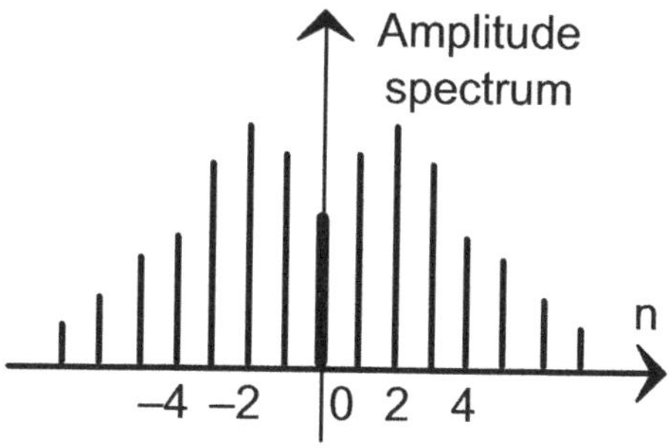


**Figure 2.1.** *Amplitude spectrum*

The previous formula can be regarded as a particular case of the Fourier transform (FT) and the second formula as an inversion formula. This allows us to move from a conventional time representation $x(t)$ to a frequency representation $X(f)$ defined by the series $C_n = X(n/T)$. The advantage of this representation is that it is discrete; it will thus be used in the discrete fourier transform (DFT). Signals $g_n(t) = e^{2i\pi nt/T}$ have the advantage of simplifying the frequency representation but they introduce the concept of negative frequencies. For any periodic signal, we obtain a decomposition into a finite or infinite sum of sinusoidal terms with a fundamental (first harmonic) and harmonics ($\upsilon_1 = 2/T$, $\upsilon_2 = 3/T$, $\upsilon_3 = 4/T$, etc.) of the second, third, fourth, etc., orders. The Fourier series allows the transition from the time domain to the frequency domain.

REMARK 2.2.– *The concept of negative frequencies has been introduced in the complex form of the Fourier series (mathematical construct):*

*– Core elements in Fourier series (sin, cos, $g_n$) are orthogonal; the signal is developed in series on an orthogonal basis;*

*– The condition for convergence for the Fourier series formula is: $x(t)$ piecewise continuous and admits a right derivative and a left derivative at any point of $[0, T]$.*

REMARK 2.3.– *We will use the properties of the Fourier series, in non-periodic cases, for the calculation of discrete spectra (sampled in $f$) by periodizing the original signal (see DFT).*

### 2.1.3. *Properties of Fourier series*

– The term $C_0 = \frac{a_o}{2}$ corresponds to the average value of the signal;

– the spectral representation is composed of lines;

– the frequency term $1/T$ is the fundamental or first harmonic;

– the frequency term $n/T$ is the harmonic of order $n$;

– if the signal $x(t)$ is real, then $C_n = C^*_{-n}$;

– If the signal $x(t)$ is even and real, then $C_n$ is real and $b_n = 0$;

– $x(t)$ is real and odd $C_n$ is imaginary and $a_n = 0$.

THEOREM 2.1.– Parseval equality.

Average (quadratic) power (over a period $T$) of the signal is equal to the sum of the square of the amplitude of all harmonics.

$$
\begin{aligned}
< x(t), x(t) > &= \frac{1}{T} \int_0^T x(t).x^*(t).dt \qquad [2.12] \\
&= \sum_{n=-\infty}^{n=+\infty} |C_n|^2 = |C_0|^2 + 2 \sum_{n=1}^{n=+\infty} |C_n|^2
\end{aligned}
$$

This last relation indicates there is conservation of energy when moving from the temporal representation to Fourier representation.

### 2.1.4. *Some examples*

EXAMPLE 2.1.– *Given a square signal of period T. Let* $\omega_o = 2/T$. *Its expansion into a Fourier series gives us:*

$$y(t) = \frac{4}{\pi}\left(\sin(\omega_o t) + \frac{1}{3}\sin(3\omega_o t) + \frac{1}{5}\sin(5\omega_o t) + \frac{1}{7}\sin(7\omega_o t) + ...\right) \quad [2.13]$$

*In order to reconstruct the signal, an infinite number of terms is necessary. If a truncation is carried out (partial sum of the* $n$ *first terms), an error occurs.*

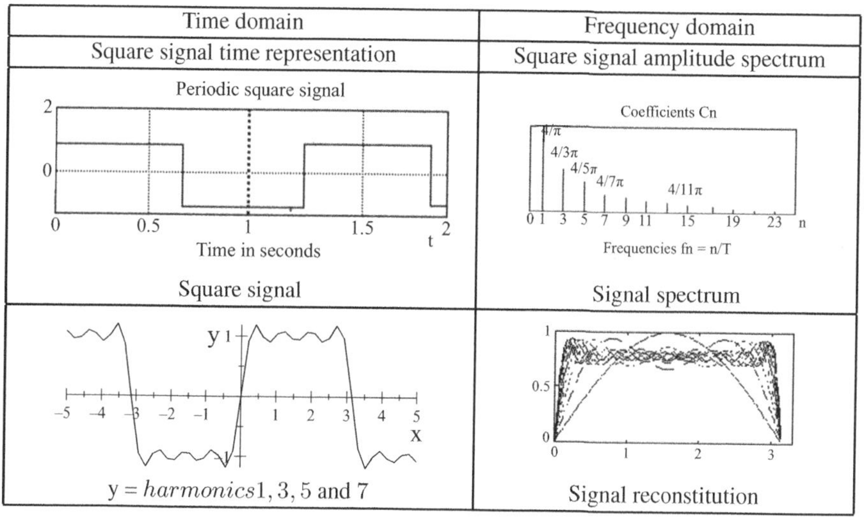


**Figure 2.2.** *Fourier series expansion of a square signal*

In the above case, there are oscillations and a significant overrun in the neighborhood of discontinuity points (Gibbs phenomenon: truncation. The reconstruction of the signal with harmonics 1,2,3 in the time domain is translated by oscillations in the frequency domain and vice versa).

EXAMPLE 2.2.– *As another example of functions, we represent as*

$x(t) = A.\sin(2\pi t/T + \varphi)$. *Its Fourier series can be written as:*

$$x(t) = A.(\sin\varphi.\cos(2\pi t/T) + \cos\varphi.\sin(2\pi t/T)) \quad [2.14]$$

*or even*

$$x(t) = A\frac{e^{i\varphi}.e^{j2\pi t/T} - e^{-i\varphi}.e^{-i2\pi t/T}}{2i} \quad [2.15]$$

*with* $C_1 = A\frac{e^{i\varphi}}{2i}$; $C_{-1} = A\frac{e^{-i\varphi}}{2i}$ *and* $C_n = 0$ *for* $n$ *different from* $1$ *and* $-1$.

The average or continuous component of the signal is zero $C_0 = a_0/2 = 0$. Thus, plotting the modulo of the coefficients of the Fourier series with respect to $n$, two lines can be obtained (Dirac at 1 and at $-1$ at height $A/2$). If we consider the same signal but limit it in time, that is we cancel it outside of the range $[0; b]$, then the spectrum becomes broader. The appearance of additional lines confirms the Gibbs phenomenon, as shown Figure 2.3.

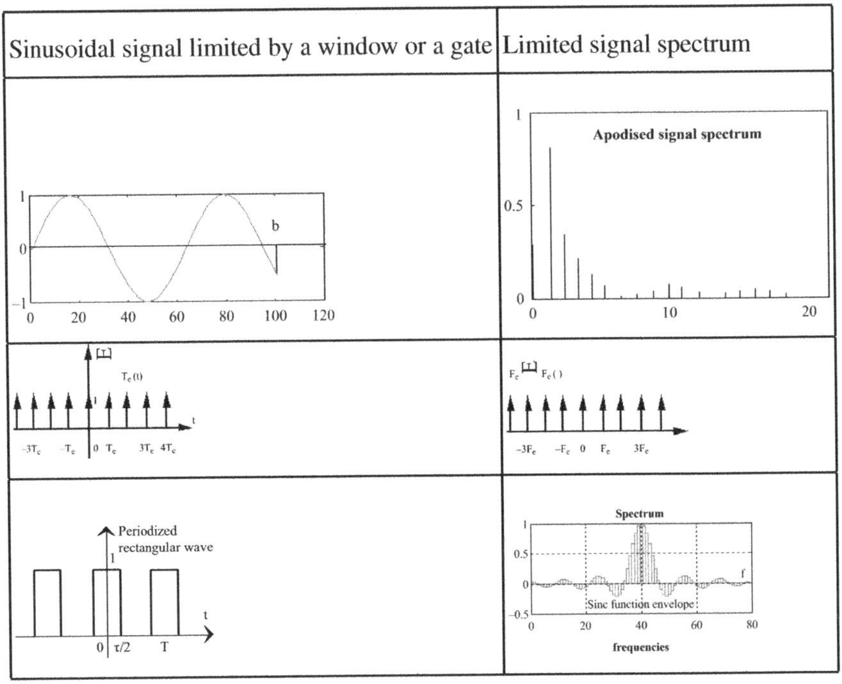


**Figure 2.3.** *Fourier series expansion of a sinusoidal signal*

EXERCISE 2.1.– *Given the Fourier series decomposition of the Dirac comb:* $\sqcup\!\sqcup_T(t) = \sum_{n=-\infty}^{n=+\infty} \delta(t - nT)$.

## 2.2. FT applied to non-periodic functions

Non-periodic signals cannot be expanded into Fourier series, but it may be considered that a non-periodic signal results from an expansion at infinity of the period $T$ of a periodic signal. The frequency of the fundamental $v_o = 1/T$ tends to zero and the spectrum then becomes a continuous function of the frequency. Thus, we will associate with the time signal an equivalent frequency by FT that we will denote as $T \cdot F$.

The FT is defined by:

$$X(f) = \widehat{x}(f) = \int_{-\infty}^{+\infty} x(t).e^{-2\pi.i.f.t}dt \qquad [2.16]$$

and the inverse FT by:

$$x(t) = \int_{-\infty}^{+\infty} X(f).e^{2\pi.i.f.t}df \qquad [2.17]$$

Often, in the literature, the FT and its inverse FT (IFT) are expressed as a function of the pulse. It is then possible to write:

$$X(\omega) = \widehat{x}(\omega) = \int_{-\infty}^{+\infty} x(t).e^{-i.\omega.t}dt \qquad [2.18]$$

$$x(t) = \frac{1}{2\pi}\int_{-\infty}^{+\infty} X(\omega).e^{i.\omega.t}d\omega \qquad [2.19]$$

where $X(\omega)$ constitutes the FT $X(f)$ in equation [2.16] up to a scale constant. In the case of a real signal $x(t)$, the real and imaginary parts are easily isolated:

$$Re(X(f)) = \int_{-\infty}^{+\infty} x(t).\cos\omega.tdt = R(X); \qquad [2.20]$$

$$Im(X(f)) = \int_{-\infty}^{+\infty} x(t).\sin\omega.tdt = I(X); \qquad [2.21]$$

The power or energy spectrum is then given by:

$$|X(f)|^2 = |R(X)|^2 + |I(X)|^2 \qquad [2.22]$$

The phase spectrum is given by:

$$\varphi(f) = Arctg\left(\frac{-I(x)}{R(X)}\right) \qquad [2.23]$$

– An important consequence is that the FT of a differential equation in $f(t)$ yields in frequency a polynomial in $f$ multiplied by $F(f)$.

| | $f(t)$ | $F(f)$ |
|---|---|---|
| Linearity | $\lambda.f(t)+\mu.g(t)$ | $\lambda.F(f)+\mu.G(f)$ |
| Parity | $f(t)$ real even | $F(f)$ real even |
| | $f(t)$ real odd | $F(f)$ real odd |
| Similarity | $f(at)$ | $(1/\lvert a\rvert)F(f/a)$ |
| Translation | $f(t-\tau_o)$ | $e^{-2\pi.i.ft_o}.F(f)$ |
| | $e^{2\pi.i.t.f_o}.f(t)$ | $F(f-f_o)$ |
| Derivation $/t$ | $\frac{d}{dt}f(t)$ | $2i.\pi.f.F(f)$ This property plays a key role in automation. |
| Derivation $/f$ | $-2i.\pi.t.f(t)$ | $\frac{d}{df}F(f)$ |
| | $(-2i.\pi.t)^m.f(t)$ | $\frac{d^m}{df^m}F(f)$ |

**Table 2.1.** *Properties of the Fourier transform*

### 2.4.3. *Plancherel theorem and convolution product*

Given $f(t)$ and $g(t)$ two square summable functions such that: $\int_{-\infty}^{+\infty}|f(t)|^2\,dt<\infty$ and $\int_{-\infty}^{+\infty}|g(t)|^2\,dt<\infty$ or a function and a distribution. If $f(t)*g(t)$ do exist, we then have the Plancherel–Parseval formula:

$$T.F.(f(t)*g(t)) = T.F.(f(t)).T.F.(g(t)) = F(f).G(f) \quad [2.34]$$

$$T.F.(f(t).g(t)) = T.F.(f(t)) * T.F.(g(t)) = F(f) * G(f) \quad [2.35]$$

The FT transforms a convolution product of two functions into a simple product of their transforms and a simple product into a convolution product. Let

$$h(t) = f(t)*g(t) \text{ we get } H(f) = F(f).G(f) \quad [2.36]$$

$$h(t) = \int_{-\infty}^{+\infty} f(t').g(t-\tau').dt' = \int_{-\infty}^{+\infty} F(f).G(f).e^{2\pi.i.f.t}.df \quad [2.37]$$

we replace $g(x)$ by $g^*(-x)$, we obtain in frequency $G^*(f)$ instead of $G(f)$, [2.37] gives us:

$$h(t) = \int_{-\infty}^{+\infty} f(t').g(t'-t).dt' = \int_{-\infty}^{+\infty} F(f).G^*(f).e^{2\pi.i.f.t}.df \quad [2.38]$$

– An important consequence is that the FT of a differential equation in $f(t)$ yields in frequency a polynomial in $f$ multiplied by $F(f)$.

| | $f(t)$ | $F(f)$ |
|---|---|---|
| Linearity | $\lambda.f(t) + \mu.g(t)$ | $\lambda.F(f) + \mu.G(f)$ |
| Parity | $f(t)$ real even | $F(f)$ real even |
| | $f(t)$ real odd | $F(f)$ real odd |
| Similarity | $f(at)$ | $(1/\|a\|)F(f/a)$ |
| Translation | $f(t-\tau_o)$ | $e^{-2\pi.i.ft_o}.F(f)$ |
| | $e^{2\pi.i.t.f_o}.f(t)$ | $F(f-f_o)$ |
| Derivation $/t$ | $\frac{d}{dt}f(t)$ | $2i.\pi.f.F(f)$ This property plays a key role in automation. |
| Derivation $/f$ | $-2i.\pi.t.f(t)$ | $\frac{d}{df}F(f)$ |
| | $(-2i.\pi.t)^m.f(t)$ | $\frac{d^m}{df^m}F(f)$ |

**Table 2.1.** *Properties of the Fourier transform*

### 2.4.3. *Plancherel theorem and convolution product*

Given $f(t)$ and $g(t)$ two square summable functions such that: $\int_{-\infty}^{+\infty} |f(t)|^2 \, dt < \infty$ and $\int_{-\infty}^{+\infty} |g(t)|^2 \, dt < \infty$ or a function and a distribution. If $f(t) * g(t)$ do exist, we then have the Plancherel–Parseval formula:

$$T.F.(f(t) * g(t)) = T.F.(f(t)).T.F.(g(t)) = F(f).G(f) \qquad [2.34]$$

$$T.F.(f(t).g(t)) = T.F.(f(t)) * T.F.(g(t)) = F(f) * G(f) \qquad [2.35]$$

The FT transforms a convolution product of two functions into a simple product of their transforms and a simple product into a convolution product. Let

$$h(t) = f(t) * g(t) \text{ we get } H(f) = F(f).G(f) \qquad [2.36]$$

$$h(t) = \int_{-\infty}^{+\infty} f(t').g(t-\tau').dt' = \int_{-\infty}^{+\infty} F(f).G(f).e^{2\pi.i.f.t}.df \qquad [2.37]$$

we replace $g(x)$ by $g^*(-x)$, we obtain in frequency $G^*(f)$ instead of $G(f)$, [2.37] gives us:

$$h(t) = \int_{-\infty}^{+\infty} f(t').g(t'-t).dt' = \int_{-\infty}^{+\infty} F(f).G^*(f).e^{2\pi.i.f.t}.df \qquad [2.38]$$

the Fourier series to non-periodic and finite energy signals. For infinite energy or finite average power signals, it is either possible to consider the FT in the sense of distributions or to restrict the signal in time to calculate the FT.

## 2.4. FT properties

### 2.4.1. *Properties*

The FT has the following properties:

– Linearity:

$$TF[ax(t) + by(t)] = aX(f) + bY(f) \qquad [2.28]$$

– Convolution:

$$TF[x(t) * y(t)] = X(f).Y(f) \qquad [2.29]$$

$$TF[x(t).y(t)] = X(f) * Y(f)$$

– Derivation:

$$TF[\frac{d^n x(t)}{dt^n}] = (2\pi j f)^n X(f) \qquad [2.30]$$

– Change in scales:

$$TF[x(at)] = \frac{1}{|a|} X(\frac{f}{a}) \qquad [2.31]$$

– Translation in time:

$$TF[x(t-\tau)] = e^{-2\pi j f \tau} X(f) \qquad [2.32]$$

– Multiplication by $t^n$:

$$TF[t^n x(t)] = (-\frac{1}{2\pi j})^n X^{(n)}(f) \qquad [2.33]$$

### 2.4.2. *Properties of the FT*

– If $f(t)$ is summable (but not necessarily continuous), then $\text{TF}(f(t)) = F(f)$ is continuous, bounded and equal to zero at infinity;

– Other properties of the FT and its inverse are given in Tables 2.2 and 2.3;

– A dilatation in $t \Longrightarrow$ a contraction in $f$ and vice versa;

## 2.3. Necessary conditions for the Fourier integral

### 2.3.1. *Definition*

The FT is an extension of the decomposition into Fourier series, but for signals not necessarily periodic. A function $x(t)$ admits as FT the function $X(f)$ if:

$$\int_{-\infty}^{+\infty} |x(t)|\, dt < \infty \qquad [2.24]$$

It is defined by:

$$X(f) = TF[x(t)] = \int_{-\infty}^{+\infty} x(t)e^{-j2\pi ft}dt \qquad [2.25]$$

The inverse FT is given by:

$$x(t) = TF^{-1}[X(f)] = \int_{-\infty}^{+\infty} X(f)e^{+j2\pi ft}df \qquad [2.26]$$

### 2.3.2. *Necessary condition*

A function $f(t)$ admits an FT $F(f)$, calculable by formula [2.16], if and only if:

1) $f(t)$ is bounded;

2) $\int_{-\infty}^{+\infty} f(t).dt$ exists;

3) the discontinuities of $f(t)$ and its extrema are in finite number.

On the other hand, these conditions are not necessarily satisfied when the transform concerns distributions. For example, $\delta(t)$ which is not bounded, has an FT that is the constant function equal to 1. $TF(\delta(t)) = 1$. The cosine function is not integrable in $]-\infty, +\infty[$ and its FT is the sum of two Dirac distributions.

$$TF(\delta(t-a)) = \int_{-\infty}^{+\infty} \delta(t-a).e^{-2\pi.i.f.t}dt = e^{-2\pi.i.f.a} \qquad [2.27]$$

The knowledge of the FT of a signal reflects all characteristic frequencies of this signal. It also provides knowledge of amplitude, power or energy and phase spectra as for the Fourier series in the case of a periodic signal.

It makes it possible to obtain a frequency representation for signals thus characterizing in another way their properties. It therefore extends the advantages of

The FT is defined by:

$$X(f) = \widehat{x}(f) = \int_{-\infty}^{+\infty} x(t).e^{-2\pi.i.f.t}dt \qquad [2.16]$$

and the inverse FT by:

$$x(t) = \int_{-\infty}^{+\infty} X(f).e^{2\pi.i.f.t}df \qquad [2.17]$$

Often, in the literature, the FT and its inverse FT (IFT) are expressed as a function of the pulse. It is then possible to write:

$$X(\omega) = \widehat{x}(\omega) = \int_{-\infty}^{+\infty} x(t).e^{-i.\omega.t}dt \qquad [2.18]$$

$$x(t) = \frac{1}{2\pi}\int_{-\infty}^{+\infty} X(\omega).e^{i.\omega.t}d\omega \qquad [2.19]$$

where $X(\omega)$ constitutes the FT $X(f)$ in equation [2.16] up to a scale constant. In the case of a real signal $x(t)$, the real and imaginary parts are easily isolated:

$$Re(X(f)) = \int_{-\infty}^{+\infty} x(t).\cos\omega.tdt = R(X); \qquad [2.20]$$

$$Im(X(f)) = \int_{-\infty}^{+\infty} x(t).\sin\omega.tdt = I(X); \qquad [2.21]$$

The power or energy spectrum is then given by:

$$|X(f)|^2 = |R(X)|^2 + |I(X)|^2 \qquad [2.22]$$

The phase spectrum is given by:

$$\varphi(f) = Arctg\left(\frac{-I(x)}{R(X)}\right) \qquad [2.23]$$

By imposing, $f(t) = g(t)$ and $t = 0$, equation [2.37] becomes:

$$\int_{-\infty}^{+\infty} |f(t')|^2 \, dt' = \int_{-\infty}^{+\infty} |F(f)|^2 \, df \qquad [2.39]$$

Equation [2.39] translates the energy conservation when moving from the time domain into the frequency domain. This relation is identical to the Parseval equality for Fourier series.

*Parseval theorem*: For non-periodic signals, the FT preserves the energy.

## 2.5. Fourier series and FT

Let $s(t)$ a periodic signal of period $T$; the FT of a periodic signal is not directly computable, however we can take its Fourier series ($\omega_o = 2\pi/T$):

$$s(t) = \sum_{n=-\infty}^{n=+\infty} C_n.g_n(t) = x(t) = \sum_{n=-\infty}^{n=+\infty} C_n.e^{in\omega_o t} \qquad [2.40]$$

In the following, we will assume a periodic signal wherefrom we extract a period to apply the FT. Then, we will compare this FT with the Fourier series of the periodic signal.

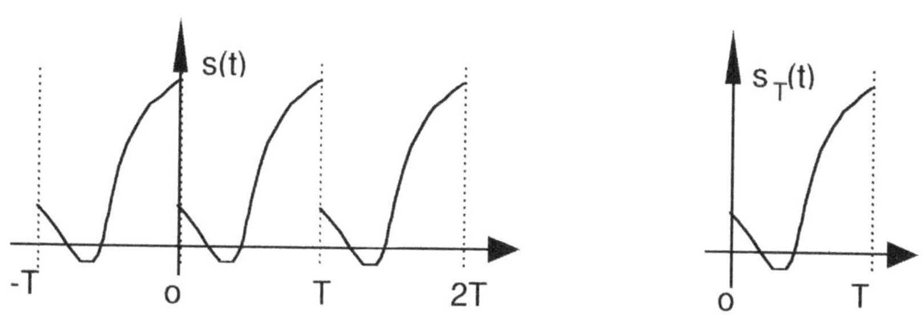


**Figure 2.4.** *Periodized signal*

Denote $s_T(t)$ the truncated signal over a period $T$. $s_T(t) = s(t)$ if $t \in [0, T]$ ; $s_T(t) = 0$ if $t \notin [0, T]$. Using the offset property of Dirac's delta function, we can write:

$$s(t) = s_T(t) * \sum_{n=-\infty}^{n=+\infty} \delta(t - nT) = s_T(t) * \sqcup\!\sqcup_T(t) \qquad [2.41]$$

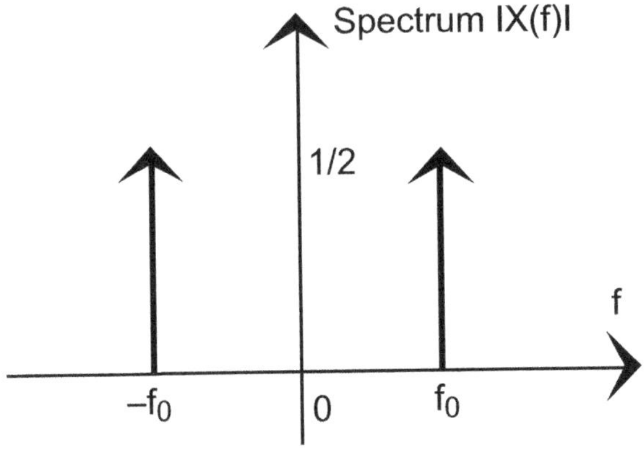


**Figure 2.7.** *Spectrum of a sinusoid*

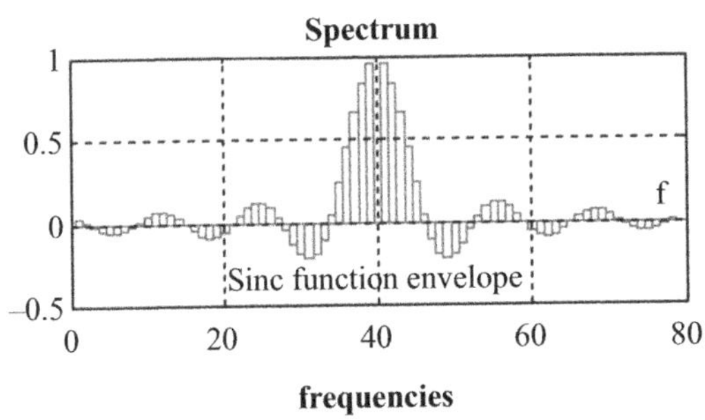


**Figure 2.8.** *Sinc function spectrum*

4) The so-called sample signal : $x(t) = \sin(2\pi f_o t)/\pi.t$. Its usefulness: theory of antennas, sampling, etc. $X(f) = \pi_{f_o}(f)$ carries in frequency;

5) The FT of a Dirac comb in time $\sqcup\!\sqcup_T(t) = \sum\limits_{n=-\infty}^{n=+\infty} \delta(t - nT)$ is a Dirac comb at frequency $\frac{1}{T}\sqcup\!\sqcup_T(t) = \frac{1}{T}\sum\limits_{n=-\infty}^{n=+\infty} \delta(f - n\frac{1}{T})$.

By imposing, $f(t) = g(t)$ and $t = 0$, equation [2.37] becomes:

$$\int_{-\infty}^{+\infty} |f(t')|^2 \, dt' = \int_{-\infty}^{+\infty} |F(f)|^2 \, df \qquad [2.39]$$

Equation [2.39] translates the energy conservation when moving from the time domain into the frequency domain. This relation is identical to the Parseval equality for Fourier series.

*Parseval theorem*: For non-periodic signals, the FT preserves the energy.

## 2.5. Fourier series and FT

Let $s(t)$ a periodic signal of period $T$; the FT of a periodic signal is not directly computable, however we can take its Fourier series ($\omega_o = 2\pi/T$):

$$s(t) = \sum_{n=-\infty}^{n=+\infty} C_n . g_n(t) = x(t) = \sum_{n=-\infty}^{n=+\infty} C_n . e^{in\omega_o t} \qquad [2.40]$$

In the following, we will assume a periodic signal wherefrom we extract a period to apply the FT. Then, we will compare this FT with the Fourier series of the periodic signal.

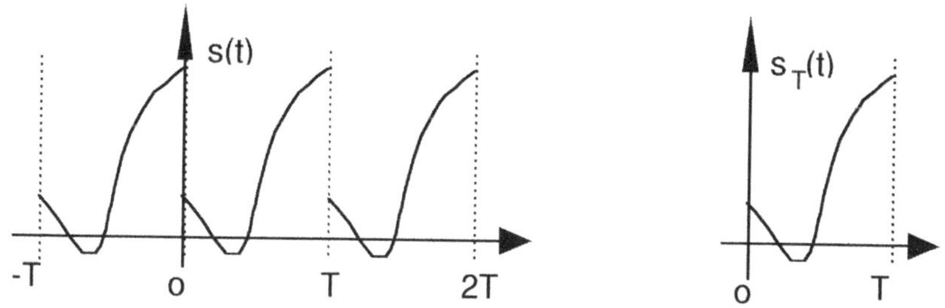


**Figure 2.4.** *Periodized signal*

Denote $s_T(t)$ the truncated signal over a period $T$. $s_T(t) = s(t)$ if $t \in [0, T]$ ; $s_T(t) = 0$ if $t \notin [0, T]$. Using the offset property of Dirac's delta function, we can write:

$$s(t) = s_T(t) * \sum_{n=-\infty}^{n=+\infty} \delta(t - nT) = s_T(t) * \sqcup\!\sqcup_T(t) \qquad [2.41]$$

$S_T(f) = TF(s_T(t))$ exists because $s_T(t)$ is of bounded support and is a continuous function of frequency $f$, (FT of a non-periodic function). Taking the FT in equation [2.41] (using Plancherel–Parseval's formula):

$$S(f) = S_T(f).\frac{1}{T}\sum_{n=-\infty}^{n=+\infty}\delta(f-n/T) \qquad [2.42]$$

The comb $\frac{1}{T}\sqcup\!\sqcup_{\frac{1}{T}}(t) = \frac{1}{T}\sum_{n=-\infty}^{n=+\infty}\delta(f-n\frac{1}{T})$ is the FT of the Dirac comb: $\sqcup\!\sqcup_T(t) = \sum_{n=-\infty}^{n=+\infty}\delta(t-nT)$.

Whence

$$S(f) = \frac{1}{T}\sum_{n=-\infty}^{n=+\infty} S_T(f).\delta(f-\frac{n}{T}) = \frac{1}{T}\sum_{n=-\infty}^{n=+\infty} S_T(\frac{n}{T}) \qquad [2.43]$$

This amount is non-zero only for $f = \frac{n}{T}$ because $\delta(f - \frac{n}{T})$ is zero everywhere except at frequency points $\frac{n}{T}$. However, since

$$Cn =< s(t), gn(t) >= \frac{1}{T}\int_0^T s(t).g_n^*(t).dt = \frac{1}{T}\int_0^T s(t).e^{-2\pi.i.n.t/T}.dt \qquad [2.44]$$

and

$$S_T(f) = \frac{1}{T}\int_0^T s(t).e^{-2\pi.i.f.t}dt \qquad [2.45]$$

In addition, $s(t)$ and $s_T(t)$ coincide between 0 and $T$, we thus obtain for $f_o = \frac{1}{T}$:

$$T.C_n = S_T(\frac{n}{T}) \qquad [2.46]$$

$$S(f) = \frac{1}{T}.S_T(f).\sum_{n=-\infty}^{+\infty}\delta(f-n/T) = \sum_{n=-\infty}^{+\infty} C_n.\delta(f-n/T) \qquad [2.47]$$

The periodic signal spectrum is discrete (evaluated at the multiple frequencies of $\frac{1}{T}$). This is the sampling of the continuous spectrum of the truncated signal $s_T(t)$ up to the constant. This property will be used for the numerical calculation of the FT: the DFT.

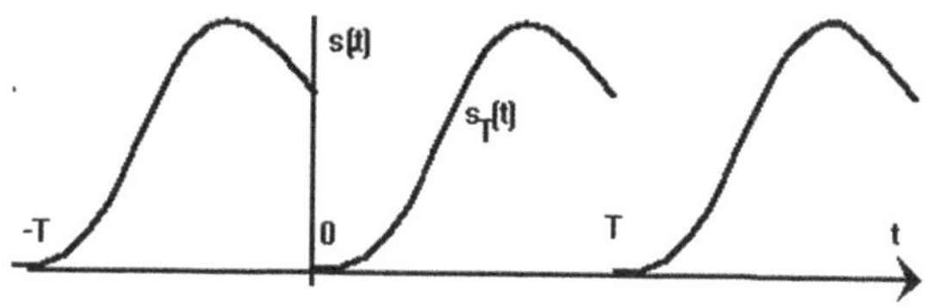


**Figure 2.5.** *Periodic signal* $s(t) = s_T(t) * \sqcup\!\sqcup_T(t)$

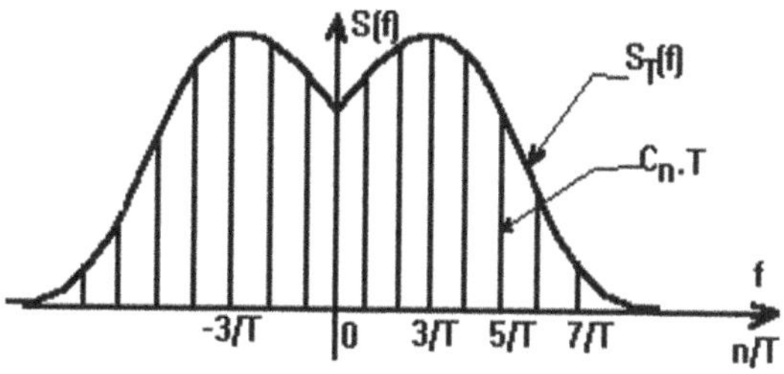


**Figure 2.6.** *Line spectrum at* $f_n = n/T$

## 2.6. Elementary signals and their transforms

A sinc function refers to a function denoted $\mathrm{sinc}(.)$ defined by $\mathrm{sinc}(x) = \sin(x)/x$.

The sine integral function is also defined by $\int_0^x (\sin(y)/y)dy$. This function is equal to $\pi/2$ for $x = \infty$.

1) Sinusoidal signal: $x(t) = a.\cos(2\pi f_o.t + \varphi)$ ($a = 1$ and $\varphi = 0$). Its spectral representation is: $X(f) = \frac{1}{2}(\delta(f - fo) + \delta(f - fo))$;

2) Gate or rectangular signal, $x(t) = \pi_\tau(t) = 1$ if $|t| \leq \tau/2$ and 0 otherwise. Its spectral representation is: $X(f) = sin(\pi f.t)/\pi.f$;

3) Periodized gate of period $T$, $\pi_{p\tau}(t) = 1$ if $|t| \bmod(T) \leq \tau/2$ and 0 otherwise; $\pi_{p\tau}(t)$ is a periodic function, its FT is:

$$TF(\pi_{p\tau}(t)) = \sum_{n=-\infty}^{+\infty} C_n \delta(f - nf_o) \text{ with } f_o = 1/T \qquad [2.48]$$

and the coefficients of the Fourier series: $C_n = \int_0^T \pi_\tau(t).e^{-2\pi.i.n.f_o.t}dt$. A line spectrum (discontinuous) is formed by distributions of masses $C_n$ (by convention each line can be drawn giving it a height $|C_n|$) amplitude-modulated by a sine function;

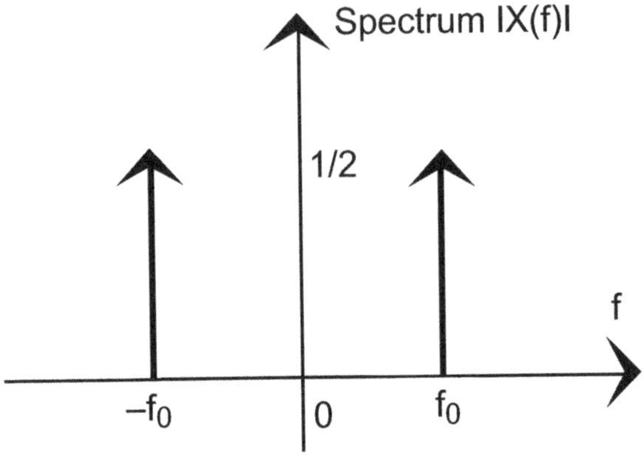


**Figure 2.7.** *Spectrum of a sinusoid*

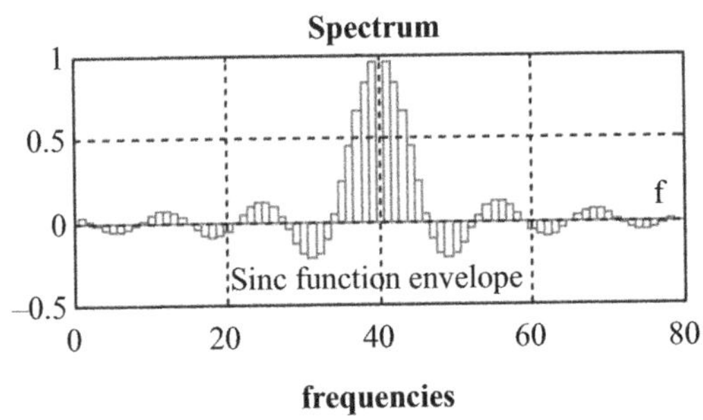


**Figure 2.8.** *Sinc function spectrum*

4) The so-called sample signal : $x(t) = \sin(2\pi f_o t)/\pi.t$. Its usefulness: theory of antennas, sampling, etc. $X(f) = \pi_{f_o}(f)$ carries in frequency;

5) The FT of a Dirac comb in time $\sqcup\!\sqcup_T(t) = \sum\limits_{n=-\infty}^{n=+\infty} \delta(t - nT)$ is a Dirac comb at frequency $\frac{1}{T}\sqcup\!\sqcup_T(t) = \frac{1}{T} \sum\limits_{n=-\infty}^{n=+\infty} \delta(f - n\frac{1}{T})$.

| Table of common Fourier transforms | |
|---|---|
| Time domain | Frequency domain |
| $\delta(t)$ | $1$ |
| $1$ | $\delta(f)$; delta function |
| $1/t$ | $-i\pi sgn(f)$; $sgn(f) = 1$ if $f > 0$, $-1$ if $f < 0$ |
| $\|t\|$ | $-1/(2(\pi f)^2)$ |
| $\exp(-\pi t^2)$ | $e^{-(\pi f)^2}$ |
| $\exp(-at^2)$ with $a > 0$ | $\sqrt{\frac{\pi}{a}}\exp(-(\frac{\pi^2}{a})f^2)$ |
| $e^{-a\|t\|}$ | $2a/(a^2+4\pi^2 f^2)$ |
| $e^{-at}.u(t)$ ; $u(t)$ the step function | $1/(a+2i\pi f)$ |
| $e^{-at}.\sin(2\pi f_o t)u(t)$ ; $u(t)$ the step | $\frac{2\pi f_o}{(a+2i\pi f)^2+(2\pi f_o)^2}$ |
| $e^{-at}.\cos(2\pi f_o t)u(t)$ ; $u(t)$ the step | $\frac{a+2i\pi f}{(a+2i\pi f)^2+(2\pi fo)^2}$ |
| $e^{\pm 2i\pi f_o t}$ | $\delta(f \pm fo)$ |
| $\cos(2\pi f_o t + \varphi)$ | $\frac{1}{2}(e^{i\varphi}\delta(f-fo)+e^{-i\varphi}\delta(f+fo))$ |
| $\text{sign}(t)$ | $1/(i\pi f)$ |
| $(-2i\pi t)^m$ | $\delta^{(m)}(f)$ |
| $\delta^{(m)}(t)$ | $(-2i\pi f)^m$ |
| $\delta(t \pm T)$ | $e^{\pm 2i\pi Tf}$ |
| $\sum_{n=-\infty}^{+\infty} \delta(t-nT)$ | $\frac{1}{T}\sum_{n=-\infty}^{+\infty} \delta(f-n/T)$ |
| $\sum_{n=-\infty}^{+\infty} C_n\delta(t-nT)$ | $\sum_{n=-\infty}^{+\infty} C_n e^{-2i\pi nfT}$ |
| $\sum_{n=-\infty}^{+\infty} C_n e^{2i\pi n f_o t}$ | $\sum_{n=-\infty}^{+\infty} C_n\delta(f-nf_o)$ |
| Centered gate: $\prod_T(t) = u(t+\frac{T}{2}) - u(t-\frac{T}{2})$ | $\frac{T.sin(\pi fT)}{\pi fT} = T\text{sinc}(\pi fT)$ |
| Triangular gate $1-\frac{\|t\|}{T}$ si $\|t\| < T$ and 0 otherwise | $T.(\frac{T.sin(\pi fT)}{\pi fT})^2 = T\text{sinc}^2(fT)$ |
| $f_o.\sin(\pi f_o t)/(f_o t)$ | $u(f+f_o/2)-u(f-f_o/2)$ |
| $y(t) = x^*(t)$ | $Y(f) = X^*(-f)$ |
| $y(t) = x(-t)$ | $Y(f) = X(-f)$ |
| $y(t) = x(t).\cos(2\pi f_o t)$ modulation | $Y(f) = (X(f+f_o)+X(f-f_o))/2$ Offset |
| $y(t) = X(t)$ | $Y(f) = x(-f)$ |
| $\sqcup\sqcup_T(t) = \sum_{n=-\infty}^{n=+\infty} \delta(t-nT)$; Dirac comb | $\frac{1}{T}\sqcup\sqcup_{\frac{1}{T}}(t) = \frac{1}{T}\sum_{n=-\infty}^{n=+\infty}\delta(f-n\frac{1}{T})$ |
| $\text{sign}(t) = t/\|t\|$ | $1/i\pi f$ |
| $u(t)$ unit step | $(\delta(f)+1/i\pi f)/2$ |
| $\prod_T(t).\cos(2\pi f_o t)$ | $(T/2)(\text{sinc}(\pi(f-fo)T)+\text{sinc}(\pi(f+fo)T))$ |

**Table 2.2.** *Common Fourier transforms*

## 2.7. Laplace transform

### 2.7.1. *Definition*

The bilateral LT of any signal $x(t)$ is given by:

$$X(p) = L[x(t)] = \int_{-\infty}^{+\infty} x(t)e^{-pt}dt \quad [2.49]$$

where $p$ is a complex variable written as $p = \sigma + j\omega$.

The LT of a causal signal $x(t)$ is defined by:

$$X(p) = L[x(t)] = \int_{0}^{\infty} x(t)e^{-pt}dt$$

A signal $x(t)$ is causal if it verifies:

$$x(t) = \begin{cases} x(t) & \text{if } t \geq 0 \\ 0 & \text{if } t < 0 \end{cases} \quad [2.50]$$

The IFT is given by:

$$x(t) = TL^{-1}[X(p)] = \frac{1}{2\pi j}\int_{\sigma-j\infty}^{\sigma+j\infty} X(p)e^{pt}dp \quad [2.51]$$

with $\sigma$ being real and called the convergence abscissa, it is chosen at a larger scale than the real part of all singular points of $X(p)$.

The LT can be considered a generalization of the FT for non-summable functions at the absolute value. To this end, functions $f(t)e^{-\sigma t}$ are introduced that enable the definition of a summable function based on a function that may not be summable, by an appropriate choice of $\sigma$, if it exists. By defining a relation for the transition from the time domain into the frequency domain, this transform will thus allow, for these functions, the simplification of the study of characteristics, the solution of differential equations and the study of systems and filters.

DEFINITION 2.1.– *If $f(t)$ is a time function of the same order of magnitude as is from $t = t_o$, we could then write $f(t).e^{-\sigma t} \leq M$ and $f(t)$ has an LT. Let $p = \sigma + i.\omega$, we define the LT by:*

$$L(f(t)) = F(p) = \int_{-\infty}^{+\infty} f(t)e^{-pt}dt \textit{ Bilateral Laplace transform} \quad [2.52]$$

$$L(f(t)) = F(p) = \int_{0}^{+\infty} f(t)e^{-pt}dt \textit{ Unilateral Laplace transform} \quad [2.53]$$

$$L(f(t)) = F(p) = F(\sigma + i.\omega) = \int_{0}^{+\infty} f(t)e^{-\sigma t}e^{-i.\omega t}dt \quad [2.54]$$

$\int_0^{+\infty} |f(t)| e^{-\sigma t}dt$ converge in a domain $D$, defined by a limit $\sigma_o$ for the real part of $p$ and here denoted as $\sigma_o$.

The unilateral transform (or monolateral) is of particular interest to us for causal systems and signals. It holds only for $Re(p) = \sigma > \sigma_o$ where the Laplace integral converges. This means that poles or singularities or singular points $F(p)$ lie, in the complex plane $p$, to the left of the axis $\sigma = \sigma_o$ (parallel to the imaginary axis). The convergence domain is thus defined ($D$ the left half plane). $F(p)$ is a complex function of a complex, analytical and uniform (holomorphic) variable in the half plane on the right of the axis $\sigma = \sigma_o$ (convergence domain of the LT); all its derivatives, with respect to $p$, exist and are continuous in this domain.

LT of a periodic function: let $f(t)$ be a periodic function of period $T$, $f(t) = f(t + nT)$ for any $n$. It can be written that: $f(t) = f_T(t) * \sum_{n=-\infty}^{n=+\infty} \delta(t - nT)$, knowing further that $L(\delta(t - nT)) = e^{-pTn}$ and that a convolution product is transformed into a simple product, we can write the LT of a periodic function:

$$L(f(t)) = F(p) = \frac{1}{1 - e^{-pT}} \int_{0}^{T} f(t).e^{-p.t}.dt \quad [2.55]$$

We also note that by imposing $\sigma = 0$ (if the imaginary axis belongs to the convergence domain), in the definition of the LT, we get that of the FT. Thus, the LT $F(p)$ can be considered as the FT of the causal function $f(t).e^{-\sigma t}$, and by IFT, it yields:

$$f(t).e^{-\sigma t} = \int_{-\infty}^{+\infty} F(\sigma + i.\omega).e^{i.\omega.t}.d\omega \quad [2.56]$$

REMARK 2.4.– *The impulse response $h(t)$ has an LT $H(p)$ referred to as transfer function of the system. If the system is stable, we then get $\int_{-\infty}^{+\infty} |h(t)|\, dt < \infty$ and consequently the poles of $H(p)$ have a negative real part (located in the half plane on the left of $\sigma = 0$).*

The inverse LT is defined on an axis in the convergence domain $\sigma > 0$ by the inversion formula (Mellin–Fourier integral):

$$L^{-1}(F(p)) = f(t) = \frac{1}{2i\pi} \int_{\sigma-i\infty}^{\sigma+i\infty} F(p).e^{p.t}.dp \qquad [2.57]$$

$$= \frac{1}{2i\pi} \int_{\sigma-i\infty}^{\sigma+i\infty} F(\sigma + i.\omega).e^{i.\omega.t}e^{\sigma t}.d\omega \qquad [2.58]$$

For $f(t)$ of the order of $e^{\alpha t}$, it is necessary that $\sigma > \alpha$. Observe that if $\alpha < 0$, we can chose $\sigma = 0$.

RESIDUALS.– $R$ is a residual of $G(p)$ with respect to a pole, the integral $R = \frac{1}{2i\pi} \oint_\gamma G(p).dp$, taken along a closed contour $\gamma$, of the complex plane, surrounding the considered pole.

THEOREM 2.2.– Given a function $G(p)$ uniform in a region $D$ where it has its poles $(p_i)$ and holomorphic on a closed contour $\gamma$ included in this region $D$. The integral of the complex function $G(p)$ along the contour $\gamma$ taken in the direct sense is equal to $2\pi i$ times the sum of the residuals of $G(p)$ relatively to the poles $p_i$ contained within the contour:

$$2i\pi\text{Residuals}(p_i) = \frac{1}{2i\pi} \oint_\gamma G(p).dp \qquad [2.59]$$

For the computation of the inverse LT ($G(p) = F(p).e^{p.t}$), we can consider the contour formed by the axis $\sigma$ parallel to the imaginary axis and a half circle of infinitely large radius. This contour encompasses all the poles of $F(p)$. If $F(p)$ tends toward 0 when $p$ tends to infinity, we then get:

$f(t) = 2i\pi \sum \text{Residual}(p_i)$;

– if $p_i$ is a simple pole, then

$$\text{Residual}(p_i) = \left((p - p_i)F(p).e^{p.t}\right)_{p=p_i} \qquad [2.60]$$

– if $p_i$ is a multiple pole of order $m$, then

$$\text{Residual}(p_i) = \frac{1}{(m-1)!} \frac{d^{m-1}}{dp^{m-1}} \{(p - p_i)^m F(p).e^{p.t}\}_{p=p_i} \qquad [2.61]$$

Generally, the residual method is used only very rarely for the calculation of inverse transforms. Practical methods are based on the use of tables of transforms for simple functions. They will be presented in the following sections.

### 2.7.2. *Properties*

The advantage of the LT lies in these properties because they make it possible to simplify the solving of differential equations. After presenting three important theorems, we will summarize the essential properties of LT (unilateral) and FT in Table 2.3.

The FT has the following properties:

– Linearity:

$$L[ax(t) + by(t)] = aX(p) + bY(p) \quad [2.62]$$

– Convolution:

$$L[x(t) * y(t)] = X(p).Y(p) \quad [2.63]$$

$$L[x(t).y(t)] = X(p) * Y(p)$$

– Derivation:

$$L[\frac{d^n x(t)}{dt^n}] = p^n X(p) - p^{n-1}x(0^+) - p^{n-2}\frac{dx(0^+)}{dt} \quad [2.64]$$

$$-p^{n-3}\frac{d^2x(0^+)}{dt^2} - \cdots - \frac{d^{n-1}x(0^+)}{dt^2} \quad [2.65]$$

– Integration:

$$L[\int_0^t x(\tau)d\tau] = \frac{X(p)}{p} \quad [2.66]$$

– Change in scales:

$$L[x(at)] = \frac{1}{|a|}X(\frac{p}{a}) \quad [2.67]$$

– Translation in time:

$$L[x(t-\tau)] = e^{-\tau p}X(p) \quad [2.68]$$

– Translation in the $s$ domain (frequency):

$$L[e^{-at}x(t)] = X(s+a) \quad [2.69]$$

THEOREM 2.3.– Initial value theorem:

$$f(0^+) = \lim_{t \to 0} f(t) = \lim_{p \to \infty} p.F(p) \qquad [2.70]$$

THEOREM 2.4.– Final value theorem:

$$f(\infty) = \lim_{t \to \infty} f(t) = \lim_{p \to 0} p.F(p) \qquad [2.71]$$

THEOREM 2.5.– Derivation of a convolution product:

$$\frac{d}{dt}(f(t) * g(t)) = f(t) * \frac{d}{dt}g(t) + f(t).g(0) = g(t) * \frac{d}{dt}f(t) + g(t).f(0) \qquad [2.72]$$

### 2.7.3. *Examples of the use of the unilateral LT*

Process step response:

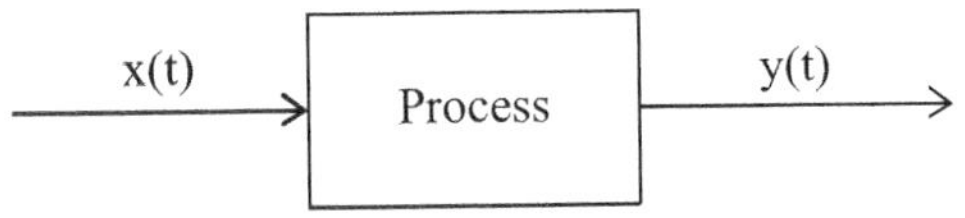


**Figure 2.9.** *Process response*

The differential equation that governs its behavior is given by:

$$\frac{d^2y(t)}{dt^2} + 3\frac{dy(t)}{dt} + 2y(t) = x(t) \qquad \text{with } y(0) = -1, \dot{y}(0) = 2 \qquad [2.73]$$

The input step has a magnitude of 5 at $t = 0$.

$$L[\frac{dy(t)}{dt}] = pY(p) - y(0) \qquad [2.74]$$

$$L[\frac{d^2y(t)}{dt^2}] = p^2Y(p) - py(0) - \dot{y}(0)$$

$$L[x(t)] = \frac{5}{p} \qquad \text{since } y = 5\Gamma(t)$$

The differential equation is transformed into an algebraic equation:

$$p^2Y(p) - py(0) - \dot{y}(0) + 3pY(p) - 3y(0) + 2Y(p) = \frac{5}{p} \qquad [2.75]$$

| Functions | $f(t), g(t)$ | $TL : F(p), G(p)$ | $TF; F(f), G(f)$ |
|---|---|---|---|
| Linearity | $a.f(t) + b.g(t)$ | $a.F(p) + b.G(p)$ | $a.F(f) + b.G(f)$ |
| Derivation/t | $\frac{d}{dt}f(t)$ | $p.F(p) - f(0^+)$ | $2i\pi f.F(f)$ |
| n derivation/t | $(\frac{d}{dt})^n f(t)$ | $pF(p) - p^{n-1}f(0) - ...$ $... - pf^{(n-2)}(0) - f^{(n-1)}(0)$ | |
| Derivation/p | $t.f(t)$ | $-\frac{d}{dp}F(p)$ | $-\frac{d}{df}F(f)/2i\pi$ |
| | $t^m.f(t)$ | $(-1)^m \frac{d^m}{dp^m}F(p)$ | $\frac{(-1)^m \frac{d^m}{df^m}F(f)}{(2i\pi)^m}$ |
| Integration/p | $f(t)/t$ | $\int_p^\infty F(u)du$ | |
| Integration/t | $\int_0^t f(u).du$ | $\frac{1}{p}F(p) + \lim_{t\to 0}\int_0^t f(u)du$ | |
| Delay | $f(t - \tau_o)$ | $e^{-p\tau_o}.F(p)$ | $e^{-2\pi.i.f\tau_o}.F(f)$ |
| Offset in f | $e^{a.t}..f(t)$ | $F(p - a)$ if $a = 2i\pi f_o$ | $F(f - f_o)$ |
| Change in scale | $f(at)$ | $\frac{1}{\lvert a\rvert}F(p/a)$ for $Re(p) > \sigma_o a$ | $\frac{1}{\lvert a\rvert}F(f/a);$ |
| Borel theorem | $f(t) * g(t)$ | $F(p).G(p)$ | $F(f).G(f)$ |
| Product* | $f(t).g(t)$ | $F(p) * G(p)$ | $F(f) * G(f)$ |
| Dirac | $\delta(t)$ | $1$ | |
| Unit step function | $u(t)$ | $1/p$ | |
| Ramp | $t.u(t)$ | $1/p^2$ | |
| | $\frac{u(t).t^{n-1}}{(n-1)!}$ | $1/p^n$ | |
| Exponential | $e^{-at}.u(t)$ | $1/(p + a)$ | |
| | $t.e^{-at}.u(t)$ | $1/(p + a)^2$ | |
| Sinusoids | $\cos(\omega t).u(t)$ | $p/(p^2 + \omega^2)$ | |
| | $\sin(\omega t).u(t)$ | $\omega/(p^2 + \omega^2)$ | |
| Sinus amories | $e^{-at}\cos(\omega_t + \varphi).u(t)$ | $\frac{(p+a)\cos(\varphi)-\omega\cos(\varphi)}{(p+a)^2+\omega^2}$ | |
| | $e^{-at}\sin(\omega t + \varphi).u(t)$ | $\frac{\omega\cos(\varphi)-(p+a)\sin(\varphi)}{(p+a)^2+\omega^2}$ | |
| | $sh(at)$ | $a/(p^2 - a^2)$ | |
| | $ch(at)$ | $p/(p^2 - a^2)$ | |
| Gate | $u(t - a) - u(t - b)$ | $\frac{e^{-ap}-e^{-bp}}{p}$ | |
| | $(t + \tau)u(t)$ | $\frac{1}{p^2} + \frac{\tau}{p}$ | |
| | $(t - \tau)u(t - \tau)$ | $\frac{1}{p^2}e^{-\tau p}$ | |

**Table 2.3.** *A few properties of the unilateral Laplace transform*

Whence we have:

$$Y(p) = \frac{-p^2 - p + 5}{p(p^2 + 3p + 2)} = \frac{-p^2 - p + 5}{p(p + 1)(p + 2)} \qquad [2.76]$$

To calculate $y(t)$, it is necessary to decompose $Y(p)$ into simple components and perform an inverse LT:

$$Y(p) = \frac{5}{2}\frac{1}{p} - \frac{5}{p + 1} + \frac{3}{2}\frac{1}{p + 2} \qquad [2.77]$$

which yields:

$$y(t) = (\frac{5}{2} - 5e^{-t} + \frac{3}{2}e^{-2t})\Gamma(t) \qquad [2.78]$$

| $X(s)$ | $x(t)$ |
|---|---|
| $1$ | $\delta(t)$ |
| $\frac{1}{p}$ | $u(t)$ |
| $\frac{1}{1-e^{-Tp}}$ | $\sum_{n=0}^{\infty}\delta(t-nT)$ |
| $\frac{1}{p^2}$ | $tu(t)$ |
| $\frac{1}{p^3}$ | $t^2u(t)$ |
| $\frac{m!}{p^{m+1}}$ | $t^m u(t)$ |

| $X(s)$ | $x(t)$ |
|---|---|
| $\frac{1}{p+a}$ | $e^{-at}u(t)$ |
| $\frac{1}{(p+a)^2}$ | $te^{-at}u(t)$ |
| $\frac{1}{p(p+a)}$ | $(1-e^{-at})u(t)$ |
| $\frac{\omega}{p^2+\omega^2}$ | $\sin(\omega t)u(t)$ |
| $\frac{\omega}{(p+a)^2+\omega^2}$ | $e^{-at}\sin(\omega t)u(t)$ |
| $\frac{p+a}{(p+a)^2+\omega^2}$ | $e^{-at}\cos(\omega t)u(t)$ |

**Table 2.4.** *Inverse Laplace transforms table*

The final value theorem can be verified as:

$$\lim_{t\to\infty} y(t) = \lim_{s\to 0} pY(p) = \lim_{p\to 0}\frac{-p^2-p+5}{(p+1)(p+2)} = \frac{5}{2} \qquad [2.79]$$

REMARK 2.5.– *The position of the poles of the LT of a signal gives information about the transient response.*

### 2.7.4. *Transfer function*

The transfer function of a system, represented by Figure 2.10, with input $x(t)$ and output $y(t)$ is the ratio of LTs as output and input ($H(p) = \frac{Y(p)}{X(p)}$), when the initial conditions are zero.

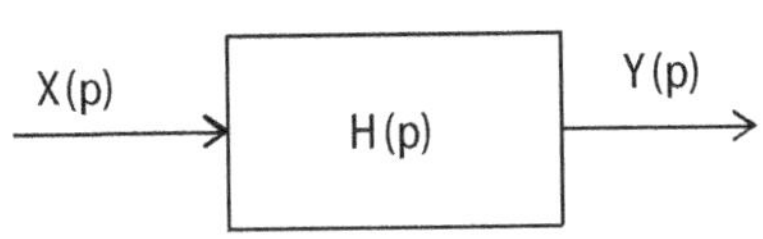


**Figure 2.10.** *Transfer function of a process*

When reconsidering the system of the previous example, we can write the transfer function of the process as:

$$H(p) = \frac{Y(p)}{X(p)} = \frac{1}{p^2+3p+2} \qquad [2.80]$$

REMARK 2.6.– *Applying an impulse to the system in the form of a Dirac peak:* $x(t) = \delta(t)$, *we will thus have* $X(p) = 1$. *Then,* $Y(p) = H(p)$. *The transfer function is therefore the* LT *of the impulse response with zero initial conditions and the poles of the transfer function* $H(p)$ *determine the free response of the system.*

*A condition for the free response of a system to tend toward zero when t tends to infinity (asymptotic stability of the system) is that the poles of the transfer function* $H(p)$ *have all a strictly negative real part.*

*The general form of a transfer function of a system is given by:*

$$H(p) = K \frac{\prod_{j=0}^{j=m} (p + z_j)}{\prod_{k=0}^{k=n} (p + p_k)} \quad [2.81]$$

*with* $z_j \in \mathbb{C}$ *the zeros of* $H(p)$ *(or the roots of its numerator) and* $p_k \in \mathbb{C}$ *the poles of* $H(p)$ *(or roots of its denominator). For a physically implementable system, we necessarily have* $m < n$.

## 2.8. FT and LT

Let $x(t)$ be a continuous signal:

– Its FT is given by:

$$X(f) = \int_{-\infty}^{+\infty} x(t) e^{-j2\pi f t} dt \quad [2.82]$$

– Its LT is given by:

$$X(p) = \int_{-\infty}^{+\infty} x(t) e^{-jp} dt \quad [2.83]$$

Comparing both equations, we find:

$$X(f) = X(p)|_{p=j2\pi f} \quad [2.84]$$

The FT is thus obtained through the LT on the imaginary axis.

## 2.9. Application exercises

EXERCISE 1.–

1) Consider the signal $f(t)$ (non-periodic) defined in Figure 2.11.

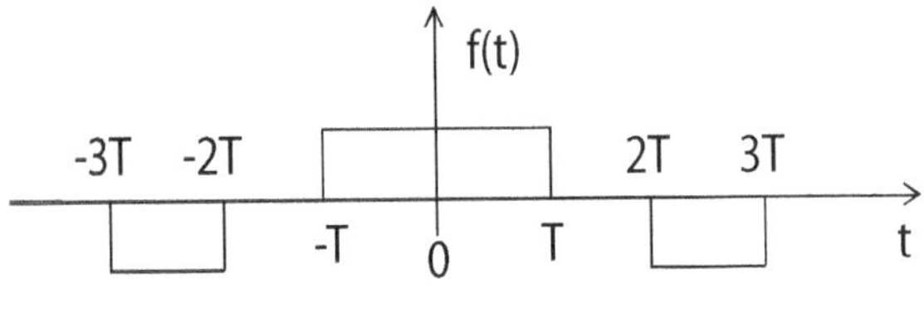


**Figure 2.11.** *Non-periodic signal*

2) a) Give the analytical expression of this signal according to the Heaviside step function $u(t)$.

b) Give the expression for the FT of its derivative with respect to time $df(t)/dt$.

c) Derive the FT of the signal $f(t)$.

3) a) Determine the spectral representation of the following periodic signals (Figure 2.12):

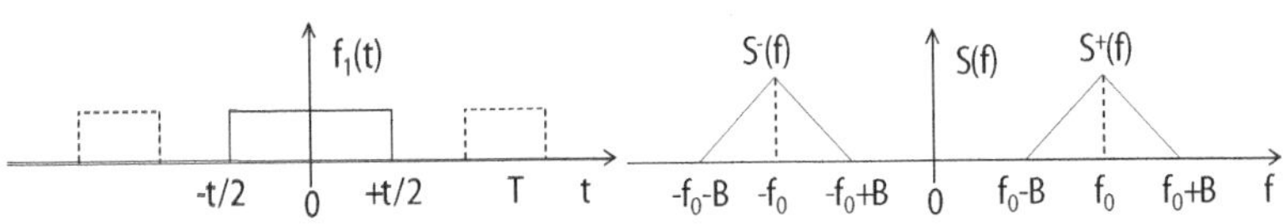


**Figure 2.12.** *Periodic signals*

b) Determine the total power of these signals and represent their average power spectral density knowing that:

$$\sum_{p=0}^{+\infty} \frac{1}{(2p+1)^2} = \frac{\pi^2}{8} \qquad [2.85]$$

c) It is assumed that $f_1(t)$ represents, for $\tau = \frac{T}{2}$, the input signal of an ideal low-pass filter.

4) Determine the bandwidth of the filter in order to retrieve an output of 98% of the power of the input signal $f_1(t)$.

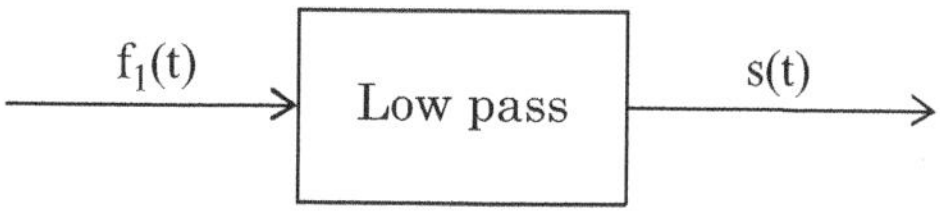


**Figure 2.13.** *Low-pass filter*

EXERCISE 2.–

1) Calculate the FTs of the signals:

$$\begin{array}{ll} \delta(t); & ⊔⊔_T(t) = \sum_{n=-\infty}^{+\infty} \delta(t-nT);\ \cos(2\pi\nu_0 t) \\ \delta(t-T_0); & g_1(t) = e^{-a|t|};\ \text{with } a>0 \\ \delta^{(n)}(t); & g_2(t) = e^{-at^2};\ \text{with } a>0 \end{array} \quad [2.86]$$

2) Calculate the IFT of:

$$X(\nu) = j\frac{[\delta(\nu+\nu_0) - \delta(\nu-\nu_0)]}{2} \quad [2.87]$$

3) Given the signal $x(t) = A\sin(2\pi at) + B\sin(2\pi bt)$ and $m_T(t) = (1/T)\pi_{T/2}(t)$:

a) Calculate $y(t) = x(t) * m_T(t)$

b) Assuming $0 < a < b$, how to chose $T$ and $A'$ to get: $y(t) = A'\sin(2\pi at)$

4) Let signal $u(t)$ defined by $u(t) = \sum_{k=-\infty}^{+\infty} v(t-kT)$ where $v(t)$ is defined by:

$$v(t) = \begin{cases} (2/T)(1-2|t|/T)\ if\ |t| < t/2 \\ 0 \qquad\qquad \text{otherwise} \end{cases} \quad [2.88]$$

5) a) Directly calculate the Fourier series coefficients of $u(t)$ explaining why this series does exist.

b) Find this result using the FT of $v(t)$.

EXERCISE 3.–

1) Give the Fourier series decomposition of the Dirac comb: $⊔⊔_T(t) = \sum_{n=-\infty}^{n=+\infty} \delta(t-nT)$

2) Calculate the FT of the gate function $\pi_\tau(t)$, which is equal to 1 in the interval $[-\tau/2, \tau/2]$ and to 0 elsewhere.

3) Derive the FTs of the following signals: $x(t) = \sin(2\pi f_o.t)/(\pi.f_o.t)$ and the triangular gate represented opposite.

4) Based on this reasoning, calculate $\int_{-\infty}^{+\infty} \sin(x)/x.dx$

5) Using previous results, find the spectral representation of the signals of Exercise 2.1.

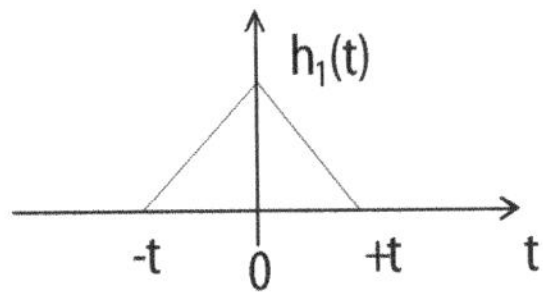


**Figure 2.14.** *Triangular gate*

EXERCISE 4.–

Model represented by the convolution product: Consider two systems with an input $x(t)$; $y(t)$ and $z(t)$ are the respective outputs of the two systems defined by:

$$y(t) = \int_{t-T}^{t} x(r).\text{dretz}(t) = \int_{t-T/2}^{t+T/2} x(r).dr \qquad [2.89]$$

i) Physically interpret these models.

ii) Show whether these systems are linear, causal, stationary and give their impulse response.

iii) Give the expression of their convolution product. Derive the frequency response (modulo and phase).

iv) To what physical systems can they correspond?

EXERCISE 5.– Laplace transform

1) Using the unilateral LT, solve the following differential equations: $\ddot{s}(t) = f(\dot{s}(t), s(t)) = -s(t) + 8.\sin(3t)$ with $s(0) = 1$ and $\dot{s}(0) = 0.5$

$$t.e^{-2t} = s(t) + \int_0^t e^{-3(t-\tau)} s(\tau)d\tau \qquad [2.90]$$

$$\ddot{s}(t) + \omega_0.s(t) = \sum_{k=0}^{+\infty} \frac{2}{\pi(2k+1)}.\sin((2k+1)2\pi t) \qquad [2.91]$$

2) Calculate the original functions of the following functions of the complex variable $p$ (obtained using the unilateral LT, with $0 < \xi < 1$): $H_1(p) = \frac{1}{p^2(p+1)}$; $H_2(p) = \frac{1}{p^2+2\xi\omega_0 p+\omega_0^2}$; $H_3(p) = \frac{p^2+2p+2}{p^2+3p+2}$; $H_4(p) = \frac{p(p+1)}{(p+2)^2}$; $H_5(p) = \frac{2.e^{-0.5p}}{p^2-6p+13} - \frac{p-1}{p^2-2p+2}$; $H_6(p) = \frac{p}{(p+1)(p+2)}$

EXERCISE 6.– The signal $f(t)$, below, is of the form $e(t) = E_o \sin\omega t$ in $[t_o, T_o]$ (for $T_o > t \geq t_o$) and elsewhere equals zero.

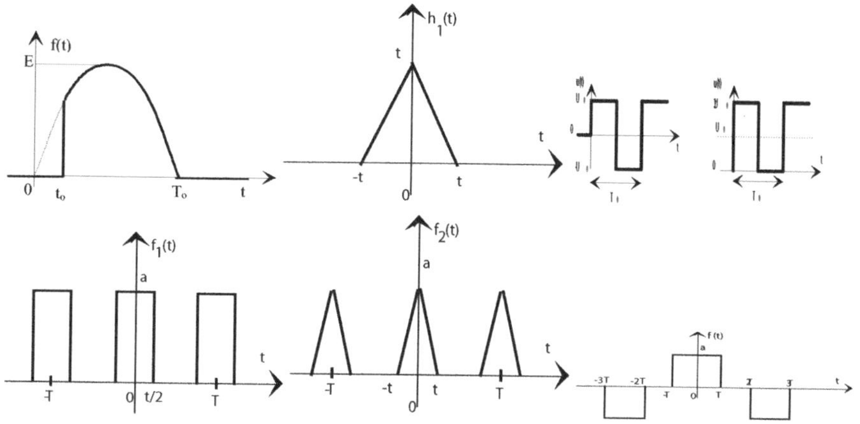

**Figure 2.15.** *Time signals*

a) Give the analytical expression of these signals with respect to time to the Heaviside step function $u(t)$. Determine the LTs of these signals. Deduce the spectral (or frequency in amplitude and phase) representation of these signals (assuming that $\tau = T/2$ and $a = 1.5$). Compare their ranges (supports) in the time and frequency domains.

b) Give the expression for the LT of the derivatives with respect to time $df(t)/dt$ of these signals.

# 3

# Spectral Study of Signals

This chapter explores correlation, weighting and the Wiener Khintchine theorem.

The focus is on the use of the tools that we have defined for the analysis of signals and systems and on a physical interpretation of the properties of time–frequency transforms. In time domain analysis, we will use intercorrelation functions that we recall in the following. Frequency analysis makes use of time–frequency transforms and frequency representations. Depending on the nature of the signal (finite energy, periodic or average finite power), we will have to choose the appropriate definitions. In the case of infinite duration signals, it is always possible to use a weighting by means of windows (apodization) for time or frequency analysis.

## 3.1. Power and signals energy

### 3.1.1. *Power and energy of random signals*

Consider a signal $x(t)$ (real or complex: $x(t) = a(t) + ib(t)$). For this signal, the following energies and powers can be defined:

– Instantaneous power:

$$P_i(t) = x(t).x(t)^* = |a(t)|^2 + |b(t)|^2 = |x(t)|^2 \quad [3.1]$$

$$P_i(t) = x(t)^2 \text{ if } x(t) \text{ is real} \quad [3.2]$$

– Instantaneous interaction power: now let $x(t)$ and $y(t)$ be two signals

$$P_{xy}(t) = x(t).y(t)^* = P^*_{yx}(t) \quad [3.3]$$

– Average power over a duration $T$:

$$P_x(t_o, T) = \frac{1}{T}\int_{t_o}^{t_o+T} x(t).x(t)^* dt \qquad [3.4]$$

– Interaction power over a duration $T$:

$$P_{xy}(t_o, T) = \frac{1}{T}\int_{t_o}^{t_o+T} x(t).y(t)^* dt \qquad [3.5]$$

– Interaction energy over a duration $T$:

$$E_{xy}(t_o, T) = \frac{1}{T}\int_{t_o}^{t_o+T} x(t).y(t)^* dt \qquad [3.6]$$

– Energy of a signal over a duration $T$:

$$E_x(t_o, T) = \frac{1}{T}\int_{t_o}^{t_o+T} x(t).x(t)^* dt = \frac{1}{T}\int_{t_o}^{t_o+T} |x(t)|^2 \cdot dt \qquad [3.7]$$

NOTES.–

– If the energy thus defined is not bounded (case of periodic signals), the notion of average power is used. The limit of the expression will be chosen in the case of non-bounded intervals (if $T$ is infinite or over a period for periodic signals):

$$P_{xy} = \frac{E_{xy}}{T} \quad or \quad P_{xy} = \lim_{T \longrightarrow \infty} \frac{E_{xy}}{T} \qquad [3.8]$$

– In the case where at least one of the two signals has bounded support (i.e. is zero outside an interval $[t_o, t_o + T]$), it is no longer possible to define any average power in $]-\infty, +\infty[$ because it would tend toward zero. We then use the notion of energy.

– Cauchy–Schwartz inequality:

$$|E_{xy}|^2 \leq E_x.E_y \qquad [3.9]$$

– For two sinusoidal signals with distinct frequencies, the interaction energy is zero and the supports of the spectra are disjoint. This property will be extended to the concept of cross-spectrum.

THEOREM 3.1 (Parseval's theorem).– Consider that $X(f)$ and $Y(f)$ are the FT of two signals $x(t)$ and $y(t)$, we then have:

$$\int_{-\infty}^{+\infty} x(t) \cdot y(t)^* \cdot dt = \int_{-\infty}^{+\infty} X(f) \cdot Y^*(f) \cdot df \qquad [3.10]$$

The interaction energy of two signals is the same whether considering the time domain or the frequency domain. In particular, if $x(t) = y(t)$, we get:

$$\int_{-\infty}^{+\infty} x(t) \cdot x(t)^* \cdot dt = \int_{-\infty}^{+\infty} X(f) \cdot X^*(f) \cdot df = \int_{-\infty}^{+\infty} |x(t)|^2 \cdot dt$$

$$= \int_{-\infty}^{+\infty} |X(f)|^2 \cdot df \qquad [3.11]$$

$|X(f)|^2$ can be interpreted as energy per frequency interval $df$, which is also known as *energy spectral density*.

## 3.2. Autocorrelation and intercorrelation

### 3.2.1. *Autocorrelation and cross-correlation in the time domain*

For the study of signals in the time domain, we can use various functions of the energy or average power type. These quantities can be defined using different methods but bear in mind that zero or infinite quantities cannot be used; then according to the nature of the signals (energy or finite average power, periodical or not), we will choose among the following definitions.

#### 3.2.1.1. *Non-periodic finite-energy signals*

Given two non-periodic finite-energy signals $x(t)$ and $y(t)$, their correlation is defined by

$$C_{xy}(\tau) = \int_{-\infty}^{+\infty} x(t)y(t-\tau)^* \cdot dt \qquad [3.12]$$

It is the interaction energy of $x(t)$ and $y(t-\tau)$ and their autocorrelation defined by

$$C_{xx}(\tau) = \int_{-\infty}^{+\infty} x(t)x(t-\tau)^* \cdot dt \quad \text{and} \qquad [3.13]$$

$$Cyy(\tau) = \int_{-\infty}^{+\infty} y(t)y(t-\tau)^* \cdot dt \qquad [3.14]$$

#### 3.2.1.2. *Non-periodic finite average power signals*

Let us take two non-periodic finite average power signals (infinite energy) $x(t)$ and $y(t)$, and their intercorrelation is defined by

$$C_{xy}(\tau) =)\frac{1}{T}\int_{t_o}^{t_o+T} x(t)y(t-\tau)^* \cdot dt \qquad [3.15]$$

This is the interaction average power of $x(t)$ and $y(t-\tau)$ and their autocorrelation is defined similarly by

$$C_{xx}(\tau) = \frac{1}{T}\int_{t_o}^{t_o+T} x(t)x(t-\tau)^* \cdot dt \qquad [3.16]$$

$$\text{and } C_{yy}(\tau) =)\frac{1}{T}\int_{t_o}^{t_o+T} y(t)y(t-\tau)^* \cdot dt \qquad [3.17]$$

#### 3.2.1.3. *Periodical signals*

Let us take two periodic signals $x(t)$ and $y(t)$ of period $T$ (same period, with average finite power and infinite energy), their intercorrelation is defined by averages of powers over a period $T$ of signals:

$$C_{xy}(\tau) = \frac{1}{T}\int_{t_o}^{t_o+T} x(t)y(t-\tau)^* \cdot dt \qquad [3.18]$$

This is the average power per interaction period of $x(t)$ and $y(t-\tau)$ and their autocorrelation is defined by

$$C_{xx}(\tau) = \frac{1}{T}\int_{t_o}^{t_o+T} x(t)x(t-\tau)^* \cdot dt$$

$$\text{and} \quad C_{yy}(\tau) = \frac{1}{T}\int_{t_o}^{t_o+T} y(t)y(t-\tau)^* \cdot dt \qquad [3.19]$$

#### 3.2.1.4. *Observation*

The correlation function can be expressed using the expression of the Fourier series:

$C_{xy}(\tau) = \frac{1}{T}\int_{t_o}^{t_o+T} x(t)y(t-\tau)^* dt$ knowing that $x(t) = \sum\limits_{n=-\infty}^{+\infty} C_{xn} g_n(t)$ and $y(t) = \sum\limits_{m=-\infty}^{+\infty} C_{ym} g_m(t)$

On the other hand, $C_{xn} =< x(t), g_n(t) >= \frac{1}{T}\int_{t_o}^{t_o+T} x(t).g_n^*(t) \cdot dt$, and $< g_n(t), g_m(t) >= 0$ if $n \neq m$ (orthogonal functions); this yields:

$C_{xy}(\tau) = \frac{1}{T}\int_{t_o}^{t_o+T} \sum\limits_{n=-\infty}^{+\infty} C_{xn} g_n(t))y(t-\tau)^* dt = \frac{1}{T}\sum\limits_{n=-\infty}^{+\infty} C_{xn}\int_{t_o}^{t_o+T}$ $g_n(t)y(t-\tau)^* dt$ (uniform convergence)

$$C_{xy}(\tau) = \frac{1}{T}\sum_{n=-\infty}^{+\infty} C_{xn}\frac{1}{T}\int_{t_o}^{t_o+T} g_n(t)\sum_{m=-\infty}^{+\infty} C_{ym}^* \cdot g_m(t-\tau)^* dt$$

$$= \sum_{n=-\infty}^{+\infty}\sum_{m=-\infty}^{+\infty} C_{xn}C_{ym}^* \cdot \frac{1}{T}\int_{t_o}^{t_o+T} g_n(t)g_m(t-\tau)^* dt; \text{ (orthogonality)}$$

$$C_{xy}(\tau) = \sum_{n=-\infty}^{+\infty} C_{xn}C_{yn}^* \cdot \frac{1}{T}\int_{t_o}^{t_o+T} g_n(t)g_n(t-\tau)^* dt$$

$$= \sum_{n=-\infty}^{+\infty} C_{xn}C_{yn}^* .e^{-2i\pi n\tau/T} \cdot \frac{1}{T}\int_{t_o}^{t_o+T} g_n(t)g_n(t)^* dt$$

$$C_{xy}(\tau) = \sum_{n=-\infty}^{+\infty} C_{xn}C_{yn}^* \cdot e^{-2i\pi n\tau/T}$$

Similarly, it can be shown that $C_{xx}(\tau) = \sum\limits_{n=-\infty}^{+\infty} |C_{xn}|^2 e^{-2i\pi nt/T}$ and $C_{xx}(0) = \sum\limits_{n=-\infty}^{+\infty} |C_{xn}|^2$

### 3.2.1.5. *Properties of the correlation functions*

$$C_{xx}(0) > |C_{xx}(\tau)|; \qquad \forall \tau \in R \qquad [3.20]$$

$$|C_{xy}(\tau)|^2 \leq C_{xx}(0) \cdot Cyy(0) \qquad \forall \tau \in R \qquad [3.21]$$

$$C_{xx}(\tau) = C_{xx}^*(-\tau) \qquad \text{if } C_{xx}(\tau) \text{ is real, then it is even} \qquad [3.22]$$

$$C_{xy}(\tau) = C_{yx}^*(-\tau) \qquad [3.23]$$

If $x(t)$ and $y(t)$ are periodic of period $T$, $C_{xy}$ has the same period.

$$C_{xy}(\tau) = \int_{-\infty}^{+\infty} X(f)^2 \cdot e^{2\pi.i.f.\tau} \cdot df \qquad [3.24]$$

Inverse FT of the power spectral density (PSD).

– The autocorrelation is the inverse FT of the power spectral density of the signal denoted as $S_x(f) = |X(f)|^2$, and for a signal with finite energy or $|X(f)|^2 /T$ for an average finite power signal (or the average power of the signal according to the Parseval relation).

$$C_{xy}(\tau) = \int_{-\infty}^{+\infty} X(f) \cdot Y(f) \cdot e^{2\pi.i.f.\tau} \cdot df \qquad [3.25]$$

$$C_{xx}(0) = \int_{-\infty}^{+\infty} X(f)^2 df = E \qquad [3.26]$$

The autocorrelation characterizes the memory (or predictability) of a signal, and its PSD characterizes a frequency range (a noise "without memory" will have a broadband PSD and an autocorrelation that decreases very quickly (close to a Dirac)).

### 3.2.2. A few examples of applications in steady state

*Force*: in classical mechanics, a mass $m$, when subjected to a force $F$, undergoes an acceleration such that $F = m\ddot{x}$. The acceleration is the second derivative of the position $x$ of the mass: $d^2x/dt^2 = \ddot{x}$. The system having input $F$ and output position $x$ has the differential equation $F = m\ddot{x}$; it is a double integrator. The acceleration is expressed in meters per seconds squared $(\mathrm{m/s^2})$ and the displacement $x$ in meters.

*Work*: the work of force $F$ during a displacement $dx$ is equal to the dot product of force and displacement $dW = Fdx$ or $W = Fl$ for a displacement of length $l$.

*Stored, supplied, dissipated energy*: the energy of a body characterizes the work it can perform.

A mass $m$ moving with velocity $v$ contains the kinetic energy $E = mv^2/2$.

Inductance $L$, with a current flowing with intensity $I$, has stored electromagnetic energy $E = LI^2/2$ (in Joules).

A capacity $C$ subject to voltage $V$ has stored an energy in electrostatic form $E = CV^2/2$.



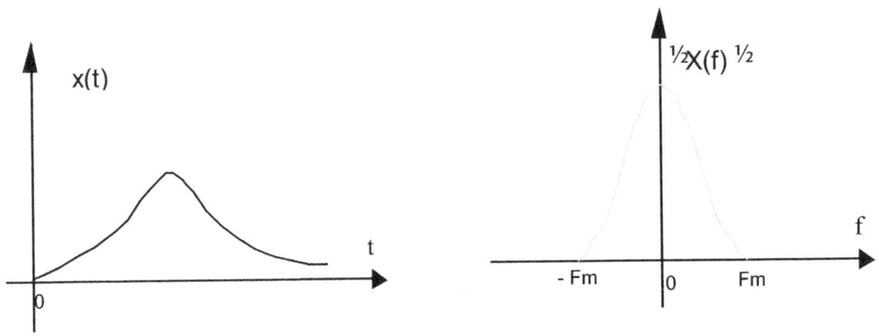


A resistance $R$ to which is applied voltage $V$, circulated by a current $I$, dissipates a power $P = RI^2 = V^2/R$.

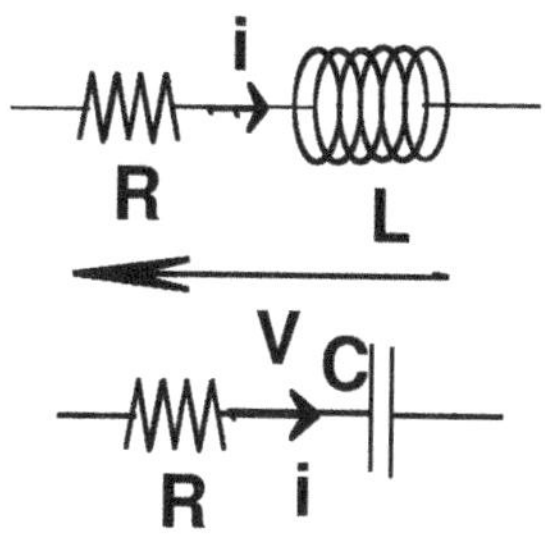


**Figure 3.1.** *RL and RC circuits*

Considering the RL circuit in Figure 3.1, its equation is written as $V = R \cdot i(t) + L \cdot di(t)/dt$. Multiplying this equation by $i(t) \cdot dt$, we obtain: $V \cdot i(t) \cdot dt = R \cdot i(t)^2 \cdot dt + L \cdot i(t) \cdot di$. The part $V \cdot i(t) \cdot dt$ is the energy delivered to the RL circuit (instantaneous supplied power $Vi(t)$). Although $R \cdot i(t)^2 \cdot dt$ is the energy dissipated by the Joule effect (instantaneous dissipated power $R \cdot i(t)^2$), $L \cdot i(t) \cdot di$ is the energy stored in inductance $L$ in electromagnetic form. Let $dE = Li(t) \cdot di$, so for a permanent current $I$, we integrate for $i(t)$ ranging from 0 to $I$ and we find $E = LI^2/2$.

For the RC circuit, the differential equation is obtained by describing $i(t) = Cdv_c(t)/dt$ and $V = R \cdot i(t) + v_c(t)$, with $v_c(t)$ the voltage at the terminals of the capacitor. Hence, the differential equation $V = R \cdot Cdv_c(t)/dt + v_c(t)$. By multiplying by $i(t)dt$, we get: $V \cdot i(t)dt = R \cdot C \cdot i(t)dv_c + v_c(t) \cdot i(t)dt = R \cdot i(t)^2 \cdot dt + C \cdot v_c(t) \cdot dv_c(t)$, where the last term is the energy stored in the capacitor in electrostatic form. Thus, for a permanent voltage $V$, this term is integrated for $v_c$ ranging from 0 to $V$ and we get: $E = CV^2/2$.

### 3.2.3. *Powers in variable state*

Let us consider, in the above examples, the case where $V$ is a variable voltage (for instance periodic of period $T$: $V(t) = A \cdot \sin(\omega t)$). In this case, the input signal has infinite energy and finite average power. The calculation of the above-mentioned quantities no longer makes sense and we instead consider average powers over a period. The instantaneous quantities are integrated over a period to obtain the average powers:

- $\frac{1}{T}\int_0^T V(t) \cdot i(t) \cdot dt$ is the average power supplied to the circuit for a period $T$;
- $\frac{1}{T}\int_0^T R \cdot i(t)^2 \cdot dt$ is the average power dissipated by the Joule effect;

- $\frac{1}{T}\int_0^T Li(t)\frac{di}{dt}dt$ is the average power stored in $L$ in electromagnetic form;

- $\frac{1}{T}\int_0^T C.v_c(t)\frac{dv_c}{dt}dt$ is the average power stored in $C$ in electrostatic form.

## 3.3. Mathematical application of the correlation and autocorrelation functions

For study in the time domain, we will use these tools that are functions based on the energy or the average power.

– Intercorrelation of two signals $x(t)$ and $y(t)$: this is the interaction energy of $x(t)$ and $y(t)$.

- Non-periodic finite energy signals

Given two non-periodic finite energy signals $x(t)$ and $y(t)$, their intercorrelation is: $C_{xy}(t) = \int_{-\infty}^{+\infty} x(t)y(t-\tau)^*.dt$.

This is the interaction energy of $x(t)$ and $y(t-\tau)$ and their autocorrelation is defined by $C_{xx}(t) = \int_{-\infty}^{+\infty} x(t)x(t-\tau)^*.dt$ and $Cyy(t) = \int_{-\infty}^{+\infty} y(t)y(t-\tau)^*.dt$.

- *Non-periodic finite average power signals*: this is the interaction average power of $x(t)$ and $y(t-\tau)$. For two non-periodic signals (infinite energy) with finite average power $x(t)$ and $y(t)$, their intercorrelation is defined by $C_{xy}(t) = \lim\limits_{T\to\infty} \frac{1}{T}\int_{t_o}^{t_o+T} x(t)y(t-\tau)^*.dt$ and their autocorrelations are similarly defined by

$C_{xx}(t) = \lim\limits_{T\to\infty} \frac{1}{T}\int_{t_o}^{t_o+T} x(t)x(t-\tau)^*.dt$ and $Cyy(t) = \lim\limits_{T\to\infty} \frac{1}{T}\int_{t_o}^{t_o+T} y(t)y(t-\tau)^*.dt$

- *Periodical signals*: let us take two periodical signals $x(t)$ and $y(t)$ of period $T$ (same period, with average finite power and infinite energy), their intercorrelation is defined by averages of powers over a period $T$ of signals: $C_{xy}(t) = \frac{1}{T}\int_{t_o}^{t_o+T} x(t)y(t-\tau)^*.dt$.

This is the average power per interaction period of $x(t)$ and $y(t-\tau)$ and their autocorrelations are defined by $C_{xx}(t) = \frac{1}{T}\int_{t_o}^{t_o+T} x(t)x(t-\tau)^*.dt$ and $Cyy(t) = \frac{1}{T}\int_{t_o}^{t_o+T} y(t)y(t-\tau)^*.dt$.

– In this case, it is possible to express the correlation function by making use of the expression of the Fourier series knowing that $x(t) = \sum\limits_{n=-\infty}^{+\infty} Cx_n g_n(t)$ and $y(t) = \sum\limits_{m=-\infty}^{+\infty} Cy_m g_m(t)$ and $< g_n(t), g_m(t) >= 0$ if $n \neq m$ (orthogonal functions):

We get: $C_{xy}(t) = \sum\limits_{n=-\infty}^{+\infty} C_{xn}C_{yn}^* e^{-2i\pi nt/T}$

Similarly, it can be shown that $C_{xx}(t) = \sum_{n=-\infty}^{+\infty} |C_{xn}|^2 e^{-2i\pi nt/T}$, and that $C_{xx}(0) = \sum_{n=-\infty}^{+\infty} |C_{xn}|^2$

– Properties of the correlation functions:

$$C_{xx}(0) \geq 0 \text{ and } C_{xx}(0) \geq |C_{xx}(t)| \; \forall t \in R \qquad [3.27]$$

$$|C_{xy}(t)|^2 \leq |C_{xx}(0)| \, . \, |Cyy(0)| \; \forall t \in R \qquad [3.28]$$

$$C_{xx}(t) = C^*_{xx}(-t) \qquad [3.29]$$

If $C_{xx}(t)$ is real, then it is even

$$C_{xy}(t) = C^*_{yx}(-t) \qquad [3.30]$$

If $x(t)$ and $y(t)$ are periodic of period $T$, $C_{xy}$ has the same period.

$$C_{xy}(t) = \int_{-\infty}^{+\infty} |X(f)|^2 . e^{2i\pi . f . t} df \qquad [3.31]$$

This is the inverse FT of the PSD.

The autocorrelation is the inverse FT of the PSD of the signal denoted as $S_x(f) = |X(f)|^2$ for a signal with finite energy or $|X(f)|^2 / T$ for an average finite power signal.

$$C_{xy}(t) = \int_{-\infty}^{+\infty} X(f) . Y^*(f) . e^{2i\pi . f . t} df \qquad [3.32]$$

$$C_{xx}(0) = \int_{-\infty}^{+\infty} |X(f)|^2 \, df = E \qquad [3.33]$$

(or the average power of the signal according to Parseval's theorem).

The autocorrelation characterizes the memory (or predictability) of a signal, and its PSD characterizes a frequency range (a noise "without memory" will have a broadband PSD and an autocorrelation that decreases very quickly (close to a Dirac)). The autocorrelation is the inverse FT of the PSD of the signal. The autocorrelation characterizes the memory or predictability of a signal, and its PSD characterizes a frequency range (noise "with no memory" will have a broadband PSD and an autocorrelation that decreases very quickly).

### 3.3.1. *Duration of a signal and its spectrum width*

Theoretically, the following relationships exist between the durations of a signal and of its spectrum:

– finite time duration signal $T \Leftrightarrow$ spectrum with unbounded support $F_o$;

– signal with bounded support spectrum $F_o \Leftrightarrow$ the duration of the signal $T$ is infinite.

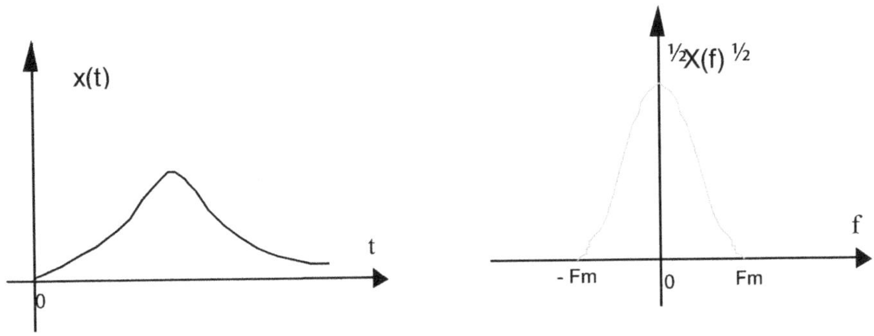


**Figure 3.2.** *Signal and its spectrum*

For example, the FT of a Dirac peak is equal to 1 for any $f$, therefore $T = \{0\}$ and $F_o = R$.

In practice, for any real signal, we can define a useful duration denoted as $T$ and a useful spectrum width $F_o$. Product $(T.F_o)$ has a lower bound (uncertainty relation).

Useful duration: let $s(t)$ be a real square-integrable signal, and its energy $E$ is finite. It is expressed by:

$E = \int_{-\infty}^{+\infty} s^2(t)dt$; its normalization can thus be derived as $1 = \int_{-\infty}^{+\infty} s^2(t)/Edt$

On this basis, we define the normalized signal $x(t) = s(t)/\sqrt{E}$ obtained by normalizing $s(t)$, we get for this signal energy:

$$E = \int_{-\infty}^{+\infty} x^2(t)dt = 1 \qquad [3.34]$$

In this case, $x(t)^2$ takes the meaning of a probability density function; this signal can then be centered around $t_o$ defined as an average of $t$ weighted by $x(t)^2$. To center

this signal, the origin of times $t = 0$ is placed at the average moment $t_o = < t >$, defined as follows.

$$< t > = \int_{-\infty}^{+\infty} t.x^2(t)dt \qquad [3.35]$$

The lifetime $T$ of the signal is then defined by the following standard deviation:

$$\frac{T^2}{2} = \int_{-\infty}^{+\infty} t^2 x^2(t)dt \qquad [3.36]$$

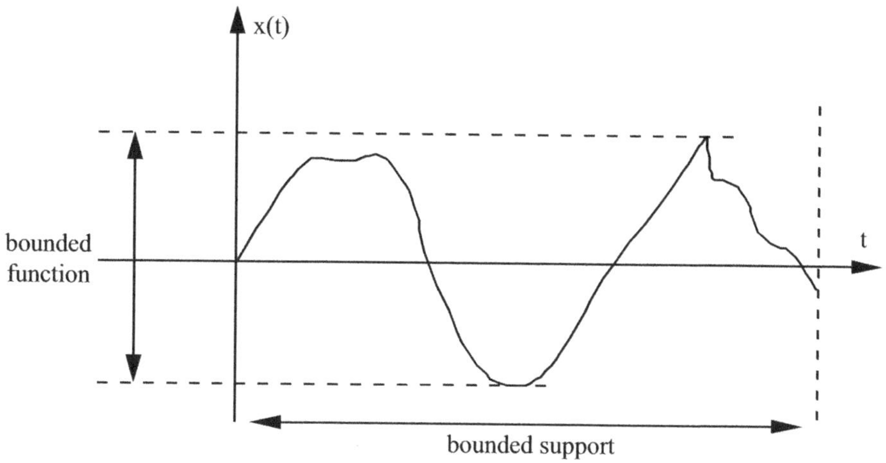


**Figure 3.3.** *Support-bounded time function*

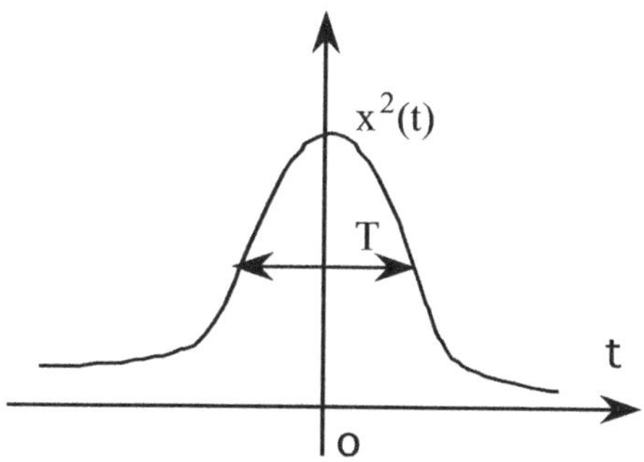


**Figure 3.4.** $x(t)^2$ *probability density*

$T$ is the duration of the signal. The same procedure can be used for the effective spectral width.

Effective spectral width: according to Parseval's theorem, we get:

$$\int_{-\infty}^{+\infty} |x(t)|^2 dt = \int_{-\infty}^{+\infty} |X(f)|^2 df$$
$$= 2\int_{0}^{+\infty} |X(f)|^2 df = 2\int_{0}^{+\infty} C^2(f) df = 1 \quad [3.37]$$

the spectral amplitude $C(f)$ of a real signal being an even function. Relation [1.8] allows us to use a reasoning similar to the one discussed previously. We then define:

$$(F_o/2)^2 = \int_{-\infty}^{+\infty} f^2 |X(f)|^2 df$$
$$= 2\int_{0}^{+\infty} f^2 |X(f)|^2 df = 2\int_{0}^{+\infty} f^2 C^2(f) df \quad [3.38]$$

$F_o$ is the effective spectral width. The uncertainty relation is given as:

$$T.F_o \geq 1/\pi \quad [3.39]$$

It can be demonstrably justified using Cauchy–Schwartz inequality.

#### 3.3.1.1. *Bounded spectrum signals*

The physical signals being considered are always supposed to be of bounded spectrum (higher frequencies are cut off by the sensor and the dynamics of the physical system). Bounded spectrum signals ( composed between $-F_{\max}$ and $+F_{\max}$) are very often used in communication techniques. For such a signal denoted $x(t)$, it can be written as:

$$x(t) = \int_{-F_{\max}}^{+F_{\max}} X(f) e^{2ipf} df \quad [3.40]$$

For this kind of signal, we have Bernstein's theorem defined as follows.

THEOREM 3.2 (Bernstein's: a bounded signal).– $|x(t)| \leq M$ with bounded spectrum, $[-F\max, +F\max]$, has all its derivatives bounded:

$$|x^{(n)}(t)| \leq (2\pi F_{\max})^n M \quad [3.41]$$

This means that a bounded signal cannot vary in an arbitrarily fast fashion and the smaller the maximal bound with frequency $F_{\max}$, the slower the signal variations.

EXAMPLE 3.1.– *For instance, let us consider an almost periodic signal defined by* $x(t) = \sum_{k=1}^{p} a_k cos(2\pi f_k t + \varphi_k)$, *assuming that* $f_k < f_p \ \forall k \neq p$ *hence* $|x(t)| \leq \sum_{k=1}^{p} a_k = M \ \forall t \in R$ *and we get by deriving this signal* $n$ *times:* $x^{(n)}(t) = \sum_{k=1}^{p} a_k (2\pi f_k)^n cos(2\pi f_k t + \varphi_k + n\pi/2)$, *where* $|x^{(n)}(t)| \leq \sum_{k=1}^{p} a_k.$ $(2\pi f_k)^n \leq (2\pi f_p)^n M$, *because we have* $f_k < f_p$.

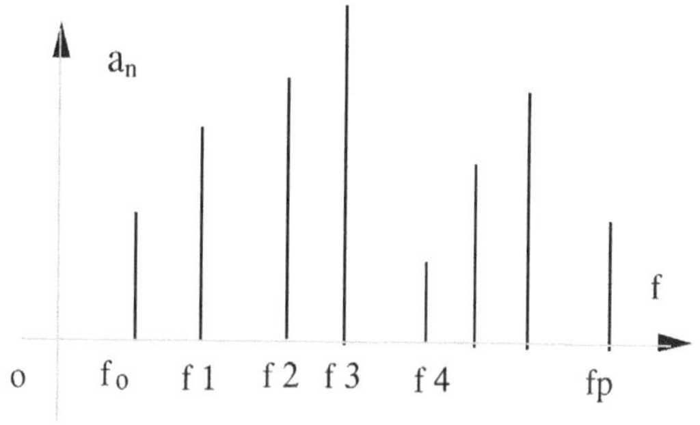


**Figure 3.5.** *Coefficient $a_n$ of the Fourier transform*

Note that the derivative of a signal represents its rate of variation, and its second derivative characterizes the acceleration of this variation and the $n$th derivative can be considered as the variation rate of the $(n-1)$th derivative of the signal. Thus, the signal represented underneath will have variation dynamics whose velocity is limited by a maximal bound depending of the spectral range of the signal.

Bernstein's theorem can also be employed in the definition of a relation between the bandwidth of a system or a filter and the speed of its dynamics (response).

#### 3.3.1.2. *Low-pass filter time constant*

Consider a physical system or a low-pass filter whose response to a step can be represented as shown in Figure 3.6. A "time constant" or the response time of such a system can be defined as being the time lapse after which the step response shifts from the value of zero to 90% of the final value in steady state or at the moment where the response reaches its first maximum overshoot. Denote by $F_c$ the cut-off frequency of the filter or system.

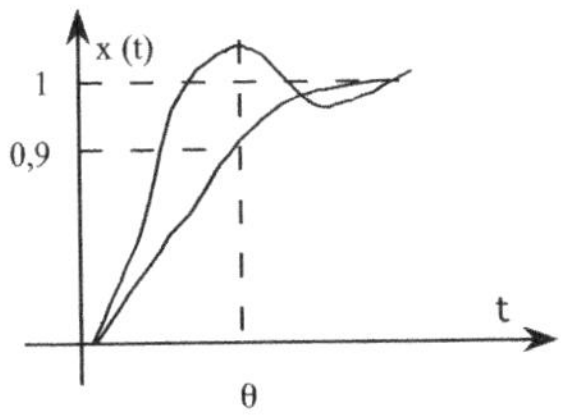


| | |
|---|---|
| $Fc$ | Cutoff frequency |
| $x(t)$ | Unit step response (with or without overshooting) |
| $\theta$ | Constant time (or response time) |

**Figure 3.6.** *Low-pass filter time constant*

Using Bernstein's theorem and finite increments, we obtain the relation $\theta \geq \frac{1}{\pi F_c}$; therefore, for the system to be fast (small $\theta$), it is necessary that the filter be broadband (large $F_c$).

*Demonstration*: let $y(t) = x(t) - x(\theta)/2$; knowing that $\theta$ is constant, we obtain by deriving $y'(t) = x'(t)$ and $|y(t)| \leq x(\theta)/2 = M$ for $0 \leq \mathrm{t} \leq \theta$ according to the finite-increment theorem, we get: $x(\theta) - x(0) = \theta.x(\theta_1)$ for $0 \leq \theta_1 \leq \theta \Rightarrow y'(\theta_1) = x'(\theta_1) = x(\theta)/\theta$ car $x(0) = 0$ according to Bernstein's theorem $|y'(t)| \leq (x(\theta)/2).2\pi F_c$ where $x(\theta)/\theta \leq x(\theta).\pi F_c$ and thus finally it yields:

$$1/\pi F_c \leq \theta \qquad [3.42]$$

### 3.3.2. *Finite or zero average power signals*

Recall that square-integrable signals are signals with finite total energy and zero average power. For these signals, we have defined an autocorrelation function:

$$C_{xx}(t) = \int_{-\infty}^{+\infty} x(t).x^*(t-\tau)dt \qquad [3.43]$$

whose FT is the spectral energy density: $\hat{C}(f) = |X(f)|^2$.

#### 3.3.2.1. *Average PSD*

We will consider here signals with finite average power and infinite total energy. These are periodic signals or more generally non-periodic signals with infinite total energy. Finite average power signals do not have common FT, but a generalized spectral analysis can be made highlighting their spectral properties.

The finite average power (non-zero) is the average time value:

$$\overline{|x(t)|^2} = \lim_{T\to+\infty} \frac{1}{T} \int_{-\frac{T}{2}}^{\frac{T}{2}} |x(t)|^2 dt \neq 0 \qquad [3.44]$$

It is admitted that this does not depend on the time origin $t$ (stationarity). We slice $x(t)$ with a rectangular window of width $T$ centered in zero $\pi_T(t)$; $x_T(t) = \pi_T(t).x(t)$

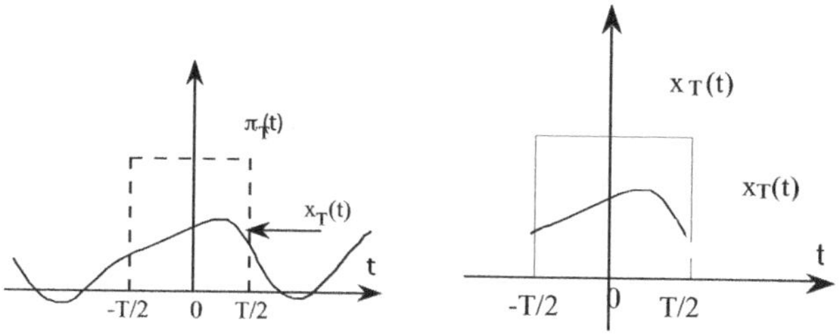


**Figure 3.7.** *Rectangular window*

$x_T(t)$ has an FT and Parseval's theorem can be applied

$$\begin{aligned} < |x(t)|^2 >_T &= \frac{1}{T}\int_{-\frac{T}{2}}^{\frac{T}{2}} |x(t)|^2 dt \\ &= \frac{1}{T}\int_{-\infty}^{+\infty} |x_T(t)|^2 dt = \frac{1}{T}\int_{-\infty}^{+\infty} |XT(f)|^2 df \end{aligned} \quad [3.45]$$

We denote by $\hat{\gamma}_T(f) = \frac{1}{T}|X_T(f)|^2$ the average PSD in the interval $[-\frac{T}{2}; \frac{T}{2}]$. As the first member of the Parseval relation has a limit equal to $|x(t)|^2$ when $T \to +\infty$, it can be concluded that:

$$\begin{aligned} \overline{|x(t)|^2} &= \lim_{T\to+\infty} \frac{1}{T}\int_{-\frac{T}{2}}^{\frac{T}{2}} |x(t)|^2 dt \\ &= \lim_{T\to+\infty} \frac{1}{T}\int_{-\infty}^{+\infty} |x_T(t)|^2 dt = \int_{-\infty}^{+\infty} \hat{\gamma}(f) df \end{aligned} \quad [3.46]$$

having for definition of the PSD $\hat{\gamma}(f)$ (this limit exists and is positive):

$$\hat{\gamma}(f) = \lim_{T\to+\infty} \hat{\gamma}_T(f) = \lim_{T\to+\infty} \frac{1}{T}|X_T(f)|^2 \quad [3.47]$$

#### 3.3.2.2. *Wiener–Khintchine theorem*

For a finite average power signal, a correlation (autocorrelation) function is defined whose FT is the average PSD.

The correlation function is defined similarly to that for finite total energy signals:

$$C_{xx}(t) = \lim_{T \to +\infty} \frac{1}{T} TF^{-1}(|X_T(f)|^2) \qquad [3.48]$$

with

$$TF^{-1}(\frac{1}{T}|X_T(f)|2) = TF^{-1}(\frac{1}{T}X_T(f).X_T^*(f)) = \frac{1}{T}x_T(\tau) * x_T^*(-\tau) \qquad [3.49]$$

and

$$\frac{1}{T}\int_{-\infty}^{+\infty} x_T(t).x_T^*(t-\tau)dt = \frac{1}{T}\int_{-\frac{T}{2}}^{\frac{T}{2}} x(t).x^*(t-\tau)dt \qquad [3.50]$$

When $T \to +\infty$ the last member admits a limit because, due to the Cauchy–Schwartz inequality, its modulo is smaller than $[\frac{1}{T}\int_{-\frac{T}{2}}^{\frac{T}{2}} |x(t)|^2.dt]$ . $[\frac{1}{T}\int_{-\frac{T}{2}}^{\frac{T}{2}} |x(t-\tau)|^2 dt]$, which has a limit. Hence, the definition of the autocorrelation

$$\gamma(\tau) = \lim_{T \to +\infty} \frac{1}{T} TF^{-1}(|X_T(f)|^2) = \overline{x_T(t).x_T^*(t-\tau)} \qquad [3.51]$$

THEOREM 3.3 (Wiener–Kintchine theorem).– $\gamma(\tau)$ is the correlation function of the signal $x(t)$ with

$$\hat{\gamma}(f) = TF(\gamma(\tau)) \qquad [3.52]$$

– Properties of $\gamma(\tau)$:

- $\gamma(0)$= $|\overline{x(t)}|^2$ is the average power of the signal;
- $|\gamma(\tau)| \leq \gamma(0)$;
- $\gamma(-\tau)$=$\gamma^*(\tau)$.

### 3.3.3. *Application for linear filtering*

A stationary linear filter is characterized by its impulse response $h(t)$.

If $u(t)$ is the input and $y(t)$ is the output of this filter, then it can be written as:

$$y(t) = \int_{-\infty}^{+\infty} h(t)u(t-\tau)dt = h(t) * u(t) \qquad [3.53]$$

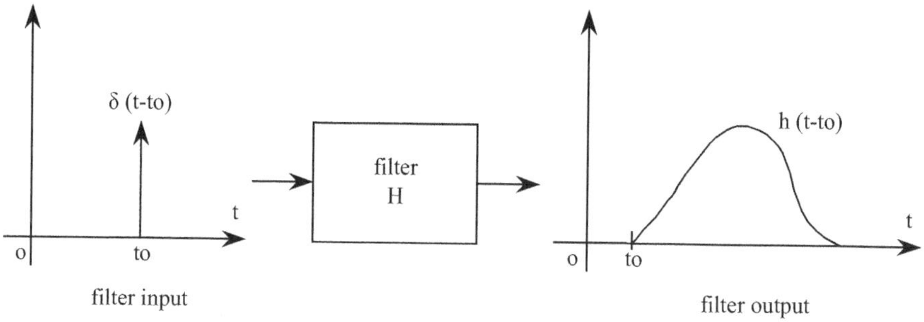


**Figure 3.8.** *Impulse response of the filter*

Furthermore, we have $Y(f) = H(f).U(f)$ with $H(f) = TF(h(t))$ the transfer function of the filter or the gain in amplitude as a function of the frequency. By definition, the power gain of the filter is given by:

$$G(f) = \frac{|Y(f)|^2}{|U(f)|^2} = |H(f)|^2 \quad [3.54]$$

Consider a unity gain filter ($G(f) = 1$) between $\alpha - \frac{\Delta}{2}$ and $\alpha + \frac{\Delta}{2}$. This filter is excited by a signal $u(t)$. The energy of output $y(t)$ can be expressed according to the properties of the FT:

$$E = \int_{\alpha-\frac{\Delta}{2}}^{\alpha+\frac{\Delta}{2}} |U(f)|^2 df \simeq |U(\alpha)|^2.\Delta \quad [3.55]$$

The value in $a$ of a signal DSP is therefore (up to the factor $\Delta$) the energy of the signal filtered through a filter of bandwidth $\Delta$ and unit gain centered in $\alpha$.

## 3.4. A few application exercises

EXERCISE 3.1.– Weighting

1) Calculate the FT of the function $f(t) = \cos(2\pi\nu_0 t)$ limited by the rectangular window $\prod_{\frac{T}{2}}(t)$.

2) Calculate the new FT if $f(t)$ is limited by the Hanning window $h_N(t)$.

$$h_N(t) = \begin{cases} \frac{1}{2}[1 + \cos(2\pi\frac{t}{T})] & \text{if } t \in [-\frac{T}{2}, \frac{T}{2}] \\ 0 & \text{elsewhere} \end{cases} \quad [3.56]$$

EXERCISE 3.2.– Modulation

1) Determine the spectral representation of amplitude-modulated signals $f_1(t)$ and $f_2(t)$, when the spectrum of $m(t)$ is bounded ($M(f) = 0$ if $\mid f \mid > f_0$ and $F_0 \gg f_0$:

$$f_1(t) = A_0 m(t)\cos(2\pi F_0 t) \qquad \text{with} \mid m(t) \mid < 1 \tag{3.57}$$

$$f_2(t) = A_0[1 + m(t)]\cos(2\pi F_0 t) \qquad \text{with} \mid m(t) \mid < 1$$

Application: $m(t) = cos(2\pi f_0 t)$.

2) The aim is to transmit on a phone line two messages $m_1(t)$ and $m_2(t)$ that occupy the band $[-B, +B]$. To this end, the signal $r(t) = u(t)m_2(t) + m_1(t)$ is created.

Determine $u(t)$ such that, after transmission, $m_1(t)$ and $m_2(t)$ are found again.

Give an overview of the system to be implemented to obtain the two messages separately.

EXERCISE 3.3.– Convolution and correlation of periodic signals

Consider the three signals:

$$f_1(t) = \cos(2\pi\nu_1 t), \quad f_2(t) = \cos(2\pi\nu_1 t + \phi_1), \quad f_3(t) = \cos(2\pi\nu_2 t + \phi_2) \tag{3.58}$$

1) Calculate the FT $F_1(\nu)$, $F_2(\nu)$ and $F_3(\nu)$.

2) Calculate convolution products $f_1(t) * f_1(t)$, $f_1(t) * f_2(t)$ and $f_1(t) * f_3(t)$.

3) Calculate correlations $C_{f_1 f_1}(\tau)$, $C_{f_1 f_2}(\tau)$ and $C_{f_1 f_3}(\tau)$.

EXERCISE 3.4.– Parseval's theorem

Calculate Parseval's theorem to solve the integral:

$$I = \int_{-\infty}^{+\infty} [\frac{\sin(at)}{t}]^2 dt \tag{3.59}$$

EXERCISE 3.5.– RC filters

Consider the RC circuit of the figure below with input $e(t)$ and output $s(t)$.

1) Write the differential equation of this system relating $e(t)$ and $s(t)$.

2) Express the impulse response of this system. Derive thereof the frequency response $H(\nu)$ of the filter. Derive thereof its impulse response $h(t)$.

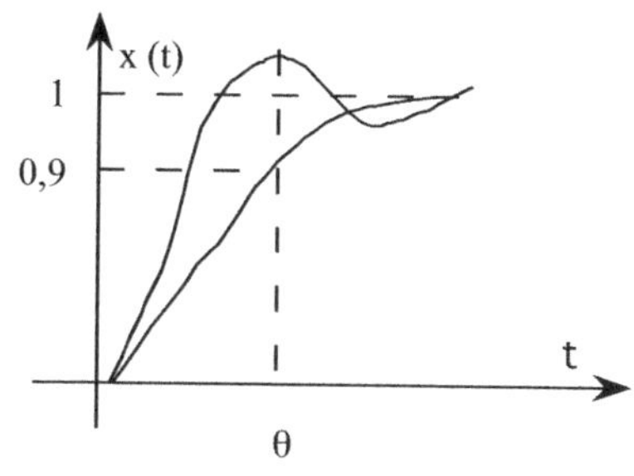


**Figure 3.9.** *RC filter*

3) Using the Laplace transform determine its transfer function $H(p) = S(p)/E(p)$ (assuming zero-valued original conditions), calculate and represent its unit step response $u(t)$ with the inverse Laplace transform $(e(t) = u(t))$.

4) Calculate $C_{hh}(\tau)$ and apply the Wiener–Khinchine theorem to determine the spectral energy density $S_{hh}(\nu)$.

5) Deduce thereof the cutoff frequency of the filter $\nu_c$.

$$\text{Notes :} \quad S_{hh}(\nu_c) = \frac{1}{2}\max(S_{hh}(\nu));$$

$$TF(e^{-at}u(t)) = \frac{1}{a + 2\pi j\nu} \quad \text{with } u(t) = \text{step} \tag{3.60}$$

EXERCISE 3.6.– Modulation–demodulation

Consider a deterministic signal $a(t)$ with zero spectrum $a(\nu)$ outside the range $[-B, +B]$.

1) Show that the spectrum of $a^2(t)$, denoted $A_2(\nu)$, is zero outside $[-2B, +2B]$.

2) Assume that $a(t) > 0$ and that it is transmitted through amplitude modulation without carrier:

$$X(t) = a(t)\cos(2\pi f_0 t) \tag{3.61}$$

To recover $a(t)$ three operations are performed:

- A quadratic operator transforms $X(t)$ into $Y(t) = X^2(t)$.

- A low-pass filter with cutoff frequency $f_c$ transforms $Y(t)$ into $Z(t)$.

- A square root operator transforms $Z(t)$ into $u(t) = k\sqrt{Z(t)}$.

What condition should $B$ and $f_0$ verify?

How should we choose $f_c$ (according to $B$ and $f_0$) so as to get $u(t) = a(t)$? Deduce thereof the value of $k$.

EXERCISE 3.7.– Consider the communication system composed as explained in the following. A message $s(t)$ is filtered through a low-pass filter $F_1$, then the output of $F_1$ (denoted $y(t)$) is multiplied by the signal $\cos(2\pi f_0 t)$ to obtain $z(t)$ that is emitted by an antenna. On receipt, an antenna picks up the signal $z(t)$ (assumed to be received without noise) that is multiplied by $(2\cos(2\pi f_0 t))$ to give $u(t)$. The signal $u(t)$ is filtered through a filter $F_2$ ($x(t)$ is the output from $F_2$). $F_1$ is an ideal low-pass filter with cut-off frequency $f_c = (4/\tau)\,\mathrm{Hz}$. For the study of this system, we consider the signal:

$$s(t) = \exp(-t^2/\tau) \qquad [3.62]$$

1) With a diagram, represent this communication system. Determine the conditions on the cutoff frequencies of $F_1$, $F_2$ and $f_o$ so that $x(t)$ and $u(t)$ may contain the spectral information of the message.

2) Calculate and represent the spectral densities of $s(t)$, $y(t)$, $z(t)$, $u(t)$ and $x(t)$.

3) Determine the bandwidth of the filter $F_2$ that makes the reconstitution of the emitted message possible. What is the minimal value that $f_o$ can take?

EXERCISE 3.8.– The objective is to study the suspension of a car wheel; to this end, the system is summarized as indicated in Figure 3.10. Consider a mass $M$ attached to a spring of stiffness $k$ and a damper of viscosity coefficient $a$; the deformation of the tire is ignored to simplify the study. The excitation produced by the road profile is denoted $x(t)$ and the reaction on the mass $M$, $y(t)$. The focus is on the dynamic motion of mass $M$.

1) Write the differential equation linking the output $y(t)$ to the input $x(t)$.

2) Derive thereof the transfer function of the suspension in $p$ and its frequency response.

3) Calculate and represent the step response (assembled on a sidewalk) of the system.

4) Study the form of the response according to the coefficients $M$, $a$ and $k$ of the system. Explain in the various cases the energy transfer between the wheel $x(t)$ and the vehicle $y(t)$.

5) Generalize the study to the case of half a vehicle then of a whole vehicle by assimilating the tires to springs of stiffness $kp$ (see Figure 3.10).

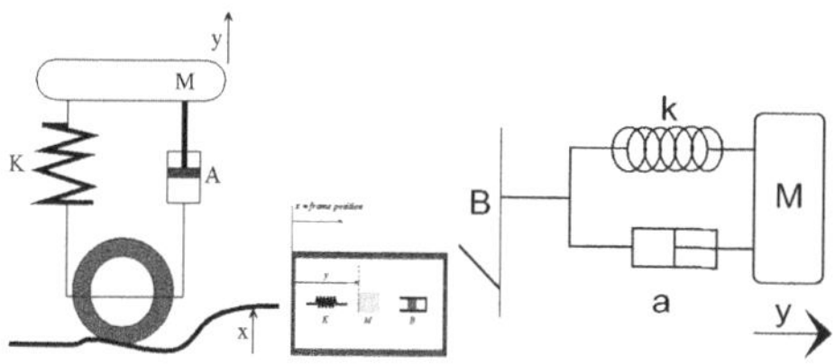


**Figure 3.10.** *Modeling of the tires of a vehicle*

EXERCISE 3.9.– Figure 3.10 represents a simple mechanical accelerometer.

1) Determine the transfer function between the input acceleration $a$ ($a = \frac{d^2x}{dt^2}$) and the output quantity $y$.

2) Calculate and represent the step response of the system. Discuss the response according to the parameters of the system.

EXERCISE 3.10.– Consider the system in Figure 3.10 of mass $M$ attached to a fixed frame by a spring of stiffness $k$ and a damper with a viscous friction coefficient $a$. We want to study the transient behavior of the mass after deviation from the equilibrium.

1) Write the differential equation translating the movement in $y$ of the mass.

2) Deduce thereof the transfer function and the frequency response.

3) Block $B$ is considered mobile and has a displacement $x$; write the differential equation linking $x(t)$ and $y(t)$; derive thereof the transfer function from $x$ to $y$.

EXERCISE 3.11.– In the following arrangements, the operational amplifiers are considered as perfect and we will define $\tau_1 = R_1C$ and $\tau_2 = R_1R_2C/(R_1 + R_2)$.

1) Determine the transfer functions of these arrangements $Vs(p)/Ve(p)$. Represent their frequency response.

2) Express their differential equation.

3) Calculate and plot their unit step response.

# 4

# Representation of Discrete (Sampled) Systems

This chapter explores Shannon sampling, the Z-transform and parametric representation.

## 4.1. Shannon and sampling, discretization methods, interpolation, sample and hold circuits

### 4.1.1. *Sampling and interpolation*

Sampling and interpolation are inseparable concepts for continuous systems considered by automatic control and signal processing. In addition, in order for signals to be usable by digital computers, quantification should be conducted (discretization of $y(\xi_n)$ at the moment at the value $\xi_n$ of the $x$-axis).

In this section, we present sampling (without quantization) associated with exact interpolation (without loss of information), then we will indicate how this theoretical principle can be applied in practice. To this end, we will begin by recalling a number of basic definitions.

DEFINITION 4.1.– *Sampling: representation of a continuous function $y(x)$ by a set of values $y(\xi_n)$ with $n$ being an integer ($n \in \mathbb{Z}$ or $\mathbb{N}$). Values $y(\xi_n)$ are extracted at intervals regularly spaced by $T_e$ ($\xi_n = nT_e$). $T_e$ is the sampling period.*

DEFINITION 4.2.– *Samples: values $y(\xi_n)$, $n \in \mathbb{Z}$ (or $\mathbb{N}$). In practice, only a finite number of values is necessary $y(\xi_n)$ (samples).*

DEFINITION 4.3.– *Interpolation: knowing samples $y(\xi_n)$ for $n = 0, 1, .., N, ..$, we should be able to recover any value $y(x)$ $\forall x \in \mathbb{R}$ using an interpolation process.*

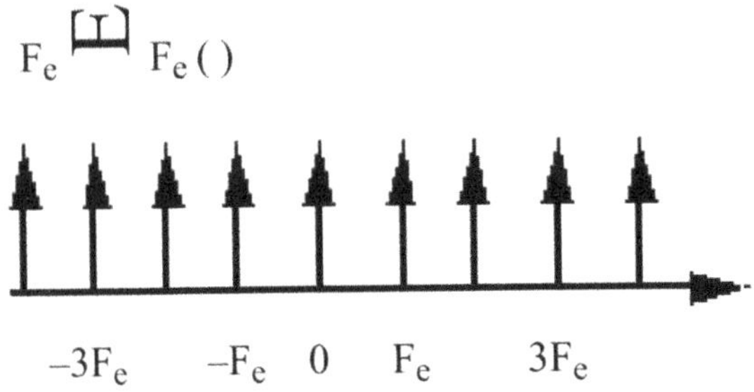


**Figure 4.2.** *Frequency comb distribution*

#### 4.1.1.2. *Ideal sampling*

Let $x(t)$ be the signal (assumed as continuous and bounded) that we want to sample. Its ideal sampling with period $T_e$ is the product $x(t).\sqcup\!\sqcup_{Te}(t) = v(t)$.

$$v(t) = x(t).\sqcup\!\sqcup_{Te}(t) = x(t) \sum_{n=-\infty}^{n=+\infty} \delta(t - nT_e) = \sum_{n=-\infty}^{n=+\infty} x(nT_e)\delta(t - nT_e) \qquad [4.7]$$

– $v(t)$ is the sum of impulse samples $x(nT_e)$ (discrete values, a countable infinity) regularly spaced of $T_e$ (the sampling period);

– $v(t)$ is a series of delta functions of mass $x(nT_e)$ spaced by $T_e$.

Assuming that $x(t)$ has a FT, the FT of its ideal sample will be: $V(f) = X(F) * TF(\sqcup\!\sqcup_{Te}(t))$

$$V(f) = \int_{-\infty}^{+\infty} X(f').\hat{\sqcup\!\sqcup}_{Te}(f - f')df' \qquad [4.8]$$

$$= \int_{-\infty}^{+\infty} X(f').(\sum_{n=-\infty}^{n=+\infty} F_e\delta(f - nF_e - f'))df'$$

We finally get:

$$V(f) = \sum_{n=-\infty}^{n=+\infty} X(f - nF_e) \qquad [4.9]$$

The Fourier transform $V(f)$ of the sampled signal $v(t)$ is obtained by periodizing, on the frequency axis, the FT $X(f)$ of the signal $x(t)$ multiplied by $F_e = \frac{1}{T_e}$ (*Te* being the sampling period).

# 4

# Representation of Discrete (Sampled) Systems

This chapter explores Shannon sampling, the Z-transform and parametric representation.

## 4.1. Shannon and sampling, discretization methods, interpolation, sample and hold circuits

### 4.1.1. *Sampling and interpolation*

Sampling and interpolation are inseparable concepts for continuous systems considered by automatic control and signal processing. In addition, in order for signals to be usable by digital computers, quantification should be conducted (discretization of $y(\xi_n)$ at the moment at the value $\xi_n$ of the $x$-axis).

In this section, we present sampling (without quantization) associated with exact interpolation (without loss of information), then we will indicate how this theoretical principle can be applied in practice. To this end, we will begin by recalling a number of basic definitions.

DEFINITION 4.1.– *Sampling: representation of a continuous function $y(x)$ by a set of values $y(\xi_n)$ with $n$ being an integer ($n \in \mathbb{Z}$ or $\mathbb{N}$). Values $y(\xi_n)$ are extracted at intervals regularly spaced by $T_e$ ($\xi_n = nT_e$). $T_e$ is the sampling period.*

DEFINITION 4.2.– *Samples: values $y(\xi_n)$, $n \in \mathbb{Z}$ (or $\mathbb{N}$). In practice, only a finite number of values is necessary $y(\xi_n)$ (samples).*

DEFINITION 4.3.– *Interpolation: knowing samples $y(\xi_n)$ for $n = 0, 1, .., N, ..$, we should be able to recover any value $y(x)$ $\forall x \in \mathbb{R}$ using an interpolation process.*

DEFINITION 4.4.– *Quantification: this is the discretization operation of values* $y(\xi_n)$ *(for* $n = 0, 1, .., N, ...$*) to match them with numeric values (easy to code with integers).*

#### 4.1.1.1. *Poisson formula*

If $X(f) = \text{TF} \cdot (x(t)), \forall T \in \mathbb{R}$, we get:

$$\sum_{n=-\infty}^{+\infty} x(nT) = \tfrac{1}{T} \sum_{n=-\infty}^{+\infty} X(\tfrac{n}{T}) \qquad [4.1]$$

PROOF.– In Chapter 1, we saw that if $S_T(f) = T.F.(s_T(t))$ we obtain the Fourier series decomposition of $s(t)$: $s(t) = \sum_{n=-\infty}^{+\infty} s_T(t + nT) = \sum_{n=-\infty}^{+\infty} s_T(t) * \delta(t - nT)$

$s(t) = \sum_{n=-\infty}^{+\infty} C_n e^{in\omega_o t} = \sum_{n=-\infty}^{+\infty} C_n e^{2i\pi n \frac{t}{T}}$ with $\omega_o = \frac{2\pi}{T}$ however we have found: $C_n = \frac{1}{T} S_T(\frac{n}{T})$ hence

$$\sum_{n=-\infty}^{+\infty} s_T(t + nT) = \tfrac{1}{T} \sum_{n=-\infty}^{+\infty} S_T(\tfrac{n}{T}) e^{2i\pi n \frac{t}{T}} \qquad [4.2]$$

by imposing $t = 0$, it yields:

$$\sum_{n=-\infty}^{+\infty} s_T(nT) = \tfrac{1}{T} \sum_{n=-\infty}^{+\infty} S_T(\tfrac{n}{T}) \qquad [4.3]$$

This holds for any couple $(S_T(f),\ s_T(t))$. Consequently, we get, due to the properties of the Fourier transform (FT):

1) if $T = 1$, then $\sum_{n=-\infty}^{+\infty} x(n) = \sum_{n=-\infty}^{+\infty} X(n)$;

2) translation: it is known that $x(t - a) \leftrightarrows X(f)e^{2i\pi af}$ from which $\sum_{n=-\infty}^{+\infty} x(n - a) = \sum_{n=-\infty}^{+\infty} X(n)e^{2i\pi na}$ is derived;

3) with $\delta(t - t_o) \iff e^{2i\pi f t_o}$, where we get $\sum_{n=-\infty}^{+\infty} \delta(n - t_o) = \sum_{n=-\infty}^{+\infty} e^{2i\pi n t_o}$.

We therefore retain the following formulas that constitute a generalization of the Fourier series.

PROPOSITION 4.1.– A periodized function with the form of a series of translates is expanded into a series of harmonics (denoted $F = \frac{1}{T}$);

$$\sum_{n=-\infty}^{+\infty} x(t+nT) = F \sum_{n=-\infty}^{+\infty} X(nF)e^{2i\pi nFt} \qquad [4.4]$$

$$\sum_{n=-\infty}^{+\infty} X(f-nF) = T \sum_{n=-\infty}^{+\infty} x(nT)e^{2i\pi nTf} \qquad [4.5]$$

##### 4.1.1.1.1. Application to a Dirac comb (or train pulse denoted $\text{Ш}_T(t)$)

The following distribution is thus composed of a series of translated Dirac impulses: $\text{Ш}_T(t) = \sum_{n=-\infty}^{n=+\infty} \delta(t-nT)$ with $n$ integer. This distribution is illustrated in Figure 4.1. It can be written by means of the Poisson formula: $\text{Ш}_T(t) = \sum_{n=-\infty}^{n=+\infty} \delta(t-nT) = F \sum_{n=-\infty}^{n=+\infty} e^{2i\pi nFt}$ because $TF(\delta(t) = 1$. Its FT is therefore

$$FT(\text{Ш}_T(t)) = F \sum_{n=-\infty}^{n=+\infty} \delta(f-nF) = F\text{Ш}_F(f) \qquad [4.6]$$

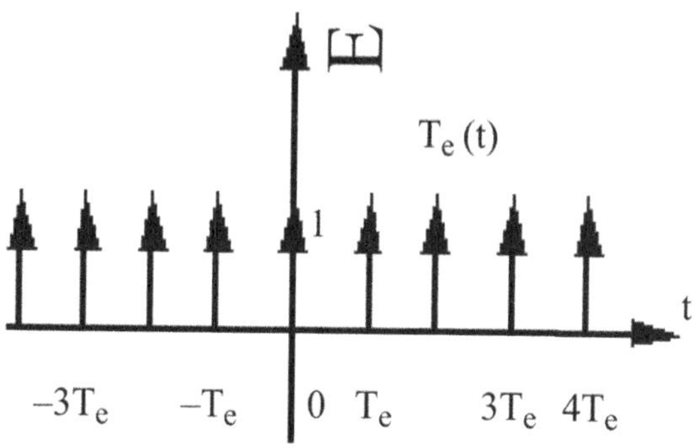


**Figure 4.1.** *Temporal comb distribution*

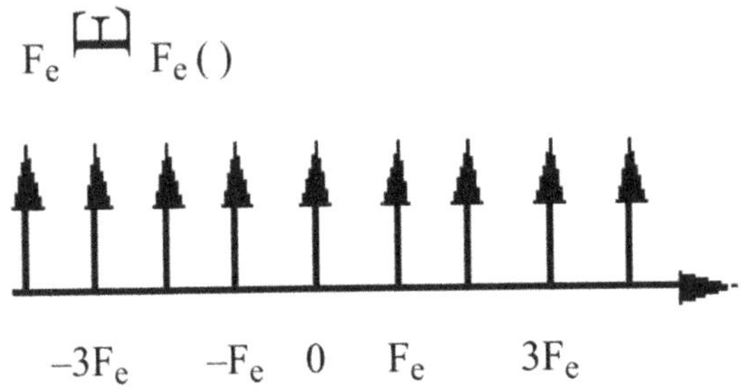


**Figure 4.2.** *Frequency comb distribution*

#### 4.1.1.2. *Ideal sampling*

Let $x(t)$ be the signal (assumed as continuous and bounded) that we want to sample. Its ideal sampling with period $T_e$ is the product $x(t).\text{Ш}_{Te}(t) = v(t)$.

$$v(t) = x(t).\text{Ш}_{Te}(t) = x(t) \sum_{n=-\infty}^{n=+\infty} \delta(t - nT_e) = \sum_{n=-\infty}^{n=+\infty} x(nT_e)\delta(t - nT_e) \qquad [4.7]$$

– $v(t)$ is the sum of impulse samples $x(nT_e)$ (discrete values, a countable infinity) regularly spaced of $T_e$ (the sampling period);

– $v(t)$ is a series of delta functions of mass $x(nT_e)$ spaced by $T_e$.

Assuming that $x(t)$ has a FT, the FT of its ideal sample will be: $V(f) = X(F) * TF(\text{Ш}_{Te}(t))$

$$V(f) = \int_{-\infty}^{+\infty} X(f').\hat{\text{Ш}}_{Te}(f - f')df' \qquad [4.8]$$

$$= \int_{-\infty}^{+\infty} X(f').(\sum_{n=-\infty}^{n=+\infty} F_e\delta(f - nF_e - f'))df'$$

We finally get:

$$V(f) = \sum_{n=-\infty}^{n=+\infty} X(f - nF_e) \qquad [4.9]$$

The Fourier transform $V(f)$ of the sampled signal $v(t)$ is obtained by periodizing, on the frequency axis, the FT $X(f)$ of the signal $x(t)$ multiplied by $F_e = \frac{1}{T_e}$ (*Te* being the sampling period).

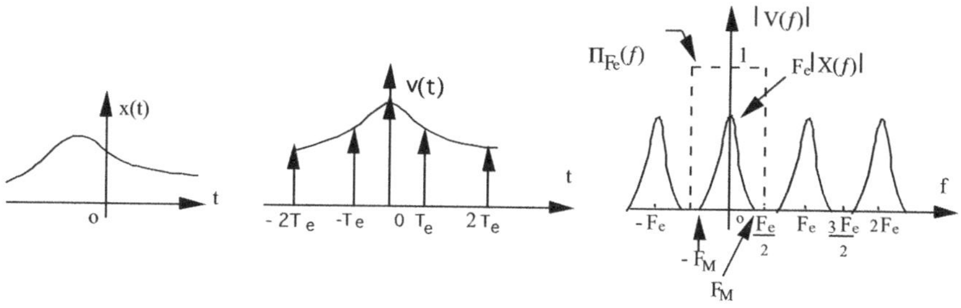


**Figure 4.3.** *Spectral representation of a sampled signal*

#### 4.1.1.3. *Sampling theorem*

##### 4.1.1.3.1. Ideal sampling (theoretical) and Shannon theorem

*Hypothesis*: the spectrum $X(f)$ of the signal is bounded in the interval $[-F_m, F_m]$ and zero outside this range. The pattern of rank 0 can then be extracted, $V_o(f)$ of $V(f)$ using an ideal low-pass filter, with transfer function $\pi_{F_e}(f)$, of cutoff frequency $f_c = \frac{1}{2}F_e \geq F_m$. This ideal low-pass filter is called a Shannon filter. □

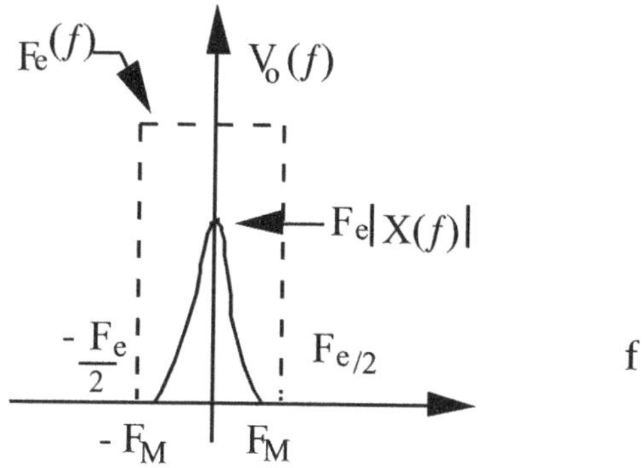


**Figure 4.4.** *Shannon filter*

We then obtain:

$$V_o(f) = \pi_{F_e}(f).V(f) = F_e X(f)$$

where applying the inverse FT: $v_o(t) = v(t) * \frac{\sin(\pi F_e t)}{\pi t} = F_e x(t)$

$$v_o(t) = \sum_{n=-\infty}^{n=+\infty} x(nT_e)\delta(t - nT_e) * \frac{\sin(\pi F_e t)}{\pi t} = F_e x(t)$$

$$v_o(t) = \sum_{n=-\infty}^{n=+\infty} x(nT_e)\frac{\sin(\pi F_e(t-nT_e))}{\pi(t-nT_e)} = F_e x(t)$$

$$v_o(t) = F_e \sum_{n=-\infty}^{n=+\infty} x(nT_e)\frac{\sin(\pi F_e(t-nT_e))}{\pi F_e(t-nT_e)} = F_e x(t) \qquad [4.10]$$

where by identifying from both results and establishing $x(pT_e) = x_p$ and

$$s_p(t) = \frac{\sin(\pi\frac{(t-pT_e)}{T_e})}{\pi(\frac{t}{T_e}-p)} = s_0(t-pT_e) \qquad [4.11]$$

we get the exact interpolation formula:

$$x(t) = \sum_{n=-\infty}^{n=+\infty} x_p s_p(t) = \sum_{n=-\infty}^{n=+\infty} x_p \frac{\sin(\pi\frac{(t-pT_e)}{T_e})}{\pi(\frac{t}{T_e}-p)} \qquad [4.12]$$

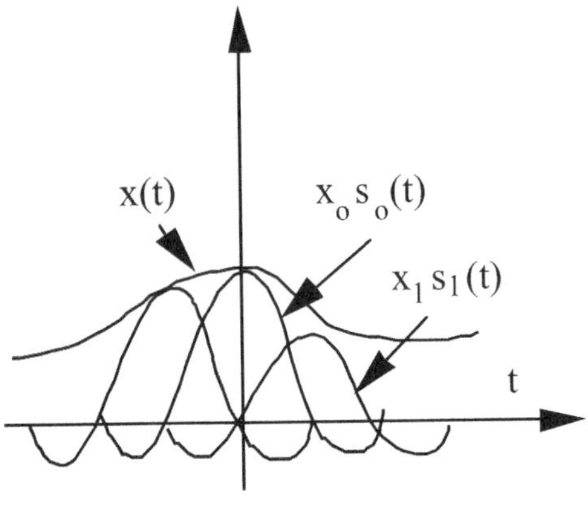


**Figure 4.5.** *Shannon interpolation*

REMARK 4.1.– *The periodization of* $X(f)$ *is obtained by convoluting* $X(f)$ *with the comb* $F_e \sqcup\!\sqcup_F(f)$.

*The frequency* $F_e$ *must be greater than twice* $F_m$ *(*$F_m$ *being the maximal frequency for which the signal has a non-zero spectrum) to avoid spectrum aliasing (overlapping) (loss of information).*

*It suffices to isolate a copy of the spectrum (* $F_e/X(f)/$ *) to recover the signal* $x(t)$ *by applying the inverse FT.*

*Similarly, when a function $x(t)$ is periodized on the time axis, this equates to frequency sampling.*

*To exactly reconstruct the signal, the infinite sum of terms $x_p.s_p(t)$ is carried out, in which $x_p$ are the samples and $s_p(t)$ functions are previously known. With $2F_m$ samples per second the signal is perfectly reconstituted.*

THEOREM 4.1.– Shannon's theorem: a continuous signal with bounded spectrum in $[-F_m, F_m]$ is completely determined, using Shannon's interpolation formula, by the values that it takes at intervals regularly spaced by $T_e = \frac{1}{F_e}$ seconds at most ($F_e = 2F_m$).

REMARK 4.2.–

*a) This formula cannot be used in real time.*

*b) It comprises an infinite number of terms.*

*c) Oversampling provides no additional information (in theory).*

*d) The functions $s_p(t)$ are orthogonal.*

*e) When it is desirable to sample at $F_e < 2F_m$, we must use a low-pass antialiasing filter to eliminate the frequencies above $F_e/2$.*

#### 4.1.1.3.2. Case of a finite number of samples

If for the reconstitution of the signal $x_r(t)$ we restrict ourselves to a finite number $N = 2P$ (with $P$ finite) of samples:

$$x_r(t) = \sum_{n=-P}^{n=P} x_p s_p(t) = \sum_{n=-\infty}^{n=+\infty} x_p \frac{\sin(\pi \frac{(t-pT_e)}{T_e})}{\pi(\frac{t}{T_e}-p)} \quad [4.13]$$

and considering that samples $x(pT_e)$ outside the time interval $[-PT_e, PT_e]$ are zero. $x_r(t)$ gives an approximate reconstruction of the signal $x(t)$ from a finite number of samples ($x_r(t)$ tends to $x(t)$ in quadratic mean in the interval $[-PT_e, PT_e]$).

#### 4.1.1.4. *Hold sampling and real periodic sampling*

Since ideal sampling involves infinitely narrow pulses, it is only approximately implementable. Technological means require the use of pulses of short but finite duration. This leads us to consider practical applications by means of the study of *average sampling*.

*Average sampling*: each sample is the average of signal $x(t)$ taken over the whole duration of a rectangular signal $\pi_\tau(t)$ of width $\tau$ and periodized with period $T_e$, $e_p(t) = \pi_\tau(t) * \sqcup\!\sqcup_{Te}(t)$. This signal is shown in Figure 4.6.

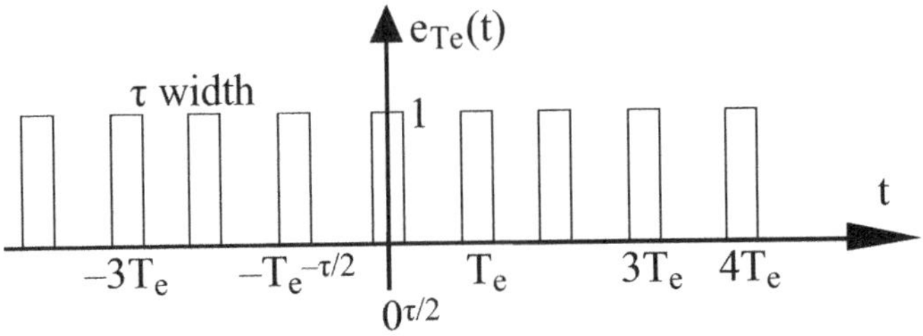


**Figure 4.6.** *Average sampling signal*

For the sample $v_p = v(pT_e)$ obtained at moment $pT_e$, we can write the averaging expression during the interval of duration $\tau$ around $pT_e$:

$$v_p = \frac{1}{\tau} \int_{pT_e - \frac{\tau}{2}}^{pT_e + \frac{\tau}{2}} x(t')dt' = \frac{1}{\tau} \int_{-\infty}^{+\infty} \pi_\tau(t' - pT_e)x(t')dt' \qquad [4.14]$$

The sample $v_p = v(pT_e)$ is therefore written as:

$$v_p = \frac{1}{\tau}(\pi_\tau(t) * x(t)).\delta(t - pT_e)$$

and from this we obtain the model of the discrete signal or distribution representative of the sampled signal:

$$v(t) = \frac{1}{\tau} \sum_{p=-\infty}^{p=+\infty} (\pi_\tau(t) * x(t)).\delta(t - pT_e) \qquad [4.15]$$

The spectrum of the sampled signal is then

$$\hat{v}(f) = V(f) = \frac{1}{\tau} \sum_{n=-\infty}^{n=+\infty} \frac{\sin(\pi\tau f)}{\pi f}.X(f) * \frac{1}{T_e}\delta(f - \frac{n}{T_e}) \qquad [4.16]$$

$$V(f) = F_e \sum_{n=-\infty}^{n=+\infty} \frac{\sin(\pi\tau(f - nF_e))}{\pi\tau(f - nF_e)}.X(f - nF_e) \qquad [4.17]$$

The multiplicative factor $\frac{\sin(\pi\tau f)}{\pi\tau f}$ expresses the deformation produced by the use of $e_p(t)$, in the real case, instead of the Dirac comb (ideal case). The spectrum of the signal to be sampled being limited in $[-\frac{F_e}{2}, \frac{F_e}{2}]$, the deformation will be small if the product $\tau\frac{F_e}{2} << 1$ or even $\tau << 2T_e$ or $\frac{F_e}{2} << \frac{1}{\tau}$.

## 4.2. Z-transform – representation of discrete (sampled) systems

### 4.2.1. *Definition – convergence and residue*

The Z-transform, for discrete signals, can be considered as a generalization of the Fourier transform. It is an indispensable tool for the analysis of discrete signals and systems.

#### 4.2.1.1. *Bilateral Z-transform*

DEFINITION 4.5.– *Let $x(k)$ $(k \in \mathbb{Z})$ a numerical sequence and $z$ a complex variable; the bilateral $z$-transform $x(k)$ is defined by:*

$$X(z) = \sum_{k=-\infty}^{+\infty} x(k)z^{-k} \tag{4.18}$$

Sufficient condition for convergence $X(z)$ exists if series $\sum_{k=-\infty}^{+\infty} \left|x(k)z^{-k}\right|$ exists (converges).

REMARK 4.3.– *The series* $\sum_{k=0}^{+\infty} u(k)$ *converges if* $\lim_{k\to+\infty} \sum_{k=-\infty}^{+\infty} |u(k)|^{1/k} < 1$.

The series in equation [4.18] can be divided into two terms:

$$X(z) = \sum_{k=0}^{+\infty} x(k)z^{-k} + \sum_{k=1}^{+\infty} x(-k)z^{k} \tag{4.19}$$

Therefore, for the ZT to exist, it suffices that:

$$\begin{cases} (a) \lim_{k\to+\infty} \left|x(k)z^{-k}\right|^{1/k} < 1 \\ (b) \lim_{k\to-\infty} \left|x(k)z^{-k}\right|^{1/k} < 1 \end{cases} \tag{4.20}$$

The previous inequalities lead to the following sufficient condition: $0 \leq \alpha < |z| < \beta \leq +\infty$ with $\alpha = \lim_{k\to+\infty} |x(k)|^{1/k}$ and $\beta = \lim_{k\to+\infty} |u(-k)|^{-1/k}$

Values $\alpha$ and $\beta$ as responses to the problem form two sets $\{\alpha\}$ and $\{\beta\}$ generally disjoint. The so-called region of convergence of the ZT is the set of points $z$ of the complex plane defined by: $r_1 = sup(\alpha)$ and $r_2 = Inf(\beta)$ and which are such that $r_1 < |z| < r_2$.

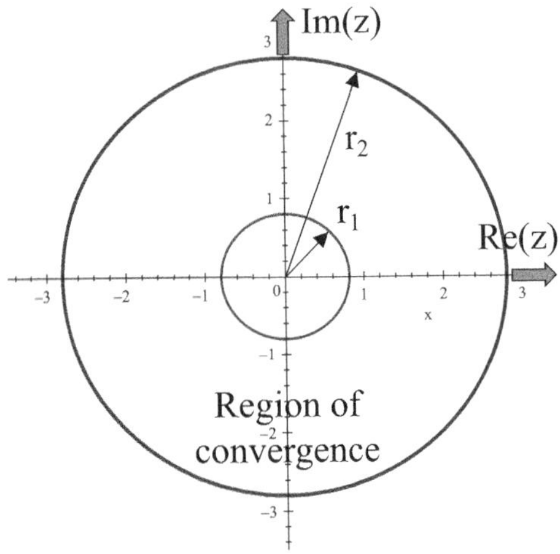


**Figure 4.7.** *Region of convergence of the bilateral ZT*

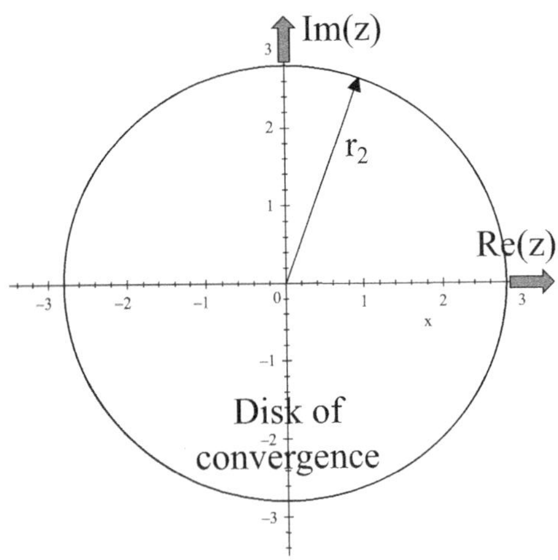


**Figure 4.8.** *Disc of convergence of the unilateral ZT*

| $x(k)$ | $X(z)$ | $(r1, r2)$ |
|---|---|---|
| $\begin{cases} x(k) = a^k & k \geq 0 \\ x(k) = 0 & k < 0 \end{cases}$ | $\frac{1}{1-az^{-1}}$ | $(\lvert a\rvert, +\infty)$ |
| $\begin{cases} x(k) = 0 \; k \geq 0 \\ x(k) = -a^k \; k < 0 \end{cases}$ | $\frac{1}{1-az^{-1}}$ | $(o, \lvert a\rvert)$ |
| $x(k) = a^k \, \forall k$ | Does not exist | (a, a) Undefined |
| $\begin{cases} x(k) = a^k \; k \geq 0 \\ x(k) = b^k \; k < 0 \end{cases}$ | $\frac{1}{1-az^{-1}} - \frac{1}{1-bz^{-1}}$ | $(\lvert a\rvert, \lvert b\rvert)$ if $\lvert a\rvert < \lvert b\rvert$ |
| $x(k) = E(k) \begin{cases} 1 \; k \geq 0 \\ 0 \; k < 0 \end{cases}$ | $\frac{1}{1-\; z^{-1}}$ | $(1, +\infty)$ |
| $x(k) = \prod_N (k - \frac{N}{2})$ | $\frac{1-\; z^{-N}}{1-\; z^{-1}}$ | $(o, +\infty)$ |

**Table 4.1.** *Some examples of the bilateral ZT*

REMARK 4.4.– *The first two examples underline the importance of the region of convergence (one must specify $r_1$ and $r_2$). If sets $\{\alpha\}$ and $\{\beta\}$ defined above are not disjoint, the ZT is not defined.*

*For a finite sequence:*

$$X(z) = \sum_{k=k_1}^{k_2} x(k) z^{-k} \qquad [4.21]$$

*the ZT is a polynomial in $z$ that absolutely converges for all non-zero $z$. We have $(r_1 = 0, r_2 = +\infty)$.*

*Linearity: $x(k) = ax_1(k) + bx_2(k)$ therefore $X(z) = a.X_1(z) + bX_2(z)$ and the convergence domain (sometimes strictly) contains the intersection of the domains of convergence of $X_1(z)$ and $X_2(z)$.*

#### 4.2.1.2. *Unilateral Z-transform*

DEFINITION 4.6.– *For causal signals ($x(k) = 0$ if $k < 0$), the unilateral $z$-transform proves sufficient because they are equal to zero for negative moments. It is defined by:*

$$X(z) = \sum_{k=0}^{+\infty} x(k) z^{-k} \qquad [4.22]$$

### **4.2.2. *Inverse Z-transform***

There is a formulation close to Fourier's using Cauchy's theorem. More often, to calculate the inverse of a ZT, practical methods are employed such as simple elements decomposition or polynomial division.

## General method

Let $X(z) = \sum_{k=-\infty}^{+\infty} x(k)z^{-k}$, the ZT of the sequence $x(k)$ with $(r_1, r_2)$ the convergence domain. Let $\Gamma$ be an oriented closed contour, as shown in Figure 4.9, surrounding the origin and located in the region of convergence. We get

$$z^n X(z) = \sum_{k=-\infty}^{+\infty} x(k)z^{n-k} \qquad [4.23]$$

Integrating equation [4.23] on the contour $\Gamma$:

$$\oint_{\Gamma} z^n X(z)dz = \sum_{k=-\infty}^{+\infty} x(k) \oint_{\Gamma} z^{n-k}dz \qquad [4.24]$$

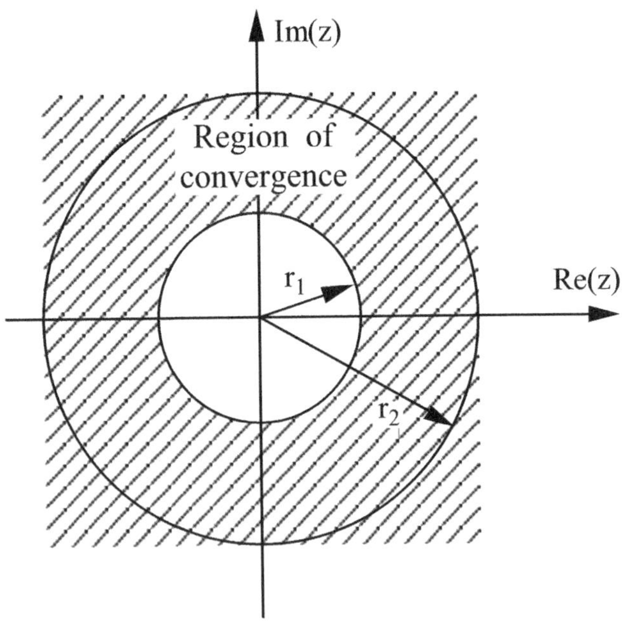


**Figure 4.9.** *Convergence of the bilateral ZT*

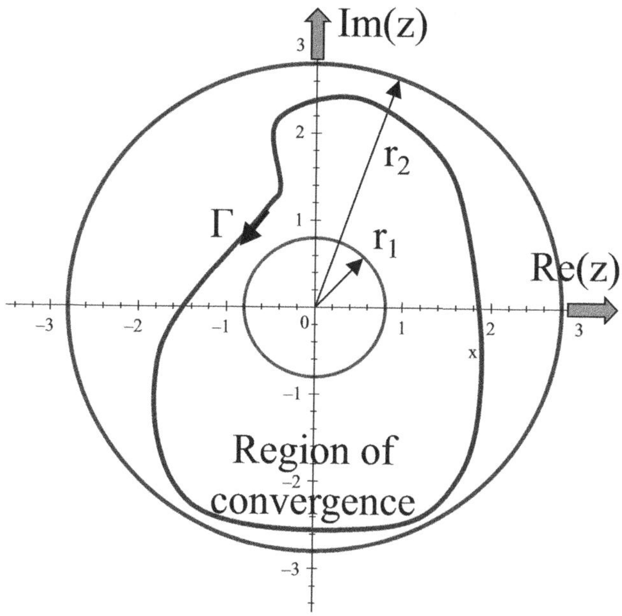


**Figure 4.10.** *Contour for the computation of the inverse ZT*

The series converges uniformly, therefore it is justified to swap symbols $\sum$ and $\oint$; moreover we know that

$$\oint_{\Gamma} z^m dz = \left\{ \begin{array}{l} 1 \text{ if } m \neq 1 \\ 2\pi j \text{ if } m = 1 \end{array} \right| \quad [4.25]$$

as a result in the second member of equation [4.24] there only remains the term $k-n = 1$ or $k = n+1$, we then obtain:

$$\oint_{\Gamma} z^m X(z) dz = 2\pi j . x(m+1) \quad [4.26]$$

$$x(k) = \frac{1}{2\pi i} \oint_{\Gamma} z^{k-1} X(z) dz \quad [4.27]$$

hence the inversion formula.

$$X(z) \text{ in } (r_1, r_2) \rightarrow x(k) = \frac{1}{2\pi i} \oint_{\Gamma} z^{k-1} X(z) dz \quad [4.28]$$

with $G$ inside the region of convergence.

EXAMPLE 4.1.–

*1)* $X(z) = \frac{1}{1-az^{-1}}$ $|z| > a. \rightarrow x(k) = \frac{1}{2\pi i}\oint\limits_{\Gamma} z^{k-1}\frac{1}{1-az^{-1}}\,dz = \frac{1}{2\pi i}\oint\limits_{\Gamma}\frac{z^k}{z-a}\,dz$

*if* $k > 0$ $\frac{1}{2\pi i}\oint\limits_{\Gamma}\frac{z^k}{z-a}\,dz = \text{Residue}(\frac{z^k}{z-a}\,, z = a) = a^k$

*if* $k > 0$ $\frac{1}{2\pi i}\oint\limits_{\Gamma}\frac{z^k}{z-a}\,dz = \text{Residue}(\frac{z^{-k}}{z-a}\,, z = 0) + \text{Residue}(\frac{z^{-k}}{z-a}\,, z = a) = a$

Recall that the residue of a function $F(z)$ with respect to a pole $a$ of multiplicity $p$ is given by the formula:

$$\begin{aligned}\text{Residue}(F(z), a) &= \text{Residue}(F(z)\,, z = a)\\ &= \frac{1}{p-1}\left[\frac{d^{p-1}}{dz^{p-1}}((z-a).F(z))\right], \text{ for } z = a\end{aligned} \qquad [4.29]$$

$$\begin{aligned}\text{Residue}(F(z), a) &= \text{Residue}(\frac{1}{z^p(z-a)}\,, z = 0)\\ &= \frac{1}{p-1}\left[\frac{d^{p-1}}{dz^{p-1}}\left(\frac{1}{(z-a)}\right)\right]_{z=0} = -a^p = -a^k\end{aligned} \qquad [4.30]$$

for $k = -p$

if $k < 0$; $x(k) = -a^k + a^k = 0$

from which finally

$$\left\{\begin{array}{l} x(k) = a^k \; k \geq 0 \\ x(k) = 0 \; k < 0 \end{array}\right| \qquad [4.31]$$

2) $X(z) = \frac{1}{1-az^{-1}}$ for $|z| < a$. The poke is no longer within the contour $G$ thus: if $k > 0$

$$\rightarrow x(k) = \frac{1}{2\pi i}\oint\limits_{\Gamma} z^{k-1}\frac{1}{1-az^{-1}}\,dz = \frac{1}{2\pi i}\oint\limits_{\Gamma}\frac{z^k}{z-a}\,dz = 0$$

if $k < 0$

$$\rightarrow x(k) = \frac{1}{2\pi i}\oint\limits_{\Gamma} z^{k-1}\frac{1}{1-az^{-1}}\,dz = \frac{1}{2\pi i}\oint\limits_{\Gamma}\frac{1}{z^{-k}(z-a)}\,dz = -a^k$$

where finally

$$\left\{ \begin{array}{l} x(k) = 0 \; k \geq 0 \\ x(k) = -a^k \; k < 0 \end{array} \right| \qquad [4.32]$$

#### 4.2.2.1. *Practical methods for the calculation of an inverse $Z$-transform*

Often, $X(z)$ is in the form of a rational fraction that can be simplified into the sum of a polynomial in $z$ and a fraction whose numerator degree is smaller than that of the denominator:

$$X(z) = q_0 + q_1 z + q_2 z^2 .... + q_l z^l + \frac{b_0 + b_1 z + b_2 z^2 .... + b_m z^m}{a_0 + a_1 z + a_2 z^2 .... + a_n z^n} \qquad [4.33]$$

It is then possible to perform a simple elements decomposition and then to expand these elementary fractions into series, taking into account the region of convergence. After the decomposition into simple elements, we may also use the tables.

EXAMPLE 4.2.– *(a)* $X(z) = \frac{5z}{1+z-6z^2} \; \frac{1}{3} < |z| < \frac{1}{2}$

$X(z) = \frac{5z}{1-2z} \frac{1}{1+3z} = \frac{A}{1-2z} \frac{B}{1+3z} \longrightarrow A = (1-2z)X(z)|_{z=1/2} = 1$ *and* $B = (1+3z)X(z)|_{z=-1/3} = -1$

$|2z| < 1 \longrightarrow$ *the first fraction is expanded into positive powers of* $2z$.

$|3z| > 1 \longrightarrow$ *the second fraction* $1/3z$ *is expanded*

$\frac{1}{1-2z} = 1 + 2z + 2z^2 + .......$

*and* $\frac{1}{1+3z} = \frac{1}{3z} \frac{1}{1+\frac{1}{3z}} = \frac{1}{3z}\left(1 - \frac{1}{3z} + \left(\frac{1}{3z}\right)^2 - ...\right)$ *hence the sequence*

$$\left\{ \begin{array}{l} x(k) = \left(-\frac{1}{3}\right)^k \;\; \textit{if} \;\; k \geq 0 \\ x(k) = \left(\frac{1}{2}\right)^k \;\; \textit{if} \;\; k < 0 \end{array} \right| \qquad [4.34]$$

EXAMPLE 4.3.– *(b)* $X(z) = e^{(z-1)}.(1+z^{-1})$.

$$X(z) = \left(\sum_{k=0}^{+\infty} \frac{z^{-k}}{k!}\right)(1+z^{-1}) = \left(\sum_{k=0}^{+\infty}(k+1)z^{-k}\right)\frac{1}{k!} = \sum_{k=-\infty}^{+\infty} x(k)z^{-k}$$

from which:

$$x(k) = \left\{ \begin{array}{lll} \frac{k+1}{k!} & \text{if} & k \geq 0 \\ 0 & \text{if,} & k < 0 \end{array} \right| \qquad [4.35]$$

#### 4.2.2.2. *Polynomial division*

The division is performed according to the increasing powers of $z^{-1}$.

EXAMPLE 4.4.–

$$X(z) = \frac{z^{-1}}{1-2z+z^2} = z^{-1} + \frac{2z^{-2} - z^{-3}}{1-2z+z^2} = z^{-1} + 2z^{-2} + \frac{\text{remainder}}{1-2z+z^2} \qquad [4.36]$$

$X(z) = z^{-1} + 2z^{-2} + z^{-3} + 0.z^{-4} - z^{-5} + .....$ *where* $x(k) = 0; 1; 2; 1; 0; -1; .....$

### 4.2.3. *Properties of the Fourier transform*

1) Linearity: $\lambda x(k) + \mu y(k) \longmapsto \lambda X(z) + \mu Y(z); \quad (r_1^{''}, r_2^{''})$ with $r_1^{''} = \max(r_1, r_1^{'}),\ r_2^{''} = \max(r_2, r_2^{'})$

2) Offset of a signal: $y(k) = x(k-k_0) \mapsto Y(z) = \sum_{k=-\infty}^{+\infty} x(k-k_0)z^{-k} = z^{-k_0}\sum_{k=-\infty}^{+\infty} x(k)z^{-k} = z^{-k_0}X(z)$; $(r_1, r_2)$ Offset of $k_o \longmapsto$ multiplication by $z^{-k_0}$.

3) Multiplication by a$^k$: $y(k) = a^k x(k) \longmapsto Y(z) = \sum_{k=-\infty}^{+\infty} a^k x(k) z^{-k} = \sum_{k=-\infty}^{+\infty} x(k)(\frac{z}{a})^{-k} = X(\frac{z}{a})$; $(|a|\, r_1, |a|\, r_2)$ if $z = p_o$ is a pole of $X(z)$ then $ap_o$ is a pole of $Y(z) = X(\frac{z}{a})$. The region of convergence is thus modified (by contraction).

4) Derivation of the ZT (or multiplication by $k$):

$$\frac{dX(z)}{dz} = \sum_{k=-\infty}^{+\infty} (-k)x(k)z^{-k-1} = -z\frac{dX(z)}{dz} = \sum_{k=-\infty}^{+\infty} kx(k)z^{-k}; (r_1, r_2)$$

$$k^n x(k) \mapsto \left(-z\tfrac{d}{dz}\right)^n X(z)\,; (r_1, r_2)$$

5) Product of transforms:

$$\text{if } X(z) = \sum_{k=-\infty}^{+\infty} x(k)z^{-k}\ ; (r_1, r_2) \text{ and } Y(z) = \sum_{k=-\infty}^{+\infty} y(k)z^{-k};$$

$$(r_1^{'}, r_2^{'}) \text{ then } X(z)Y(z) = \sum_{k=-\infty}^{+\infty}\left(\sum_{l=-\infty}^{+\infty} x(l)y(k-l)\right) z^{-k}\, (max(r_1, r_1^{'}), max(r_2, r_2^{'}))$$

6) Correlation of two signals:

$$C_{xy}(k) = \sum_{l=-\infty}^{+\infty} x(l)y(k+l)$$

– $ZC$ denotes the ZT of $C_{xy}(k)$ we define $m = l + k$:

$$ZC_{xy}(z) = \sum_{k=-\infty}^{+\infty}\left(\sum_{l=-\infty}^{+\infty} x(l)y(k+l)\right) z^{-k} = \sum_{m=-\infty}^{+\infty}\sum_{l=-\infty}^{+\infty} x(l)y(m)z^{-m}z^{+l}$$

$$ZC_{xy}(z) = \sum_{m=-\infty}^{+\infty} y(m)z^{-m} \sum_{l=-\infty}^{+\infty} x(l)z^{+l} = X(\tfrac{1}{z})Y(z)$$

EXAMPLE 4.5.–

$$x(k) = y(k) = rect(k) = \Pi_N(k - \frac{N}{2}) = \left\{ \begin{array}{ll} 1 & if\ k \geq 0\, and\, k < N \\ 0 & if\ k < 0\ or\ k \geq N \end{array} \right| \quad [4.37]$$

$$\begin{aligned} ZC_{xx}(z) &= X(\frac{1}{z})X(z) = \frac{1-z^N}{1-z}\frac{1-z^{-N}}{1-z^{-1}} \\ &= \frac{z(1-z^{-N})(1-z^N)}{(z-1)^2}; \quad (0;+\infty) \end{aligned} \quad [4.38]$$

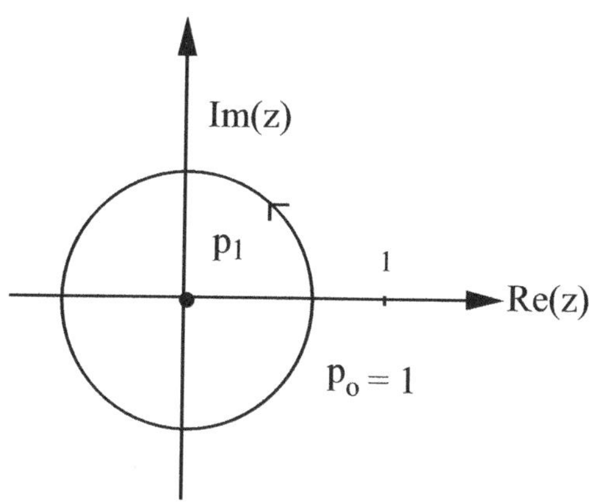


**Figure 4.11.** *Unit circle*

In the inversion, we use the residue theorem. We have a double pole $p_o = +1$.

We are restricting ourselves to the case in which $k > 0$, because it is known that $C_{xx}(k)$ is even.

a) If $0 \leq k < N$, $z^{k-1}ZC_{xx}(z) = \frac{(z^k - z^{k-N})(1-z^N)}{(z-1)^2}$ has a pole in $p_1$ at 0 of order $N-k$ and $p_o = +1$. A contour $\Gamma$ is chosen to avoid the pole $p_o = +1$. The residue can be calculated in an $N-k$ order of $z^{k-1}ZC_{xx}(z)$ and yields: $N-k = I$, thus $C_{xx}(k) = N-k$, if $|k| \leq N$.

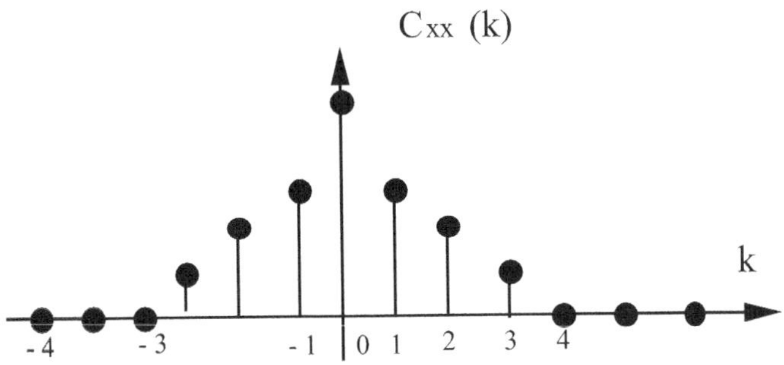


**Figure 4.12.** *Correlation*

b) If $k \geq N$, $z^{k-1}ZC_{xx}(z)$ has no pole inside $\Gamma$ therefore $k \geq N \Longrightarrow C_{xx}(k) = 0$. As a conclusion:

$$C_{xx}(k) = \left\{ \begin{array}{cc} 0 & if \;\; |k| \geq N \\ N-|k| & if \;\; |k| < N \end{array} \right| \qquad [4.39]$$

– Parseval's theorem: given $x(k)$ a square-integrable sequence $\sum_{k=-\infty}^{+\infty} |x(k)|^2 < \infty$ we calculate $I = \sum_{k=-\infty}^{+\infty} |x(k)|^2$ if $X(z) = \sum_{k=-\infty}^{+\infty} x(k)z^{-k}$ in $(r_1, r_2)$ we then have $x(k) = \frac{1}{2\pi i} \oint_\Gamma z^{k-1}X(z)dz$ and $\Gamma$ is a closed contour, oriented in the region of convergence $(r_1, r_2)$.

$$I = \sum_{k=-\infty}^{+\infty} |x(k)|^2 = \frac{1}{2\pi i} \sum_{k=-\infty}^{+\infty} x(k) \oint_\Gamma z^{k-1}X(z)dz$$

$$= \frac{1}{2\pi i} \oint_\Gamma \sum_{k=-\infty}^{+\infty} x(k)z^k X(z) z^{-1} dz \qquad [4.40]$$

however $\sum_{k=-\infty}^{+\infty} x(k)z^k = X(z^{-1})$ converges in the region $(\frac{1}{r_2}, \frac{1}{r1})$.

THEOREM 4.2.– If the two regions $(r_1, r_2)$ and $(\frac{1}{r_2}, \frac{1}{r1})$ have a non-empty intersection, we obtain Parseval's theorem in $z$:

$$\sum_{k=-\infty}^{+\infty} |x(k)|^2 = \frac{1}{2\pi i} \oint_{\Gamma} X(z)X(z^{-1})dz \qquad [4.41]$$

where $\Gamma$ is a closed path contained in the regions $(r_1, r_2)$ and $(\frac{1}{r_1}, \frac{1}{r_2})$.

EXAMPLE 4.6.– *Take the signal*

$$\left\{ \begin{array}{ll} x(k) = a^k & k \geq 0 \\ x(k) = 0 & k < 0 \end{array} \right| \qquad [4.42]$$

*we have*

$$\sum_{k=-\infty}^{+\infty} |x(k)|^2 = \sum_{k=0}^{+\infty} a^{2k} = \frac{1}{1-a^2} \qquad [4.43]$$

*The contour $\Gamma$ lies in the region $((a, f(1; a))$. Only the singularity is inside the contour therefore:*

– $Z$-transform of a convolution product: let $x(k)$ and $y(k)$ be two numerical sequences; $X(z)$ $(r_1, r_2)$ and $Y(z)$ $(r_1^{'}, r_2^{'})$ their ZT. We calculate the ZT of the convolution product.

EXAMPLE 4.7.– *If it exists, the region of convergence of the product is the intersection of the two regions of convergence of $X(z)$ and $Y(z)$.*

### 4.2.4. Representation and modeling of signals and discrete systems

DEFINITION 4.7.– *A discrete system is an operator or an algorithm that transforms an input sequence $x(n)$ into an output sequence $y(n)$.*

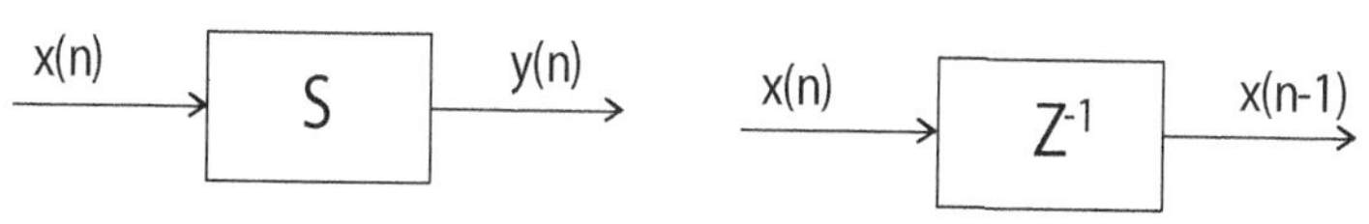


**Figure 4.13.** *Discrete systems*

#### 4.2.4.1. *Impulse response representation*

Let $h(n)$ be the response of a discrete system to the discrete impulse $\delta(n)$. Signal $h(n)$ is the impulse response of the discrete system (it is sometimes referred to by an abuse of language as the weighting sequence). It characterizes the system, because if $h(n)$ is known, the response $y(n)$ of the system can be determined for any input $x(n)$:

$$x(n) = \sum_{m=-\infty}^{+\infty} x(m)\delta(n-m) \quad [4.44]$$

the system being linear, we get

$$y(n) = S(x(n)) \quad [4.45]$$

$$y(n) = \sum_{m=-\infty}^{+\infty} x(m)S(\delta(n-m)) \quad [4.46]$$

and as $S(\delta(n-m)) = h(n-m)$, it yields

$$y(n) = \sum_{m=-\infty}^{+\infty} x(m)h(n-m) = \sum_{k=-\infty}^{+\infty} h(k)x(n-k) \quad [4.47]$$

(by establishing $n = m + k$)

This is the discrete convolution equation of $x(n)$ and $h(n)$; it only makes sense if the series thus defined does exist (converges).

$$y(n) = k(n) * x(n) = \sum_{m=-\infty}^{+\infty} x(m)h(n-m) = \sum_{k=-\infty}^{+\infty} h(k)x(n-k) \quad [4.48]$$

#### 4.2.4.2. *Causality and stability*

DEFINITION 4.8.– *A system is known as causal if and only if* $h(n) = 0$ *for* $n < 0$.

The output of the system is written as:

$$y(n) = k(n) * x(n) = \sum_{k=0}^{+\infty} h(k)x(n-k) \quad [4.49]$$

DEFINITION 4.9.– *A system is stable if for any bounded input there is a corresponding bounded output. A necessary and sufficient condition for a discrete system to be stable is that* $\sum_{k=+\infty}^{+\infty} |h(k)| < \infty$.

REMARK 4.5.– *This condition can be deduced from the conditions for the existence of the convolution product.*

EXAMPLE 4.8.– *Let the discrete system with impulse response* $h(n) = 2^n.E(n)$ *with the step signal defined in the discrete domain by*

$$E(n) = \begin{cases} 1 \text{ if } n \geq 0 \\ 0 \text{ if } n < 0 \end{cases} \qquad [4.50]$$

*This is a causal system because* $h(n) = 0$ *for* $n < 0$. *To consider stability, we evaluate* $\sum_{k=+\infty}^{+\infty} |h(k)| = \sum_{k=0}^{+\infty} 2^k$, *the series thus diverges and therefore the system is unstable.*

EXAMPLE 4.9.– *Now consider the discrete system of impulse response:* $h(n) = 2^{-n}.E(n)$. *This is a causal system because* $h(n) = 0$ *for* $n < 0$. *To assess stability, we evaluate* $\sum_{k=+\infty}^{+\infty} |h(k)| = \sum_{k=0}^{+\infty} 2^{-k} = \frac{1}{1-\frac{1}{2}} = 2$, *the series converges and the system is therefore stable.*

$$H(z) = TZ(h(k) = \sum_{k=0}^{+\infty} h(k)z^{-k} = \sum_{k=0}^{+\infty}(2z)^{-k} = \frac{1}{1-\frac{1}{2z}} = \frac{1}{1-0.5z^{-1}} \qquad [4.51]$$

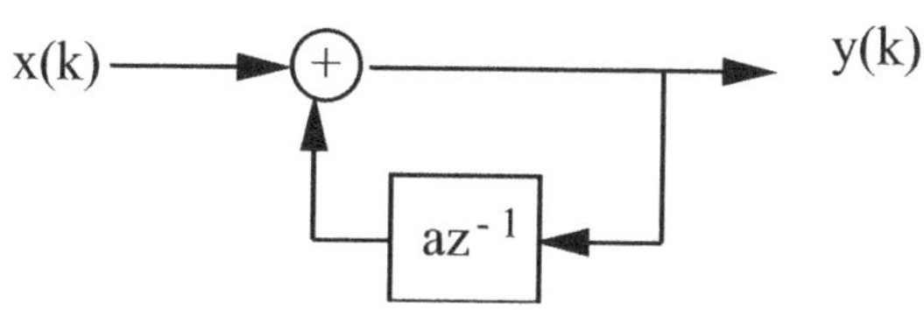


**Figure 4.14.** *Input/output system with feedback*

### 4.2.4.3. *Frequency response representation*

This is simply inferred from the relation of the time convolution by Fourier transform (the latter only makes sense if the system is stable). $y(k) = h(k) * x(k) \rightleftarrows Y(f) = H(f).X(f)$ with the frequency response of the system defined by the equation giving the FT of the discrete impulse response

$H(f) = \sum_{k=+\infty}^{+\infty} h(k)e^{-2i\pi fk}$. The name harmonic response is justified when $x(k)$ is a harmonic signal: $x(k) = e^{-2i\pi fk} \longmapsto X(f) = \delta(f - f_o)$ (the frequency is such that $0 \leq f \leq 0.5$ which means implicitly that $F_e = 1$) where $Y(f) = H(f_o).\delta(f - f_o)$. The output is also a harmonic signal whose modulus and phase are characterized by $H(f_o)$.

$H(f)$ is a continuous function of $f$, periodic of period $1/(F_e)$ and has all the properties of the Fourier transforms of absolutely summable and square-integrable sequences (discrete Fourier transform).

EXAMPLE 4.10.– *Averaging filter:*

$$h(n) = \begin{cases} 1/N \; if \; N > n \geq 0 \\ 0 \quad otherwise \end{cases} \qquad [4.52]$$

*its frequency response is*

$$H(f) = \frac{1}{N}\sum_{n=0}^{N-1} e^{-2i\pi fn} = e^{-i\pi(N-1)f}\frac{1}{N}\frac{\sin(\pi Nf)}{\sin(\pi f)} \qquad [4.53]$$

*It is a low-pass filter and especially so when $N$ is large. The static gain of the filter is defined by*

$$H(0) = \sum_{n=0}^{N-1} h(n) = 1 \qquad [4.54]$$

EXAMPLE 4.11.– *Given the causal filter with infinite length impulse response (RII): $h(n) = \beta^n$, $n \geq 0$ with $0 < \beta < 1$; we have*

$$H(f) = \sum_{n=0}^{+\infty} \beta^n e^{-2i\pi fn} = \frac{1}{1-\beta e^{-2i\pi f}} = \frac{1}{1-\beta\cos(2\pi f)+\beta i\sin(2\pi f)} \qquad [4.55]$$

*which also gives us a low-pass filter $|H(f)|^2 = \frac{1}{1+\beta^2-2\beta\cos(2\pi f)}$ with static unit gain $H(0) = \frac{1}{1-\beta}$.*

### 4.2.5. *Transfer function in Z and representation in the frequency domain*

#### 4.2.5.1. *Transmittance in Z-parametric representation*

Let us take a given system and the transmittance of the discrete system $y(z) = H(z)X(z)$; if the system is stable, it has a sense and $H(z)$ exists on the unit circle that thereby belongs to the region of convergence.

A causal discrete system is stable if $H(z)$ has all its singularities (poles) inside the unit circle (singularities $p_k$ are such that $p_k < 1$).

We may consider the $m$ zeros $z_m$ of $H(z)$ and the poles of $H(z)$ denoted $p_k$ and write:

$p_k = a < 1$ stable system.

We shall see how to associate a point of the unit circle with each frequency to evaluate the Fourier transform $H(\ddot{A})$ of $h(n)$.

Hence, assuming $z = e^{2j\pi nf}$, we obtain the FT from the $z$-transform.

Recalling the previous example: $H(z) = f(z; z - a)$ ;

$H(\ddot{A}) = (e2^1 j\ddot{A}) = f(e2^1 j\ddot{A}; e2^1 j\ddot{A} - a)$ and $\acute{U}H(\ddot{A})\acute{U} = \acute{U}e2j^1\ddot{A}\acute{U}/\acute{U}e2j^1\ddot{A} - a\acute{U}$, this is the ratio of the moduli of two vectors, equal to zero ($z1 = 0$) at $e2j^1\ddot{A}$ of the complex plane and pole ($p_1 = a$) at $e2j^1\ddot{A}$.

The ratio of the two vectors passes through a minimum when approaching a zero (maximum attenuation) and a maximum when approaching a pole (maximum amplification).

Spectrum $\acute{U}H(\ddot{A})\acute{U}$ will have a peak for $\ddot{A} = 0$ (if $Re(a) > 0$, at the poles frequencies) and a minimum for $\ddot{A} = \pm f(1; 2)(Z = -1)$ (at the frequencies of the zeros).

NOTE.– For the symmetry of the spectrum, the frequencies are varied from

$\ddot{A} = -f(1; 2)(\backslash f(-Fe; 2))$ to $f = +f(1; 2)(\backslash f(+Fe; 2))$

#### 4.2.5.2. *Parametric representation*

When we aim to simulate a system, it is advantageous to express its ZT as a rational fraction, especially, if $h(k)$ is of infinite length.

The simulation becomes possible because $Y(z) = H(z)X(z)$ is written as follows:

– the term by term identification of both series gives the recurrence equation of the system

$$y(k) = -a_1 y(k-1) - ...a_n y(k-n) + b_0 x(k) + ... + b_m x(k-m) \quad [4.56]$$

The output at moment $k$ is calculated as a function of the previous inputs and outputs.

EXAMPLE.–

Given the system defined by the recurrence equation:

– we calculate the ZT member by member:

– hence the transmittance in $Z$.

It is therefore possible to analyze the system, derive therefrom its frequency response and calculate poles and zeros.

Digital filters, controllers and digital systems can be analyzed by means of the tools presented above.

NOTE.–

For the unilateral ZT, we get: $Z(x(k-ko)) = z - ko[X(z) + x(-1)z + ... + x(-ko)z - ko]$ and $Z(x(k+ko)) = zko[X(z) - x(o) - x(1)z - 1 - ... - z - ko + 1x(-ko)zko]$.

EXERCISE.–

Determine the inverse ZT of a digital correction element is characterized by its equation

$$y(k+1) = y(k) - kx(k) \quad [4.57]$$

1) Express its transfer function in $Z$.

2) Determine its response to a unit step function.

3) What function can this correction achieve?

### 4.2.6. *Z-domain transform, Fourier transform and Laplace transform*

#### 4.2.6.1. *Relationship with the Laplace transform*

Let $xa(t)$ be an analog signal. Its bilateral Laplace transform is: $xe(t) = T(t).xa(t)$ a version sampled in period $T$ of $xa(t)$.

The Laplace transform of a sampled signal is none other than the ZT of this signal for $Z = e^{-st}.f(Xe(Z); z) = Xe(S)$.

#### 4.2.6.2. *Relationship with the Fourier transform*

Let $Z = rexp(io)$, a representation using the polar coordinates of $Z(q = 2^1 \ddot{A})$.

It is possible to interpret the transform in $Z(X(Z))$ of $x(k)$, as the Fourier transform of this signal multiplied by an exponential signal $G - k.x(k)$.

In paticular for $G = 1(\acute{U}Z\acute{U} = 1)$, the ZT is similar to the FT. The Fourier transformation can be evaluated by taking the ZT on the unit circle.

#### 4.2.6.3. *Application of the S-plane to the Z-plane*

The relation $Z = e^{sT}$ establishes the application (transition relation) of the $S$ complex plane to the $Z$ complex plane (discrete continuous): $s(s + i2^1 \ddot{A})Z = eln\,|Z| + iArg(Z)$.

This application is not one-to-one. The inverse relationship is given by:

– the argument $Z(arg(Z)$ is determined modulus $2^1$;

– the left-half plane of the $S$-plane is applied inside the unit circle. The argument of $Z$ is periodic of period $f(1;T)$. The right-half plane of $S$ corresponds to the interior of the unit circle.

#### 4.2.6.4. *Final and initial signal values*

Initial value theorem:

$$f(0^+) = \lim_{t \to 0} f(t) = \lim_{p \to \infty} p.F(p) \qquad [4.58]$$

Final value theorem:

$$f(\infty) = \lim_{t \to \infty} f(t) = \lim_{p \to 0} p.F(p) \qquad [4.59]$$

NOTE.– For the unilateral ZT, $ZT(x(k + ko)) = z + ko[X(z) - x(o) - z - 1.x(1) - ... - z - ko + 1.x(ko - 1)]$.

## 4.3. A few application exercises

EXERCISE 1.– Let $S_{xx}(f)$ be the spectral power of signal $x(t)$. The spectrum $S_{xe}(f)$ obtained after sampling is shown in Figure 4.15.

Determine the sampling rate $F_e$ being used. Has the signal been properly sampled?

- According to Shannon, what is the minimum sampling frequency?

- At what value can $s(t)$ be downsampled?

*Digital application*: $F_0 = 1$ MHz, $\tau = 1$ ms.

EXERCISE 6.– Let $X_e(t)$ be a sampled signal. What relation links the Fourier transform and the $Z$-transform of $X_e(t)$?

EXERCISE 7.– The objective is to implement a digital integrator based on the recurrence equation:

$$y(k) = y(k-1) + T\frac{x(k) + x(k-1)}{2} \quad [4.62]$$

– Calculate the $Z$-transform of its impulse response.

– Provide a diagram for a possible implementation.

EXERCISE 8.– Calculate the $Z$-transform of sequences:

$u(n)$ =Heaviside function; $\delta(n) = 1$ for $n = 0$ and 0 elsewhere; $x(n) = a^n u(n)$;

$$x(n) = \begin{cases} 0 & n < 0 \\ 1 & n = 4p \text{or } n = 4p + 1 (p > 0) \\ 0 & n = 4p + 2 \text{or } n = 4p + 3 (p > 0) \end{cases} \quad [4.63]$$

EXERCISE 9.– Calculate the inverse $Z$-transform of:

$$X_1(z) = \frac{1}{1 - 3z^{-1} + 2z^{-2}} \qquad X_2(z) = \frac{0.25z}{z^2 - 1.5z + 0.5} \quad [4.64]$$

EXERCISE 10.– The impulse response of a filter is given by:

$h(0) = 1$; $h(1) = 0.75$; $h(2) = 0.5$; $h(3) = 0.25$ and $h(n) = 0$ elsewhere.

– Determine $H(z)$.

– What is the recurrence equation?

EXERCISE 11.– A discrete signal $x(k)$ has for $Z$-transform:

$$X(z) = \frac{1}{(1 - z^{-1})(1 - 0.5z^{-1})} \quad [4.65]$$

Find $x(k)$ using partial fraction decomposition.

#### 4.2.6.2. *Relationship with the Fourier transform*

Let $Z = rexp(io)$, a representation using the polar coordinates of $Z(q = 2^1 \ddot{A})$.

It is possible to interpret the transform in $Z(X(Z))$ of $x(k)$, as the Fourier transform of this signal multiplied by an exponential signal $G - k.x(k)$.

In paticular for $G = 1(\acute{U}Z\acute{U} = 1)$, the ZT is similar to the FT. The Fourier transformation can be evaluated by taking the ZT on the unit circle.

#### 4.2.6.3. *Application of the S-plane to the Z-plane*

The relation $Z = e^{sT}$ establishes the application (transition relation) of the $S$ complex plane to the $Z$ complex plane (discrete continuous): $s(s + i2^1 \ddot{A})Z = eln\,|Z| + iArg(Z)$.

This application is not one-to-one. The inverse relationship is given by:

– the argument $Z(arg(Z)$ is determined modulus $2^1$;

– the left-half plane of the $S$-plane is applied inside the unit circle. The argument of $Z$ is periodic of period $f(1;T)$. The right-half plane of $S$ corresponds to the interior of the unit circle.

#### 4.2.6.4. *Final and initial signal values*

Initial value theorem:

$$f(0^+) = \lim_{t \to 0} f(t) = \lim_{p \to \infty} p.F(p) \qquad [4.58]$$

Final value theorem:

$$f(\infty) = \lim_{t \to \infty} f(t) = \lim_{p \to 0} p.F(p) \qquad [4.59]$$

NOTE.– For the unilateral ZT, $ZT(x(k + ko)) = z + ko[X(z) - x(o) - z - 1.x(1) - \ldots - z - ko + 1.x(ko - 1)]$.

### 4.3. A few application exercises

EXERCISE 1.– Let $S_{xx}(f)$ be the spectral power of signal $x(t)$. The spectrum $S_{xe}(f)$ obtained after sampling is shown in Figure 4.15.

Determine the sampling rate $F_e$ being used. Has the signal been properly sampled?

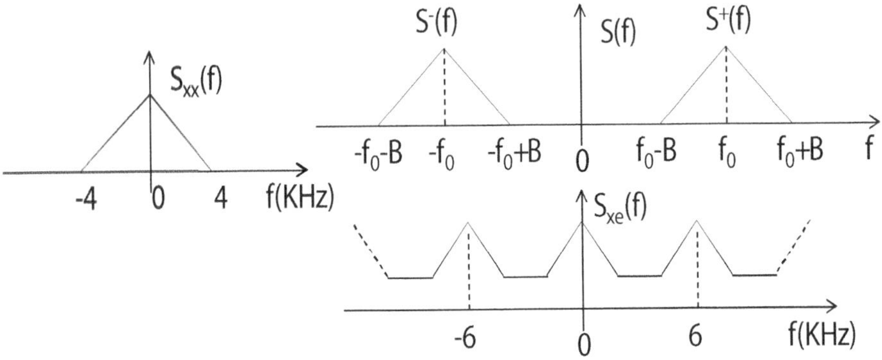


**Figure 4.15.** *Spectrum after sampling*

EXERCISE 2.– We sample the signal $x(t) = 5\cos(500 * 2\pi t)\cos^2(1000 * 2\pi t)$ at a frequency $F_e = 4.5\,\text{kHz}$. The sampled signal is filtered by a low-pass filter of cutoff frequency $F_c = 2.6\,\text{kHz}$. Give the analytical expression of the output signal.

EXERCISE 3.– Let $X(f)$ be the spectrum of the signal $x(t)$. The aim is to compare two types of sampling.

1) *Ideal sampler*

Let $T_e$ be the sampling period and $x_{ie}(t)$ be the sampled signal.

- Determine $x_{ie}(t)$ according to $x(t)$ and to the Dirac comb of width $T_e$. Derive thereof $X_{ie}(f)$ according to $X(f)$.

- Show that it is mathematically possible to reconstruct the signal using a low-pass filter with cutoff frequency $\frac{F_e}{2}$.

2) *Average sampler*

Sampling is now achieved with a sequence of impulses of finite duration $\theta$ spaced by a width $T_e$. The obtained sample $x_{me}(kT_e)$ is equal to the average of $x(t)$ during the pulse:

$$x_{me}(kT_e) = \frac{1}{\theta}\int_{kT_e-\theta/2}^{kT_e+\theta/2} x(t)dt \qquad [4.60]$$

- Show that $x_{me}(kT_e)$ can be expressed as the convolution product of $x(t)$ and the rectangular function of width $\theta$ taken at moment $kT_e$.

- Derive thereof $x_{me}(t)$ and $X_{me}(f)$. Verify that: $\lim_{\theta\to 0}(X_{me}(f)) = X_{ie}(f)$.

- Let $F_c$ be the cutoff frequency of the signal. Assuming $\theta = \lambda T_e$ ($\lambda < 1$) and $F_e = 2\alpha F_c$ ($\alpha \geq 1$). Determine the ratio $\frac{\lambda}{\alpha}$ so that the effect of filtering be lower than $1\%$ up to the cutoff frequency $F_c$. Determine $\lambda$ for the limit value of $\alpha$.

EXERCISE 4. Given a signal $s(t)$ resulting from the sum of two signals $s_1(t)$ and $s_2(t)$. It is thus assumed that the spectrum is zero outside the bands $[-f_1 - B, -f_1 + B]$, $[f_1 - B, f_1 + B]$, $[-f_2 - B, -f_2 + B]$ and $[f_2 - B, f_2 + B]$ with $f_1 < f_2$ and $B \ll f_1$.

How can the sampling frequency $F_e$ be chosen so as to avoid the aliasing phenomenon of the spectrum?

EXERCISE 5.– The idea is to sample a signal $s(t)$ whose spectrum is represented in Figure 4.16.

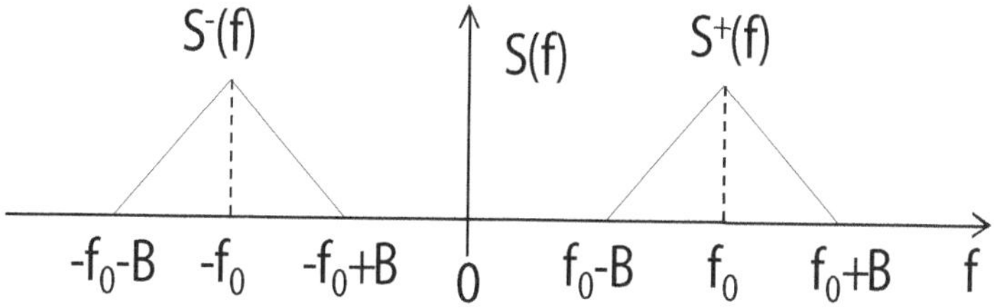


**Figure 4.16.** *Spectrum of a given signal*

1) According to Shannon, what is the minimum sampling frequency?

2) Show that $s(t)$ can be downsampled without distorting $s^+(f)$ and $s^-(f)$. Provide a range for the sampling frequency $F_e$ based on $F_0$, $B$ and $k$ ($k$ = number of spectrum translations).

3) Give a condition of $k$ so that, after sampling and filtering by a low-pass filter of cutoff frequency $\frac{F_e}{2}$, $s^+(f)$ is composed of the band $[0, \frac{F_e}{2}]$ and that $s^-(f)$ is composed of the band $[-\frac{F_e}{2}, 0]$.

*Application*: Let a signal be defined by:

$$y(t) = e^{-\frac{t}{T}} u(t) \qquad u(t)\text{= Heaviside function.} \qquad [4.61]$$

1) Filtering with an ideal low-pass filter of cutoff frequency $B$. Determine $B$ according to $\tau$, so that the filtered signal $y_F(t)$ contains $95\%$ of the energy of $y(t)$.

2) Assuming $s(t) = y_F(t)\cos(2\pi F_0 t)$:

- Represent the spectrum of $s(t)$.

- According to Shannon, what is the minimum sampling frequency?
- At what value can $s(t)$ be downsampled?

*Digital application*: $F_0 = 1$ MHz, $\tau = 1$ ms.

EXERCISE 6.– Let $X_e(t)$ be a sampled signal. What relation links the Fourier transform and the $Z$-transform of $X_e(t)$?

EXERCISE 7.– The objective is to implement a digital integrator based on the recurrence equation:

$$y(k) = y(k-1) + T\frac{x(k) + x(k-1)}{2} \quad [4.62]$$

– Calculate the $Z$-transform of its impulse response.

– Provide a diagram for a possible implementation.

EXERCISE 8.– Calculate the $Z$-transform of sequences:

$u(n)$ =Heaviside function; $\delta(n) = 1$ for $n = 0$ and 0 elsewhere; $x(n) = a^n u(n)$;

$$x(n) = \begin{cases} 0 & n < 0 \\ 1 & n = 4p \text{or } n = 4p+1 (p > 0) \\ 0 & n = 4p+2 \text{or } n = 4p+3 (p > 0) \end{cases} \quad [4.63]$$

EXERCISE 9.– Calculate the inverse $Z$-transform of:

$$X_1(z) = \frac{1}{1 - 3z^{-1} + 2z^{-2}} \qquad X_2(z) = \frac{0.25z}{z^2 - 1.5z + 0.5} \quad [4.64]$$

EXERCISE 10.– The impulse response of a filter is given by:

$h(0) = 1$; $h(1) = 0.75$; $h(2) = 0.5$; $h(3) = 0.25$ and $h(n) = 0$ elsewhere.

– Determine $H(z)$.

– What is the recurrence equation?

EXERCISE 11.– A discrete signal $x(k)$ has for $Z$-transform:

$$X(z) = \frac{1}{(1 - z^{-1})(1 - 0.5z^{-1})} \quad [4.65]$$

Find $x(k)$ using partial fraction decomposition.

Calculate $x(0)$ $x(1)$ and $x(2)$ using polynomial long division.

EXERCISE 12.– Calculate the inverse $Z$-transform of functions:

$$F(z) = \frac{2z}{(z-2)(z-1)^2} \qquad G(z) = \frac{z}{(z-1)(z-0.4)(z-0.5)} \tag{4.66}$$

– Give the step response of the digital filter defined by:

$$F(z) = \frac{0.5}{z+0.5} \tag{4.67}$$

– Give the step response of the filter:

$$F(z) = \frac{T_e z^{-1}}{1-z^{-1}} \tag{4.68}$$

What system does this filter realize?

– Calculate the inverse $Z$-transform of function:

$$F(z) = e^{z^{-1}}(1+z^{-1}) \tag{4.69}$$

EXERCISE 13.– Given a first-order system

$$H(p) = \frac{G}{1+Tp} \tag{4.70}$$

1) Give its differential equation, and a continuous simulation diagram.

2) Discretize this system in the period $T_e$, the derivative is approximated by:

$$[\frac{y((k+1)T_e) - y(kT_e)}{T_e}] \tag{4.71}$$

Give its recurrence equation, its transfer function in $Z$, the step response of the continuous system and that of the discrete system.

EXERCISE 14.– The following recurrence equations are provided:

a) $y_k = y_{k-1} + Ku_{k-1}$

b) $y_k - ay_{k-1} = Ku_{k-1} - bu_{k-2}$

c) $y_k - 1.5y_{k-1} + 0.5y_{k-2} = K(0.5u_{k-2} + u_{k-3})$

Give the transfer functions in $Z$ and calculate $K$ giving a static unit gain.

Calculate their step and impulse response.

EXERCISE 15.–

1) Determine the $Z$-transform of $s(k)$:

$$s(k) = [\frac{e^{ak} - e^{bk}}{j\omega}]u(k) \quad \text{with } u(k)\text{the unit step} \qquad [4.72]$$

2) A digital filter has the recurrence equation:

$$y(k) = 2\cos(\frac{T_e}{\sqrt{LC}})y(k-1) - y(k-2) + \frac{1}{\sqrt{LC}}\sin(\frac{T_e}{\sqrt{LC}})e(k-1) \quad [4.73]$$

where $y(k)$ is the output of the filter, $e(k)$ input, $T_e$ the sampling period and $L$, $C$ two constants.

- Give the transfer function $H(z)$ and the impulse response $h(k)$.
- Propose a digital implementation of this filter.

3) We assume $h_e(t) = h(k)\delta(t-k)$. Determine a function $h(t)$ such that:

$$h_e(t) = h(t)\delta(t-k) \qquad [4.74]$$

Determine the Laplace transform $h(t)$. Give an analog implementation of the filter.

EXERCISE 16.– Calculate the $Z$-transforms of the following signals ($u_k$= Heaviside):

1) $y_k = u_{k-l}$

2) $y_k = (\frac{1}{2})^k u_k$

3) $y_k = (\frac{1}{3})^k u_{k-l}$

4) $y_k = e^{-\alpha k}u_k$ with $\alpha > 0$

5) $y_k = -a_1 y_{k-1} - a_2 y_{k-2} + b u_{k-1}$

6) Determine the recurrence equation for the signal $y_k = cos(\omega_0 k)$.

7) Give the $Z$-transform of signal $y_k = 2cos(\omega_0)y_{k-1} - y_{k-2} + e_k$ knowing that $E(z)$ is the transform of $e_k$.

8) Give the $Z$-transform of signal $y_k = (e^{-ak} + e^{-bk})u_k$.

EXERCISE 17.– Study the step response with respect to $a$ of the following filter:

$$F(z) = \frac{Kz^{-1}}{1 - az^{-1}} \qquad [4.75]$$

Calculate $K$ such that the static gain be equal to unity.

EXERCISE 18.– Consider the filter in Figure 4.17.

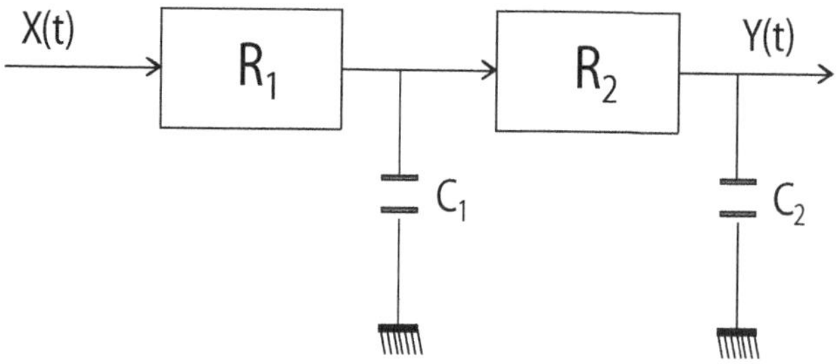


**Figure 4.17.** *Given filter*

It will be assumed that $C_2 \ll C_1$.

1) Prove that the transfer function $H(j\omega)$ can be written as:

$$H(j\omega) = \frac{1}{1 + (R_1C_1 + R_2C_2)j\omega - R_1R_2C_1C_2\omega^2} \quad [4.76]$$

Define $R_1 = 5R_2$ and $C_1 = 5C_2$.

- Calculate and represent its impulse response.

- Is the filter causal?

2) We want to implement this filter in digital form taking $T = \frac{R_2C_2}{2}$ as the sampling period.

- Calculate $H(z)$.

- Derive thereof the recurrence equation. Give a possible digital implementation of the sampled system.

*Application*: $R_2 = 10\,\Omega$ and $C_2 = 0.1$ mF.
Represent output $Y(n)$ if $x(0) = 1$ and $x(n) = 0$ for $n \neq 0$.

*Note*: $\mathrm{TF}(e^{-\alpha t}u(t)) = \frac{1}{\alpha+2\pi j\nu}$ with $u(t)$:unit step function.

EXERCISE 19.– A sampled signal $x(k)$ is applied on a filter input whose output $y(k)$ verifies

$$y(k) + ay(k-1) = x(k) + bx(k-1) + cx(k-2) \quad [4.77]$$

with $a \neq 0$ and $c \neq 0$.

1) Using the transfer function in $Z$, $H(z)$, give the relation between $a$, $b$ and $c$ so that the above equation can be written as:

$$y(k) = x(k) + dx(k-1) \quad [4.78]$$

including the parameter $d$ to be specified.

2) Recover the previous result without using $H(z)$.

EXERCISE 20.– Given a filter characterized by its transfer function $H(z)$:

$$H(z) = \frac{1 - z^{-12}}{1 + z^{-2}} \quad [4.79]$$

1) Determine its recurrence equation.

2) Give his complex gain $H(f)$ and represent $\mid H(f) \mid$.

3) Give a non-recursive $H(z)$ and deduce therefrom a new recurrence equation.

4) Determine and represent impulse and step responses.

EXERCISE 21.– A filter is characterized by its transfer function:

$$H(z) = \frac{1}{9}\frac{(1 - z^{-3})^2}{(1 + z^{-1})^2} \quad [4.80]$$

1) Determine the recurrence equation of the filter.

2) Find a non-recursive form, give the new recurrence equation.

3) Determine and represent impulse and step responses.

4) Express the complex gain, represent its modulus and its phase.

EXERCISE 22.– The recurrence equation of a filter is given by:

$$y(k) = -ay(k-1) - by(k-2) + e(k) \quad [4.81]$$

where $e(k)$ is a centered white noise of variance $\sigma^2$.

We define the autocorrelation by $C_y(n) = E[y(k)y(k-n)]$. Show that if

$E[y(k-n)e(k)] = 0$, for $n \neq 0$, we have the relation:

$$x(n) = \begin{pmatrix} C_y(0) & C_y(1) & C_y(2) \\ C_y(1) & C_y(0) & C_y(1) \\ C_y(2) & C_y(1) & C_y(0) \end{pmatrix} \begin{pmatrix} 1 \\ a \\ b \end{pmatrix} = \begin{pmatrix} \sigma^2 \\ 0 \\ 0 \end{pmatrix} \quad [4.82]$$

EXERCISE 23.– The transfer function of a digital filter is:

$$H(z) = 1 + z^{-1} + z^{-2} + z^{-3} + \ldots + z^{-l} \qquad [4.83]$$

1) What is the recurrence equation implemented by the computer? Specify the number of required operations.

2) Give a recursive form to the previous filter. Derive thereof the recurrence equation and the number of necessary operations in this case.

3) Determine and represent the modulus of the complex gain $H(j\omega)$.

4) Calculate the ratio in decibels between the amplitude of the main lobe and the first secondary lobe.

5) The sampling frequency is $1,000$ Hz. Based on the previous results, we propose to implement a low-pass filter whose attenuation is greater than 13 dB for $f > 100$ Hz. Which recurrence equation must the computer execute?

6) What are the approximate bandwidths for $0.1$ and $1$ dB.

7) Calculate the attenuation for $f = 50$ Hz.

NOTE.–

$$\sum_{f=0}^{n} r^i = \frac{1 - r^{n+1}}{1 - r}; \; \sin(\theta) \simeq \theta - \frac{\theta^3}{3!} + \ldots \qquad [4.84]$$

EXERCISE 24.– Let a discrete filter be characterized by the recurrence equation:

$$y(k) = y(k-3) + \alpha(u(k) + 3u(k-1) + 3u(k-2) + u(k-3)) \qquad [4.85]$$

1) Determine its transfer function.

2) Represent the impulse response. Demonstrate that it presents a periodic phenomenon.

3) Derive therefrom the step response without calculations.

4) What is the nature of this filter?

5) Calculate its complex gain. What happens to this gain for low frequencies? Derive the value of $\alpha$ to get the equivalent with the corresponding analog filter $H_I(p)$.

EXERCISE 25.– We aim to study the filter defined by the following recurrence equation:

$$y(k) = y(k-2) + \frac{T}{3}(u(k) + 4u(k-1) + u(k-2)) \;\; (T\text{: sampling period}) \qquad [4.86]$$

1) Determine the step response for each step and plot it.

2) Determine the transfer function and plot the frequency response ($H_R(j\omega)$).

3) What continuous transfer function ($H_I(j\omega)$) is this filter likely to achieve?

4) Study the ratio:

$$\rho(\omega) = \frac{H_R(j\omega)}{H_I(j\omega)} \quad [4.87]$$

EXERCISE 26.– It is desirable to perform the synthesis of an analog filter characterized by the transfer function

$$H(p) = \frac{1 - \tau p}{1 + \tau p} \quad [4.88]$$

The sampling period $T_e$ is equal to $\tau/10$. Then determine the recurrence equation of the impulse response of the filter when we use:

1) a transform method;

2) a step invariance;

3) an idealization of the step response by two straight lines (previously represent the step response of the analog filter) with step invariance. Therefrom, derive a non-recursive form and its impulse response.

EXERCISE 27.–

1) Determine the transfer function $H_1(z)$ that realizes the equivalent of the filter:

$$H_1(p) = \frac{1 + T_1 p}{1 + T_2 p} \quad [4.89]$$

using the results of Exercise 5.

2) Give in canonical form the recurrence equations of $H_1(z)$.

3) Give $H_1(z)$ using bilinear transform.

4) Compare impulse responses if $T_1 = 10T$ and $T_2 = 3T$ of $H_1(p)$ and $H_1(z)$.

EXERCISE 28.– Same questions as Exercise 4 for the following filters:

$$y(k) = y(k-1) + Tu(k) \quad [4.90]$$

$$y(k) = y(k-1) + \frac{T}{2}(u(k) + u(k-1)) \quad [4.91]$$

$$y(k) = -y(k-1) + \frac{2}{T}(u(k) - u(k-1)) \qquad [4.92]$$

EXERCISE 29.– Non-recursive digital filtering is an operation that, in a sampled signal $x(t)$, associates a sampled signal $y_l(t)$ whose samples are given by the relation

$$y_{l,m} = \sum_{n=-N}^{+N} b_n x_{m-n} \qquad [4.93]$$

$x_m = x(mT_e), x(mT_e)$ being the *m*th sample of the signal $x(t)$ taken at moment $mT_e$, $T_e$ is the sampling period of the signal. Variables $b_n$ are characteristic coefficients of the filter. $N$ can *a priori* take all values from $-\infty$ to $+\infty$.

1) Demonstrate that this filtering operation is linear and stationary.

2) What can be said of coefficients $b_n$ so this filter can be implementable? (We assume next that the filtering is not necessarily "feasible", which is often the case when the signal is processed offline on a computer).

3) Determine the transfer function of this filter according to coefficients $b_n$. To this end, we will use the property of a filter to transform a signal $e^{2\pi j\nu t}$ into a signal of the same frequency but different phase and amplitude.

4) What is the essential property of this transfer function?

NOTES.–

i) A filter is said to be "implementable" if the response depends only on previous inputs;

ii) If $x(t) = e^{2\pi j\nu t}$, then $y(t) = H(\nu)x(t)$.

5) *Application*:

– Give the transfer function associated with the filtering:

$$y_{l,m} = \frac{1}{2N+1} \sum_{n=-N}^{+N} x_{m-n} \qquad [4.94]$$

and especially those associated with the case in which $N = 1$.

– Draw the graph of the transfer function for the case where $N = 1$. What effect does such a filter have upon the function $x(t)$?

The signal obtained in the case $N = 1$ is again filtered $p$ times under the same conditions, such that:

$$y_{2,m} = \frac{1}{3} \sum_{n=-1}^{+1} y_{1,m-n} \quad [4.95]$$

$$y_{p,m} = \frac{1}{3} \sum_{n=-1}^{+1} y_{p-1,m-n} \quad [4.96]$$

– Give the transfer function associated with these $p$ filters. What is their effect when $p$ becomes large?

EXERCISE 30.– Let $x_e(t)$ be a sampled signal. What relation links the Fourier transform and the $Z$-transform of $X_e(t)$.

EXERCISE 31.– The objective is to implement a digital integrator based on the recurrence equation

$$y(k) = y(k-1) + T[x(k) + x(k-1)]/2 \quad [4.97]$$

– Calculate the $Z$-transform of its impulse response.

Provide a diagram for a possible implementation.

EXERCISE 32.– Calculate the $Z$-transforms of sequences: $u(n)$ Heaviside function; $d(n) = 1$ for $n = 0$ and 0 elsewhere; $x(n) = anu(n)$;

EXERCISE 33.– Calculate the inverse $Z$-transform of:

$$X_1(z) = f(1; 1 - 3z - 1 + 2z - 2); X_2(z) = f(0, 5z; z - 2 - 1.5z - 1 + 0.5) [4.98]$$

EXERCISE 34.– The impulse response of a filter is given by: $h(0) = 1$; $h(1) = 0.75$; $h(2) = 0.5$; $h(3) = 0.25$ and $h(n) = 0$ otherwise.

1) Determine $H(z)$.

2) What is the recurrence equation?

EXERCISE 35.– A discrete signal $x(k)$ has the $Z$-transform:

$$X(z) = 1/[(1 - z - 1)(1 - Az - 1/2)] \quad [4.99]$$

1) Recover $x(k)$ using partial fraction decomposition.

2) Calculate $x(0)$, $x(1)$, and $x(2)$ using polynomial long division.

EXERCISE 36.– Consider a low-pass RIF filter, with linear phase whose frequency response is represented in Exercise 18.

1) Show that the system can be considered as two systems placed in series that will be determined as $G(z)$.

2) The second system represented by $Gb(z)$ is decomposed into a battery of recursive filters of the first and second order, assembled in parallel. Determine the $Z$-transform of each of these filters.

3) Write the equation of recurrence of the global system according to $X_k$ and to $X_{k-1}$.

EXERCISE 37.– Non-recursive digital filtering is an operation which, at a sampled signal x(t), combines a sampled signal $y1(t)$ whose samples are given by the relation

$$x_m = x(mT_e), x(mT_e),$$

such that the *n*th sample of the signal $x(t)$ is taken at moment $mT_e$, $T_e$, and the sampling period of the signal is $x_{m-n} = x((m-n)T_e)$.

Coefficients $b_n$ are characteristic coefficients of the filter.

*N* can take, *a priori*, any values between - $\infty$ and + $\infty$.

1) Demonstrate that this filtering operation is linear and stationary.

2) What can be said of coefficients $b_n$ in order for this filter to be "implementable"? (In the following, we will assume that the filtering is not necessarily "implementable", which is often the case when the signal is processed offline on a computer).

3) Determine the transfer function according to the coefficients $bn$ of this filter. To this end, we will use the property of a filter to transform a signal $eiwt$ into a signal of the same frequency but with different phase and amplitude.

4) What is the essential property of this transfer function?

NOTES.–

i) A filter is said to be "implementable" if the response depends only on previous inputs.

ii) If $x(t) = e^{2\pi j\nu t}$ then $y(t) = H(n)s(t)$.

*Application*:

5) Give the transfer function associated with the filtering and more specifically that associated with the case where $N = 1$.

6) Draw the graph of the transfer function for case 5 ($N = 1$).

What effect does such a filter have upon the function $s(t)$?

7) The signal obtained in case 5 ($N = 1$) is again filtered $p$ times under the same conditions, such that we have:

i) Give the transfer function associated with these $p$ filters.

ii) What is their effect when $p$ is large?

EXERCISE 38.– A sampled signal $s(k)$ is applied on a filter input whose output $y(k)$ verifies

$$y(k) + ay(k-1) = x(k) + bx(k-1) + cx(k-2) \quad [4.100]$$

with $a_0$ and $c_0$.

1) Using the transfer function in $Z$ $H(z)$, give the relation that must link $a$, $b$ and $c$ so that the above equation can be written as: $y(k) - x(k) + dx(k-1)$ with a $d$ that will be specified.

2) Recover the previous result without using $H(z)$.

EXERCISE 39.– Given a filter characterized by its transfer function $H(z)$

$$H(z) = \frac{1 - z - 12}{1 + z - 2} \quad [4.101]$$

1) Determine its recurrence equation.

2) Give this complex gain $H(f)$ and represent $ÚH(f)Ú$.

3) Give a non-recursive form of $H(z)$ and derive therefrom a recurrence equation.

4) Determine and represent impulse and step responses.

EXERCISE 40.– Let $S_{xx}(f)$ be the spectral power of signal $x(t)$. The spectrum $S_{xe}(f)$ obtained after sampling is represented in Figure 4.18.

Determine the sampling rate $F_e$ being used. Has the signal been properly sampled?

EXERCISE 41.– We sample the following signal:

$$x(t) = 5\cos(500 * 2\pi t)\cos^2(1,000 * 2\pi t) \quad [4.102]$$

at a frequency $F_e = 4.5$ kHz. The sampled signal is filtered by a low-pass filter having a cutoff frequency $F_c = 2.6$ kHz. Give the analytical expression of the output signal.

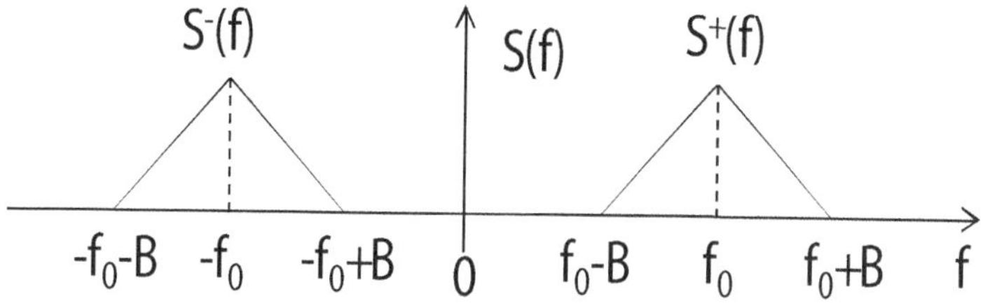


**Figure 4.18.** *Spectrum after sampling*

EXERCISE 42.– Let $X(f)$ be the spectrum of signal $x(t)$

1) *Ideal sampler*

Let $T_e$ be the sampling period and $x_{ie}(t)$ the sampled signal.

- Determine $x_{ie}(t)$ according to $x(t)$ and the Dirac comb of width $T_e$. Derive thereof $X_{ie}(f)$ according to $X(f)$.

- Show that it is mathematically possible to reconstruct the signal using a low-pass filter with cutoff frequency $\frac{F_e}{2}$.

2) *Average sampler*

Sampling is now achieved with a sequence of impulses of finite duration $\theta$ spaced by a width $T_e$. The obtained sample $x_{me}(kT_e)$ is equal to the average of $x(t)$ during the pulse:

$$x_{me}(kT_e) = \frac{1}{\theta}\int_{kT_e-\theta/2}^{kT_e+\theta/2} x(t)dt \qquad [4.103]$$

- Show that $x_{me}(kT_e)$ can be expressed as the convolution product of $x(t)$ and the rectangular function of width $\theta$, taken at moment $kT_e$.

- Derive thereof $x_{me}(t)$ and $X_{me}(f)$. Verify that: $\lim_{\theta\to 0}(X_{me}(f)) = X_{ie}(f)$.

- Let $F_c$ be the cutoff frequency of the signal. Assume $\theta = \lambda T_e$ ($\lambda < 1$) and $F_e = 2\alpha F_c$ ($\alpha \geq 1$). Determine the ratio $\frac{\lambda}{\alpha}$ so that the effect of filtering be lower than 1% up to the cutoff frequency $F_c$. Determine $\lambda$ for the limit value of $\alpha$.

EXERCISE 43.– Take a signal $s(t)$ resulting from the sum of two signals $s_1(t)$ and $s_2(t)$. It is thus assumed that the spectrum is zero outside the bands $[-f_1 - B, -f_1 + B]$, $[f_1 - B, f_1 + B]$, $[-f_2 - B, -f_2 + B]$ and $[f_2 - B, f_2 + B]$ with $f_1 < f_2$ and $B \ll f_1$.

How can the sampling frequency $F_e$ be chosen so as to avoid the aliasing phenomenon of the spectrum?

EXERCISE 44.– We aim to sample a signal $s(t)$ whose spectrum is represented in Figure 4.19.

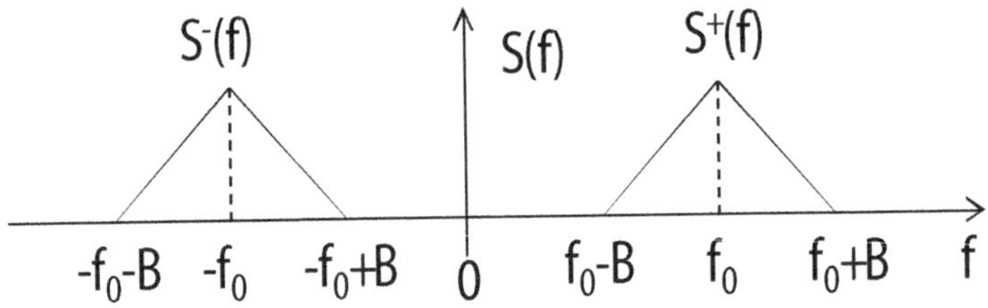


**Figure 4.19.** *Spectrum of a given signal*

1) According to Shannon, what is the minimum sampling frequency?

2) Show that $s(t)$ can be downsampled without distorting $s^+(f)$ and $s^-(f)$. Provide a range for the sampling frequency $F_e$ based on $F_0$, $B$ and $k$ ($k$ = number of spectrum translations).

3) Give a condition of $k$ so that, after sampling and filtering using a low-pass filter of cutoff frequency $\frac{F_e}{2}$, $s^+(f)$ is composed of the band $[0, \frac{F_e}{2}]$ and that $s^-(f)$ is composed of the band $[-\frac{F_e}{2}, 0]$.

*Application*: Let a signal be defined by

$$y(t) = e^{-\frac{t}{\tau}}u(t) \qquad u(t)\text{= Heaviside function.} \tag{4.104}$$

1) Filtering with an ideal low-pass filter of cutoff frequency $B$. Determine $B$ with respect to $\tau$, so that the filtered signal $y_F(t)$ contains 95% of the energy of $y(t)$.

2) Assuming $s(t) = y_F(t)cos(2\pi F_0 t)$:

- Represent the spectrum of $s(t)$.
- According to Shannon, what is the minimum sampling frequency?
- At what value can $s(t)$ be downsampled?

*Digital application*: $F_0 = 1.5$ MHz, $\tau = 1.5$ ms.

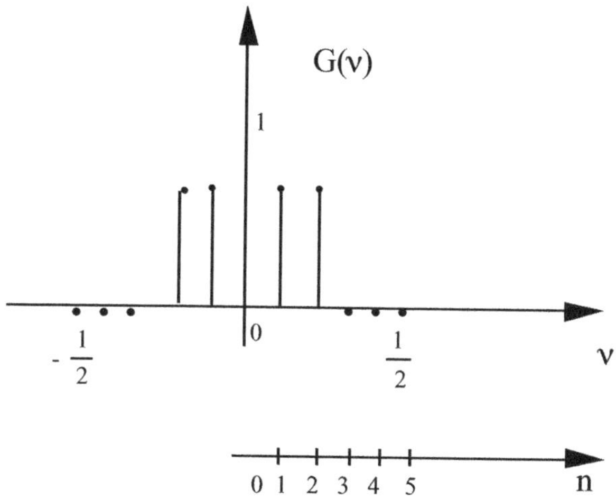


**Figure 4.20.** *Discrete Fourier transform of a filter*

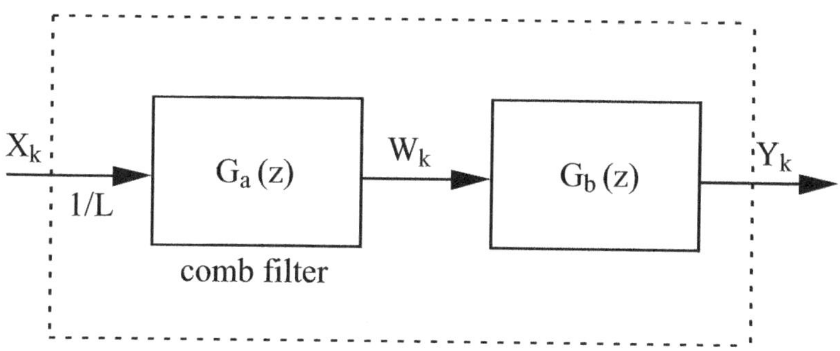


**Figure 4.21.** *Comb filter*

# 5

# Representation of Signals and Systems

This chapter explores the representation, modeling and analysis of linear systems. Modeling is an important step in the study of automated systems.

## 5.1. Introduction to modeling

In system analysis and control, it is necessary to have a model available representing its dynamic behavior. The analysis of behavior means that a strategy can be devised for system control, taking into account the performance imposed by requirements and by the physical limitations of the real system. Thus, a first study of the behavior or the performance of the control law being applied may be done in simulation. When the obtained simulation results are considered very satisfactory, the application of the control to the real system can be envisaged. If however the results are not acceptable, modeling and synthesis control should be reviewed.

To illustrate this approach, we propose the organization chart in Figure 5.1.

Modeling is achieved by writing the physical equations that describe the behavior of the system. These equations, in general, are in the continuous case ordinary differential equations (ODEs) and in the discrete case, recurrence equations. We apply transforms to these equations (Laplace transform: continuous; $Z$-transform: discrete) to shift from the time domain to the frequency domain where system behavior analysis is easier and more interesting.

In the case where the system cannot be described by physical equations, it is always possible that an approximate model be proposed, which will be adjusted by identification. This model should best describe the behavior (input/output) of the system. It can be obtained based on identification by analogy to known systems by using the responses to test signals or an identification that includes the optimization of the error criterion. In this case, the algorithm uses the input and output data of the system. These data must contain enough information to ensure the stimulation of the dynamics of the system. The schematic diagram of an identification is given in Figure 5.2.

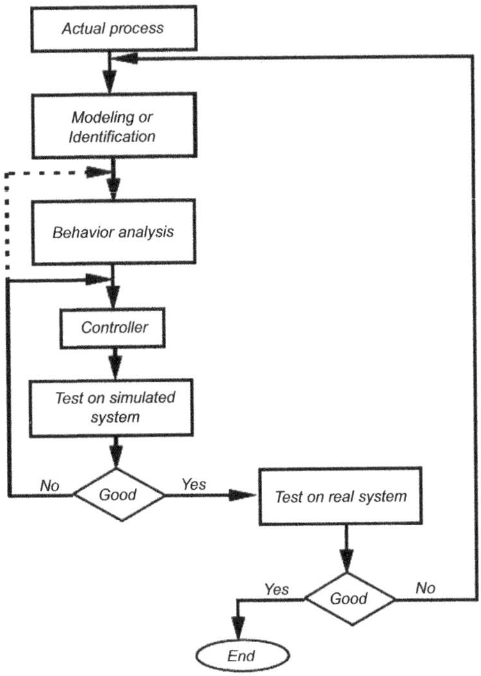


**Figure 5.1.** *Organizational chart*

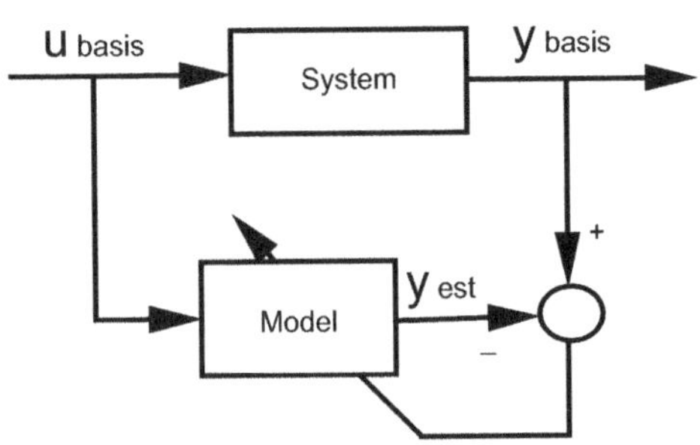


**Figure 5.2.** *Approximating model adapted by identification*

In automation control, the classical diagram of a servo system is given by the following figure:

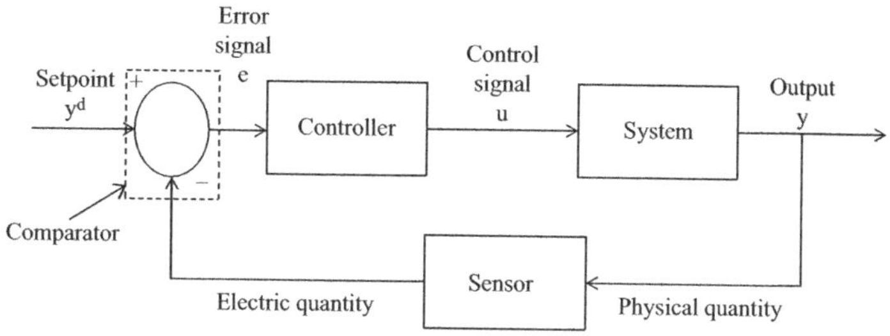


**Figure 5.3.** *Servo systems*

– $y^d$: the setpoint, is an electrical quantity that represents the desired output value of the system;

– $\varepsilon$: the error signal between the setpoint and the actual output of the system;

– $u$: the control signal generated by the controller;

– $y$: a physical quantity that represents the system output.

The output is measured with a sensor that translates it into an electric quantity $y$. Through the comparer, this electric quantity is compared to the setpoint (also in electrical form, in this case).

A system is defined by a relation or an operator $S$ connecting the input $u(t)$ and the output $y(t)$ of the system, when these time functions are vectorial; this form is then referred to as multivariable system, we will write $y = S(u)$. If the variable $t$ is discrete, the system is known as discrete or sampled. The properties of this operator will be those of the system.

Several types of systems can be distinguished:

– linear systems (they can be described by linear differential equations). The system is linear if the relation $y(t) = S(u(t))$ is linear, otherwise it is said to be nonlinear[1];

– time-invariant systems (described by differential equations with constant coefficients) and time-variable systems (described by differential equations with time-variable coefficients). The system is invariant if $y(t-\tau) = S(u(t-\tau))\ \forall \tau \in R$;

1 In this book, we will only consider the study of time-invariant linear systems or that could be approximated by LTI systems.

– the system is causal if output $y(t)$ at the moment $t$ depends only on values $u(\tau)$ for moments $\tau < t$. In other words, its impulse response is causal $(h(t) = 0\, \forall t < 0)$;

– $S$ is univariate if $u(t)$ and $y(t) \in R$ are scalar time functions;

– $S$ is multivariate if $u(t) \in R^m$ and $y(t) \in R^{\,p}$ are vector time functions ($m$ inputs and $p$ outputs);

– continuous systems for which all measured quantities are continuous. $S$ is continuous if its inputs/outputs and its impulse response are continuous;

– discrete systems for which all quantities are measured only at moments regularly spaced by $T_e$ (continuous or discrete variables). $S$ is discrete if its state and its inputs/outputs are discrete variables;

– $S$ is said to be sampled, when it results from the discretization of a continuous system, these are then referred to as sampled or digital systems;

– the order of the system $n$ is the degree or the order of the differential equation that represents it. The number of poles or roots of its denominator is $n$ (the order of the system);

– $S$ is deterministic if its state can be characterized (determined) by the knowledge of its model and its past or initial state;

– nonlinear systems (described by nonlinear differential equation). It is often possible as a first approximation to linearize nonlinear systems based on an operation point (equilibrium), considering small variations around this point;

– it is a distributed parameter system if its dimension is infinite, otherwise it is a lumped parameter system.

Recall that a linear system can be represented by its impulse response. If $h(t)$ is the impulse response of the system, it can be written as:

$$\begin{aligned} y(t) = h(t) * u(t) &= \int_{-\infty}^{+\infty} h(\tau).u(t-\tau)d\tau \\ &= \int_{-\infty}^{+\infty} u(\tau).h(t-\tau)d\tau = u(t) * h(t) \end{aligned} \quad [5.1]$$

Note also that $Y(p) = H(p).U(p)$ using Laplace transform. $H(p) = TL(h(t))$ is the transfer function of the system.

In linear systems control, modeling is a very important step and in order to correctly control a system, a good model must be known. For example, to drive a car, the better its dynamic behavior or model (by training) are known, the better it can be controlled at high speed and as a result it performs better and is easier to drive. The dynamic model is acquired by learning or by system identification. Most of the time, system modeling is carried out using ODEs, but bear in mind that, in general, physical phenomena are infinite dimensional and governed by *partial differential equation* (PDE).

### 5.1.1. *Signal representation using polynomial equations*

The signal is defined by a polynomial in $t$ or a function of $t$ such as exponential functions. These signals are generally of the same type as sinusoid or exponential signals.

EXAMPLE 5.1.–

*a)* $y(t) = \frac{K}{(a^2+t^2)^n}$

*b)* $y(t) = \cos(\omega t) = (e^{j\omega t} + e^{-j\omega t})/2$

### 5.1.2. *Representation of signals and systems by differential equations*

A signal or the response of a system to an input $u(t)$ can be defined by differential equations of which they are solutions. For example, for a signal:

$$\rho^n y(t) + a_{n-1}.\rho^{n-1} y(t) + ... + a_1.\rho y(t) + a_0.y(t) = 0 \qquad [5.2]$$

with $\rho y(t) = dx(t)/dt$ defining the differential operator and for a system of input $u(t)$ and output $x(t)$, we get:

$$\rho^n y(t) + a_{n-1}.\rho^{n-1} y(t) + ... + a_1.\rho y(t) + a_0.y(t) = b_m.\rho^m u(t) + ... \\ + b_1 \rho u(t) + b_0 u(t) \qquad [5.3]$$

In order to introduce the different types of modeling, we are going to study a few examples.

EXAMPLE 5.2.– *RLC electric circuit*

*The differential equations describing the behavior of this circuit are written as:*

$$C\frac{de_c(t)}{dt} = i(t) \quad \text{and} \qquad [5.4]$$

$$e(t) = Ri(t) + L\frac{di(t)}{dt} + e_c(t) \qquad [5.5]$$

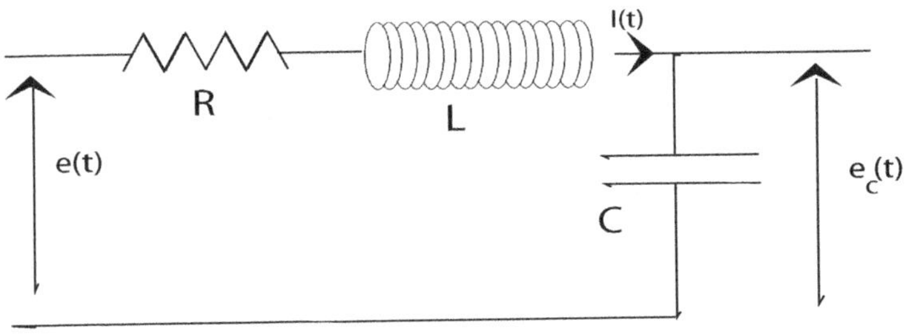


**Figure 5.4.** *RLC circuit*

*Its differential equation is of the second order:*

$$e(t) = Ri(t) + L\frac{di(t)}{dt} + \tfrac{1}{C}\int_0^t i(\tau)d\tau \qquad [5.6]$$

*or*

$$e(t) = RC\frac{de_c(t)}{dt} + LC\frac{d^2i(t)}{dt^2} + e_c(t) \qquad [5.7]$$

## 5.2. Representation using system state equations

A continuous system can have two representations: the first is called external representation or transfer function, and only takes into account the input–output behavior of the system. On the other hand, the second is called state–space representation, which considers the internal state of the system. It gives the dynamic behavior between input, the state and output. It is an interesting representation insofar as it informs us of the internal state of the process to be controlled. Shifting from a representation to the other is possible by means of the Laplace transform.

### 5.2.1. *State variables and state representation definition*

Consider the model of the dynamic system of equation [5.3]. Any dynamic system of order $n$, having $m$ inputs and $p$ outputs, can be described by a set of $n$ differential equations of the form:

$$\frac{dx_i}{dt} = f_i(x_1, x_2, ..., x_n, u_1, u_2, ..., u_m) \text{ for } i = 1, .., n \qquad [5.8]$$

where the $x_i$ are the state variables and the $u_j$ are the input variables. The dynamics of the system is therefore described by as many differential equations as there are state variables. The number of state variables is equal to the *order* of the system ($n$). The variables of the system state are the minimum set of variables $x_1(t)$, $x_2(t)$, ..., $x_n(t)$ such that if the values of these variables are known at an instant $t_0$, and if the external stresses (input signal) applied to the system at any moment $t > t_0$ are also known, the state of the system can be derived at any instant $t$.

If the system is free (without external stresses or with zero-valued inputs), state variables and their initial values completely determine the evolution of the system. The choice of a set of state variables, for a given system, is not unique.

To represent the state of a system, we define a vector $X \in R^n$ (state vector) of dimension $n$ (minimal representation for an $n$-order system), the knowledge of which along with the initial state makes it possible that the state of the system be determined at any time by using the equation of the system.

REMARK 5.1.– *State variables are not necessarily measurable quantities. Input variables can be grouped into two categories: control variables (correction action) and disturbances or exogenous inputs that may be measurable or not.*

A system described by a differential equation of degree $n$ is said to be of order $n$. Its behavior can be described by a vector with $n$ components, the state vector that will be denoted as $X(t)$.

DEFINITION 5.1.– *The minimal dimension of the state of a system (with $m$ inputs and p outputs) corresponds to the number of initial conditions necessary to integrate its differential equation (or evolution equation). For an ordern system (degree of its differential equation), this dimension is equal to $n$. The first-order differential vector equation describing the dynamics of the state vector is a state–space representation of the system.*

*State equation of the system:*

$$\dot{X}(t) = \frac{dX(t)}{dt} = f(X(t), u(t), t) \qquad [5.9]$$

*Observation or output equation:*

$$y(t) = h(X(t), u(t), t) \qquad [5.10]$$

The problem with modeling and temporal representation is thus reduced to determining an appropriate state vector to represent the system as well as functions $f(.)$ and $h(.)$ describing the dynamic behavior of the system. In general, this is achieved based on differential equations.

Often the system can be written as:

$$\dot{X}(t) = f(X(t)) + g(X(t))u(t) \quad [5.11]$$

$$y(t) = h(X(t), u(t)) \quad [5.12]$$

#### 5.2.1.1. *Linear time-variant systems*

*Linear time variant (LTV)*: When the functions $f(.)$, $g(.)$ and $h(.)$ are linear with respect to state $X(t) \in R^n$ and input $u(t) \in R^m$, the system is LTV.

$$\dot{X}(t) = A(t).X(t) + B(t)u(t) \quad \text{and} \quad y(t) = C(t).X(t) + D(t).u(t) \quad [5.13]$$

$y(t) \in R^p$, $u(t) \in R^m$ and $X(t) \in R^n$

where

- $A(t) \in R^n \times R^n$ is the matrix defining the system dynamics;
- $B(t) \in R^n \times R^m$ is the input matrix of the system (there are $m$ inputs);
- $C(t) \in R^p \times R^n$ is the output matrix of the system (there are $p$ outputs);
- $D(t) \in R^m \times R^p$ is the direct input matrix.

#### 5.2.1.2. *Linear time-invariant systems*

*Linear time invariant (LTI)*: If functions $f_i()$, defined in equation [5.8] are linear combinations, with constant coefficients, state variables and variable inputs, the model under consideration is therefore an LTI model. It is then possible to put the set of state equations in matrix form by defining the state vector $X(t)$ as the vector whose components are the $n$ state variables and the input vector $u(t)$ as the vector whose components are the $m$ input variables. When matrices $A$, $B$, $C$ and $D$ are time independent, the system is said to be LTI. In the linear case, the state equation of a system derives from rewriting the original differential equation connecting the input to the output into a first-order vector differential equation. It assumes the following form for a system whose input is $u(t)$ and output is $y(t)$.

*State equation of the system:*

$$\dot{X}(t) = A.X(t) + Bu(t) \quad [5.14]$$

Matrix $A$ is called the evolution matrix of dimension $(n, n)$ and $B$ is the control matrix of dimension $(n, m)$. These are time-independent matrices.

To provide a complete representation of the system, we must accompany the state equation by the output equation or observation equation that connects the vector $y(t)$ of output variables to the state variables.

*Observation equation*:

$$y(t) = Cx(t) + Du(t) \qquad [5.15]$$

with $y(t)$ the output signal of dimension $(p, 1)$, $C$ the output matrix of dimension $(p, n)$ and $D$ the direct transmission matrix of dimension $(p, m)$.

– the first equation [5.14] is the evolution equation of the state;

– the second equation [5.15] is the output or observation equation.

In the case of a signal, it suffices to cancel the input $u$ in the above state equation ($u(t) = 0$ or $B = 0$). In a first step and in the context of this chapter, we shall only consider single-input single-output cases (called SISO) which makes $p = m = 1$.

In summary, a continuous time-invariant linear model of a system is completely described by state and output equations:

$$\begin{aligned} \dot{x}(t) &= Ax(t) + Bu(t) \\ y(t) &= Cx(t) + Du(t) \end{aligned} \qquad [5.16]$$

For a system described by these state equations, a functional diagram is shown in Figure 5.5.

The set of all states that the system can take is called state space or phase space.

EXAMPLE 5.3.– Electric RLC Circuit

*We write* $X(t) = \begin{pmatrix} e_c(t) \\ i(t) \end{pmatrix}$, *the state vector of the system of equation [5.6]. This vector defines the internal state of the system at time t. This allows us to obtain the system state representation, which is a temporal representation. In this case, matrix A and the state–space representation are defined by:*

$$\dot{X} = \begin{pmatrix} 0 & \frac{1}{C} \\ -\frac{1}{L} & -\frac{R}{L} \end{pmatrix} \begin{pmatrix} e_c(t) \\ i(t) \end{pmatrix} + Bu \qquad [5.17]$$

$$e_c = (1; 0)X = (1; 0)\begin{pmatrix} e_c(t) \\ i(t) \end{pmatrix}; \qquad [5.18]$$

*with* $A = \begin{pmatrix} 0 & \frac{1}{C} \\ -\frac{1}{L} & -\frac{R}{L} \end{pmatrix} \qquad B = \begin{pmatrix} 0 \\ \frac{1}{L} \end{pmatrix} \qquad C = \begin{pmatrix} 1 & 0 \end{pmatrix}$

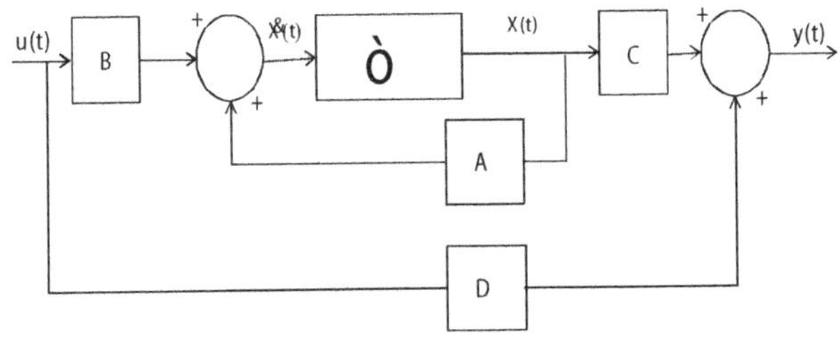


**Figure 5.5.** *Functional diagram*

EXAMPLE 5.4.– Butterworth filter case

*A normalized low-pass, fourth-order Butterworth filter, with a cut-off pulse* 1 *rad/s, whose input is* $u(t)$ *and output* $y(t)$*, has a differential equation as:*

$$(\rho^4 + 2.6131\rho^3 + 3.4142\rho^2 + 2.6131\rho + 1)y(t) = u(t) \text{ where } \rho x(t) = \frac{dx(t)}{dt} \quad [5.19]$$

*We can write* $\rho^4 y(t) = -2.6131\rho^3 y(t) - 3.4142\rho^2 y(t) - 2.6131\rho y(t) - y(t) + u(t)$*, and define* $\dot{x}_1 = \rho^4 y(t)$, $\dot{x}_2 = \rho^3 y(t) = x_1$, $\dot{x}_3 = \rho^2 y(t) = x_2$, $\dot{x}_4 = \rho y(t) = x_3$ *and* $x_4 = y(t)$.

*The system can thus be described in state form:*

$$\dot{X} = AX + Bu; \qquad y = CX \quad [5.20]$$

*with*

$$A = \begin{pmatrix} -2.6131 & -3.4142 & -2.6131 & -1 \\ 1 & 0 & 0 & 0 \\ 0 & 1 & 0 & 0 \\ 0 & 0 & 1 & 0 \end{pmatrix} \quad B = \begin{pmatrix} 1 \\ 0 \\ 0 \\ 0 \end{pmatrix} \quad C = \begin{pmatrix} 0 & 0 & 0 & 1 \end{pmatrix} \quad [5.21]$$

EXAMPLE 5.5.– Oscillator or pendulum

*The equation of a simple pendulum of mass $m$ hanging by a thread (supposedly rigid) of length $l$, under the effect of a couple $\tau$, is obtained by applying the fundamental principle of dynamics. We can write:*

$$ml^2\ddot{\theta} = \tau - mgl\sin(\theta) \qquad [5.22]$$

*If $\omega = \dot{\theta}$ denotes its angular velocity, we get:*

$$X = \begin{pmatrix} \dot{\theta} \\ \dot{\omega} \end{pmatrix} = \begin{pmatrix} \omega \\ \frac{\tau - mgl\sin(\theta)}{ml^2} \end{pmatrix} = f(X) + g(X)\tau \qquad [5.23]$$

*Limiting oneself to small movements ($\omega$ and $\theta$ close to zero), the model can be linearized assuming that $\sin(x) \approx x$; it then yields:*

$$\begin{pmatrix} \dot{\theta} \\ \dot{\omega} \end{pmatrix} = \begin{pmatrix} 0 & 1 \\ -\frac{mgl}{ml^2} & 0 \end{pmatrix} \begin{pmatrix} \theta \\ \omega \end{pmatrix} + \begin{pmatrix} 0 \\ \frac{1}{ml^2} \end{pmatrix} \tau \qquad [5.24]$$

*and*

$$\theta = \begin{pmatrix} 1 & 0 \end{pmatrix} \begin{pmatrix} \theta \\ \omega \end{pmatrix} \qquad [5.25]$$

*With $\tau = 0$, we obtain the representation of a free oscillator.*

REMARK 5.2.– *Note that here an approximation is valid only for small movements and that the possibility of such approximation depends on the characteristics of the system (vicinity of a steady-state and uniform convergence, etc.).*

*The previous type of approximation is not applicable to the two following systems, for instance: $\dot{x} + x^3 = 0$ and $\dot{x} - x^3 = 0$. These systems would give the same linear system despite their fundamentally different behaviors.*

EXAMPLE 5.6.– A fixed flexible bar

*A flexible bar fixed at one end, with length $l$ is subject to deflection $u$ on the other end. Deflection $y(x,t)$ is measured at distance $x$ (x-axis) of the fixed endpoint, at time $t$. The dynamics of its evolution is governed by the PDE:*

$$k\frac{\partial^4 y(x,t)}{\partial x^4} + \frac{\partial^2 y(x,t)}{\partial t^2} = 0 \quad \text{with} \quad y(l,t) = u(t) \qquad [5.26]$$

*$k$ is a constant that depends on the material of the bar. To determine the system state, the state of $y(x,t)$ must be known at any point $x$; it is a function of $x$ and not*

*a vector of finite dimension. This example is used to show that many physical systems are intrinsically of infinite dimension; they are governed by the PDEs. The space discretization of these equations allows the solution of ODEs. However, it should be known that the accuracy of the modeling leads to a high order concerning the number of elements or equations.*

### 5.2.2. State–space representation for discrete linear systems

Discrete systems being described by an order $n$ recurrence equation, the minimal dimension of the state vector is $n$, for a minimal representation. We have the state vector $X_k \in R^n$, inputs $u_k \in R^m$ and outputs $y_k \in R^p$; whence the state–space representation:

$$X_{k+1} = f(X_k, u_k, k) \qquad [5.27]$$

$$y_k = h(X_k, u_k, k) \qquad [5.28]$$

– *LTV* case: when functions $f(X_k, u_k, k)$ and $h(X_k, u_k, k)$ are linear with respect to state $X_k \in R^n$ and input $u_k \in R^m$, the system is discrete, linear and time varying:

$$X_{k+1} = A_k.X_k + B_k u_k \quad \text{and} \quad y_k = C_k.X_k + D_k.u_k. \qquad [5.29]$$

– *LTI* case: when matrices $A$, $B$, $C$ and $D$ are time independent, the system is said LTI:

$$X_{k+1} = AX_k + Bu_k \quad \text{and} \quad y_k = C.X_k + D.u_k. \qquad [5.30]$$

REMARK 5.3.– *To conclude with state–space representation, it should be noted that it provides a means for addressing LTV systems; in this case, harmonic analysis and the Laplace transform are not applicable. It is also used for the analysis of stability in system control and non-stationary or adaptive filtering and presents an advantage for the simulation and representation of systems in the phase plane.*

*The input variables sometimes include exogenous inputs. This leads us to separate these two types of inputs by designating them by means of different notations, the notation $u$ being traditionally used for control variables. We write for example:*

$$\dot{X}(t) = AX(t) + Bu(t) + Fw(t) \qquad [5.31]$$
$$y(t) = CX(t) + Du(t)$$

*It is always possible, by elimination, to transform the $n$ state and output equations into a single differential equation of order $n$ connecting output variables to input variables. That is the reason why the number of state variables, for a minimal representation, is called system order.*

*The state–space representation is denoted $(A, B, C, D)$ and is also called system realization. This representation is not unique.*

*State–space representation is the most appropriate representation for the simulation of a system or for programming a filter.*

## 5.3. Transfer functions

### 5.3.1. *Transfer function: external representation*

The transfer function of an SISO system is the ratio of the Laplace transform of the output to the Laplace transform of the input ($H(p) = \frac{Y(p)}{U(p)}$), assuming all initial conditions zero ($y^{(n)}(0) = u^{(m)}(0) = ... = y(0) = 0$).

Furthermore, knowing that the output is obtained by convolution of the impulse response and the input, $y(t) = h(t) * u(t)$, we obtain $Y(p) = H(p).U(p)$ using Laplace transform, which means that the transfer function is also the Laplace transform of its impulse response $h(t)$: $H(p) = \mathrm{TL}(h(t))$.

### 5.3.2. *Transfer function and state–space representation shift*

#### 5.3.2.1. *Shifting from a differential equation to an external description or transfer function*

The transfer function can be inferred from the differential equation that relates the input of a system to its output. Consider the following differential equation:

$$\frac{d^n y(t)}{dt^n} + a_{n-1}\frac{d^{n-1}y(t)}{dt^{n-1}} + \cdots + a_0 y(t) = b_m \frac{d^m u(t)}{dt^m} + b_{m-1}\frac{d^{m-1}u(t)}{dt^{m-1}} + \cdots + b_0 u(t) \quad [5.32]$$

$$\rho^n y(t) + a_{n-1}.\rho^{n-1}y(t) + ... + a_1.\rho y(t) + a_0.y(t) = b_m.\rho^m u(t) + ... + b_1 \rho u(t) + b_0 u(t) \quad [5.33]$$

or even $A(\rho)y(t) = B(\rho)u(t)$ with $A(\rho)$ and $B(\rho)$ polynomials in $\rho$ (the differential operator). Physical realizability imposes $m < n$ and it is assumed, with no loss of generality, that $a_n = 1$. The transfer function is obtained by applying the Laplace transform, member by member, to the equations above and assuming zero initial conditions. We apply the Laplace transform to both members assuming that all initial conditions are zero. This is tantamount to substituting the symbolic Laplace variable to the operator $\rho$ that yields:

$$p^nY(p) + a_{n-1}p^{n-1}Y(p) + \cdots + a_0Y(p) = b_mp^mU(p) + b_{m-1}p^{m-1}U(p) + \cdots + b_0U(p) \quad [5.34]$$

Hence, the transfer function is written as:

$$H(p) = \frac{Y(p)}{U(p)} = \frac{b_mp^m + b_{m-1}p^{m-1} + \cdots + b_0}{p^n + a_{n-1}p^{n-1} + a_{n-2}p^{n-2} + \cdots + a_0} = \frac{B(p)}{A(p)} \quad [5.35]$$

DEFINITION 5.2.– *Poles and zeros of a system: The numerator of $H(p)$ has $m$ roots (denoted $z_i$) and its denominator has $n$ roots (denoted $p_j$). The $p_j$ are the poles of the system and the $z_i$ the zeros of the system.*

*The equation $A(p) = 0$ gives the poles of the system, where $A(p)$ is the characteristic polynomial of the system.*

The transfer function can be factorized using poles and zeros and therefore written as:

$$H(p) = K\frac{\prod_{i=1}^{m}(p + z_i)}{\prod_{j=1}^{n}(p + p_i)} \quad [5.36]$$

Remember also that the transfer function is similar to the Laplace transform of the impulse response of the system.

#### 5.3.2.2. *Moving from a transfer function to a state–space representation*

It can be shown that the previous system described by the transfer function of equation [5.35] may be expressed in the state form (equation [5.37]) with $b_{m+i} = 0$ for $i > 0$; to prove it, we can calculate $\rho y$, $\rho^2 y$, $\rho^3 y$, ...$\rho^n y$ from the state–space representation.

$$\dot{X} = \begin{bmatrix} -a_{n-1} & 1 & 0 & & 0 \\ -a_{n-2} & 0 & 1 & 0 & \\ -a_{n-3} & 0 & 0 & 1 & \\ & & & & 0 \\ & & & & 1 \\ -a_o & & 0 & & 0\,0 \end{bmatrix} X + \begin{bmatrix} b_{n-1} \\ b_{n-2} \\ b_{n-3} \\ \\ b_o \end{bmatrix} u \quad \text{and} \quad y = \begin{bmatrix} 1 \; 0 \; .. \; 0 \end{bmatrix} X \quad [5.37]$$

It can also be written as: $\rho X = AX + Bu(t) \Rightarrow (\rho I - A)X = Bu(t) \Rightarrow y(t) = C(\rho I - A)^{-1}Bu(t)$, which can identify with the differential equation of the system.

#### 5.3.2.3. *Moving from a state representation to an external description (transfer function)*

In previous sections, we have defined state–space representation and observed that it could be inferred from the differential equations of the system. We consider a system defined by its state–space representation and we determine its transfer function.

$$\dot{X} = A.X + Bu(t) \quad [5.38]$$

$$y(t) = C.X + Du \quad [5.39]$$

Its function or transfer matrix is obtained by applying the Laplace transform to the previous equations considering zero initial conditions:

$$pX(p) - x(0) = A.X(p) + B.U(p) \Longrightarrow (pI - A)X(p) = B.U(p) + x(0) \quad [5.40]$$

$$\Longrightarrow X(p) = (pI - A)^{-1}(B.U(p) + x(0)) \quad [5.41]$$

where

$$Y(p) = (C.(pI - A)^{-1}B + D).U(p) + (C.(pI - A)^{-1}x(0) \quad [5.42]$$

DEFINITION 5.3.– *The transfer function is also defined by*

$$H(p) = \frac{Y(p)}{U(p)} = C.(pI - A)^{-1}B + D \quad [5.43]$$

*It is obtained considering zero initial conditions during the Laplace transform. It describes the behavior of the forced system ignoring the effect of initial conditions; it is called external representation as opposed to the state–space representation that makes use of internal states.*

DEFINITION 5.4.– *Equation* $\det(\lambda I - A) = 0$ *gives the eigenvalues of the system, and it is the characteristic equation of the system.*

REMARK 5.4.– *The denominator of the transfer function is the determinant of matrix* $(pI - A)$. *Consequently, the denominator of the transfer function is the characteristic polynomial of the system and the poles of the transfer function are the roots of the characteristic equation (which also identify with the eigenvalues of matrix A).*

*The eigenvalues of matrix A, for a given physical system, are independent of the choice of state variables. They are also found in the external description (transfer function), they characterize the system, regardless of its representation.*

#### 5.3.2.4. *Moving from a recurrence equation to a transfer function in Z (discrete case)*

For the discrete case, we use the Z-transform of its recurrence equation to derive the transfer function $H(z)$ or the transmittance in Z.

#### 5.3.2.5. *Moving from a discrete state–space representation to a transfer function in Z*

For the discrete case, we use the Z-transform of the state–space representation to obtain the transfer function $H(z)$ or transmittance in Z. Consider the system defined by

$$X_{k+1} = A.X_k + Bu_k \quad \text{and} \qquad y_k = C.X_k + Du_k \tag{5.44}$$

Its function or transfer matrix is obtained by applying the Z-transform to the state–space representation:

$$zX(z) - x(0) = A.X(z) + B.U(z) \Longrightarrow (zI - A)X(z) = B.U(z) + x(0) \tag{5.45}$$

$$\Longrightarrow X(z) = (zI - A)^{-1}(B.U(z) + x(0)) \tag{5.46}$$

where

$$Y(z) = (C.(zI - A)^{-1}B + D).U(z) + (C.(zI - A)^{-1}x(0) \tag{5.47}$$

DEFINITION 5.5.– *The transfer function in Z is also defined by*

$$H(z) = \frac{Y(z)}{U(z)} = C.(zI - A)^{-1}B + D \tag{5.48}$$

### 5.3.3. *Properties of transfer functions*

#### 5.3.3.1. *Causality of a system or of a transfer function*

A system is known as causal if its impulse response (response to an impulse) is zero for negative times ($h(t) = 0$ if $t < 0$). This means that the system does not anticipate the input for the calculation of the response

$$y(t) = \int_{-\infty}^{+\infty} h(\tau).x(t-\tau)d\tau = \int_{0}^{+\infty} h(\tau).x(t-\tau)d\tau \tag{5.49}$$

#### 5.3.3.2. *System or transfer function stability*

DEFINITION 5.6.– *A system is said stable if for any bounded input* $e(t)$ *it maps a correspondence to a bounded output* $s(t)$*. This notion of stability is called "BIBO stability" (bounded input, bounded output).*

If $h(t)$ is the impulse response of the system, the following can be stated:

$$y(t) = \int_{-\infty}^{+\infty} x(\tau).h(t-\tau)d\tau = x(t) * h(t) \qquad [5.50]$$

Knowing that $x(t)$ is bounded, $x(t) \leq M \in \mathbb{R}$, we then get:

$$|y(t)| \leq \int_{-\infty}^{+\infty} |x(\tau)| \,.\, |h(t-\tau)|\, d\tau \leq M. \int_{-\infty}^{+\infty} |h(\tau)|\, d\tau \qquad [5.51]$$

hence the system is stable if $h(t)$ its impulse response is an absolutely summable function. A sufficient condition for stability is that $h(t)$ be summable ($\int_{-\infty}^{+\infty} |h(\tau)|\, d\tau$ is finite).

A transfer function is stable if all its poles have a real negative part. It can be shown that this condition is equivalent to the previous ($h(t)$ summable).

#### 5.3.3.3. *Minimum phase system*

DEFINITION 5.7.– *A transfer function is a minimum phase transfer function if all its zeros have real negative parts.*

It should be noted that the following transfer functions all have the same modulus and different phases, $a$ and $b$ being positive constants:

$$H_1(p) = \frac{B(p)}{A(p)},\ H_2(p) = \frac{B(p)}{A(p)}\cdot\frac{p-a}{p+a},\ H_3(p) = \frac{B(p)}{A(p)}\cdot\frac{p-b}{p+b},\ H_4(p) = \frac{B(p)}{A(p)}\cdot\frac{p-a}{p+a}\cdot\frac{p-b}{p+b}$$

There is equality of moduli: $|H_1(f)| = |H_2(f)| = |H_3(f)| = |H_4(f)|$, but arguments are different. The transfer function that has the smallest phase is the one that does not comprise a pure phase shifter ($\frac{p-a}{p+a}$).

#### 5.3.3.4. *Physical realizability of a transfer function*

The general form of a system transfer function is given by:

$$H(p) = \frac{B(p)}{A(p)} = K\frac{\prod_{i=1}^{m}(p+z_i)}{\prod_{j=1}^{n}(p+p_i)} = TL(h(t)) \qquad [5.52]$$

The impulse response $h(t)$ has its transfer function $H(p)$ as a Laplace transform: for a physically implementable system; we necessarily have $m < n$. This is justified by the fact that a physically implementable system has a bounded impulse response ($h(t) < \infty\ \forall t \in \mathbb{R}_+$). In particular, $h(0^+)$ is bounded and by applying the initial value theorem (for the Laplace transform), the following result is obtained.

$$h(0^+) = \lim_{t \longrightarrow 0} h(t) = \lim_{p \longrightarrow \infty} pH(p) = \lim_{p \longrightarrow \infty} p\frac{B(p)}{A(p)} < \infty \qquad [5.53]$$

In order for $h(0^+)$ to be finite, the following condition is necessary

$$\deg(pB(p)) = \deg(B(p)) + 1 \leq \deg(A(p)) \qquad [5.54]$$

hence the following definition is proposed.

DEFINITION 5.8.– *A transfer function is proper or physically realizable if we have $m < n$. It is said to be proper in the broad sense if $m = n$.*

REMARK 5.5.– *An integrator ($H(p) = \frac{1}{p}$ proper function) is physically realizable; on the other hand, a pure derivator ($H(p) = p$ non-proper function own) is not. An approximate realization can be obtained by the proper transfer function in the broad sense $H(p) = \frac{p}{1+\tau p}$.*

#### 5.3.3.5. *Relations between transfer functions, transition matrix and state–space representation*

It is possible to switch from a state–space representation to an external description (transfer function), which only considers relations between system input and output variables.

Applying the Laplace transform to the state equation given by:

$$\dot{X}(t) = AX(t) + Bu(t) \qquad [5.55]$$

we obtain:

$$X(p) = (pI - A)^{-1}X(0^+) + (pI - A)^{-1}BU(p) \qquad [5.56]$$

The general solution of the state equation is:

$$X(t) = e^{At}X(0^+) + \int_0^t e^{A(t-\tau)}Bu(\tau)d\tau \qquad [5.57]$$

Therefrom it results that:

$$\Phi(t) = e^{At} = TL^{-1}[(pI - A)^{-1}] \qquad [5.58]$$

We have seen that the eigenvalues of matrix $A$ play a fundamental role in the dynamics of a system. Nonetheless, the solutions of the equation:

$$\det(pI - A) = 0 \qquad [5.59]$$

are nothing else than the eigenvalues of matrix $A$. The eigenvalues of matrix $A$ play an important role in the external description. As a matter of fact:

$$(pI - A)^{-1} = \frac{adj(pI - A)}{\det(pI - A)} \qquad [5.60]$$

where the adjoint matrix is the transposed cofactors matrix of $A$. The denominator $\det(pI - A)$ is the *characteristic polynomial* of matrix $A$. The degree of this polynomial is equal to the order of the system. The equation $\det(pI - A) = 0$ is called *characteristic equation* of the system. These roots, which are the eigenvalues of $A$, are called *poles* of the system.

To determine the transfer function, it is assumed that initial conditions are zero, where:

$$X(p) = (pI - A)^{-1}BU(p) \qquad [5.61]$$

The output vector is given by:

$$Y(p) = CX(p) + DU(p) \qquad [5.62]$$

The relation between the inputs Laplace transform and the outputs Laplace transform is thus given by:

$$Y(p) = [C(pI - A)^{-1}B + D]U(p) \qquad [5.63]$$

The matrix $[C(pI - A)^{-1}B + D]$ is called the transfer matrix of the variables. If the system has $m$ inputs and $p$ outputs, the transfer matrix is of dimensions $(p, m)$. For a system of the SISO type, the matrix is reduced to a scalar, which is none other than the transfer function $H(p)$.

$$H(p) = C(pI - A)^{-1}B + D \qquad [5.64]$$

where $C$ is of dimension $(1, n)$, $B$ is a matrix of dimension $(n,1)$ and $D$ is a scalar.

### 5.3.4. *Associations of functional diagrams*

We can always replace a diagram, consisting of several transfer blocks, regardless of its complexity, by a single-block functional diagram or with a unit feedback loop. Only the information on the internal signals of the system is lost. We give in what follows some simple examples of functional diagrams reduction. These examples relate to cascading associations in parallel and in negative feedback.

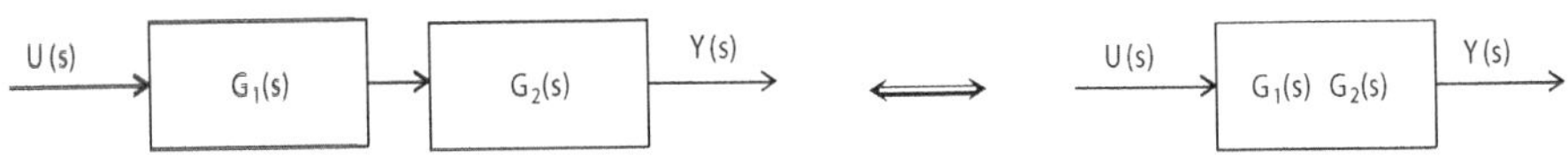


**Figure 5.6.** *Cascade diagram*

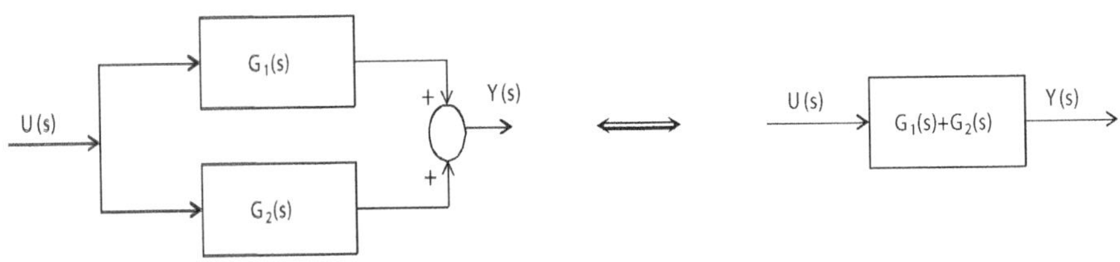


**Figure 5.7.** *Parallel diagram*

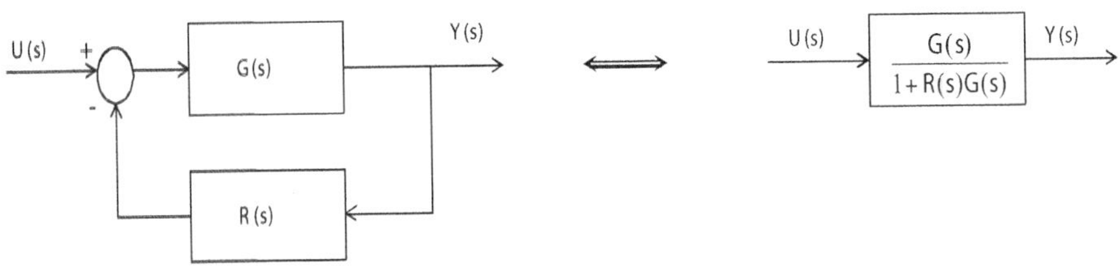


**Figure 5.8.** *Diagram 1 in parallel in negative feedback*

## 5.4. Change in representation and canonical forms

In general, for the state representation, a state vector $X$ is chosen whose components have a physical meaning (this is not an obligation). However, in most cases these quantities are not all measurable and in some cases a representation may prove to be more convenient (for manipulations) than others. This is the case of canonical forms for observation and control; they allow more easily computing a compensator, an observer or state estimator.

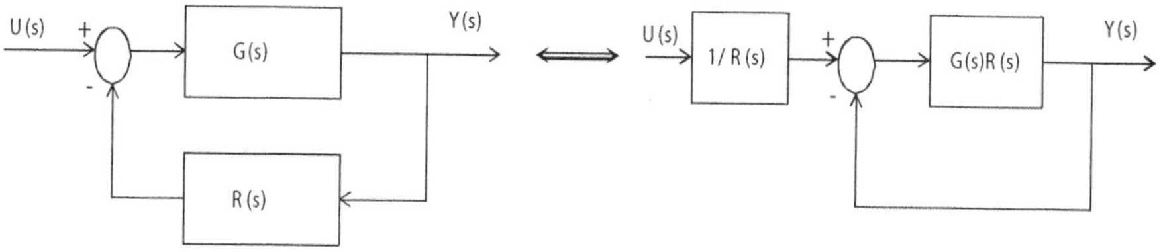


**Figure 5.9.** *Diagram 2 in parallel in negative feedback*

In this section, we present some methods that can be used to derive a state–space representation for an order $n$ univariate system defined by its transfer function. These methods can address discrete cases as well as continuous ones:

$$H(p) = \frac{b_1 p^{n-1} + b_2 p^{n-2} + ..... + b_{n-1} p + b_n}{p^n + a_{n-1} p^{n-1} + ..... + a_1 p + a_0} \quad [5.65]$$

$$H(z) = \frac{b_1 z^{n-1} + b_2 z^{n-2} + ..... + b_{n-1} z + b_n}{z^n + a_{n-1} z^{n-1} + ..... + a_1 z + a_0} \quad [5.66]$$

### 5.4.1. *Controllable canonical form*

#### 5.4.1.1. *Companion form (controllable canonical form)*

Consider a transfer function of the form:

$$H(p) = \frac{Y(p)}{U(p)} = \frac{1}{p^n + a_{n-1} p^{n-1} + a_{n-2} p^{n-2} + \cdots + a_0} \quad [5.67]$$

By reducing to the same denominator and taking the inverse Laplace transform (with zero initial condition), we have:

$$\frac{d^n y(t)}{dt^n} + a_{n-1} \frac{d^{n-1} y(t)}{dt^{n-1}} + a_{n-2} \frac{d^{n-2} y(t)}{dt^{n-2}} + \cdots + a_0 y(t) = u(t) \quad [5.68]$$

REMARK 5.6.– *We also find here that the poles of the transfer function are nothing else than the roots of the characteristic equation of the differential equation that connects the input and the output.*

A state–space representation consists of writing:

$$\begin{aligned}
\dot{x}_1 &= x_2 \\
\dot{x}_2 &= x_3 \\
&\vdots \\
\dot{x}_{n-1} &= x_n \\
\dot{x}_n &= -a_0 x_1 - a_1 x_2 - \cdots - a_{n-1} x_n + u \\
y &= x_1
\end{aligned} \qquad [5.69]$$

The matrices of the state equation are therefore:

$$A = \begin{pmatrix} 0 & 1 & 0 & \cdots & 0 \\ 0 & 0 & 1 & \cdots & 0 \\ \vdots & \vdots & \vdots & \vdots & \vdots \\ 0 & 0 & 0 & \cdots & 1 \\ -a_0 & -a_1 & -a_2 & \cdots & -a_{n-1} \end{pmatrix}; \; B = \begin{pmatrix} 0 \\ 0 \\ \vdots \\ 0 \\ 1 \end{pmatrix}; \; C = \begin{pmatrix} 1 \, 0 \cdots 0 \end{pmatrix}; \; D = 0 \quad [5.70]$$

REMARK 5.7.– *This form of state–space representation is a controllable canonical form.*

EXAMPLE 5.7.– *Let* $H(p) = \frac{1}{p^3+a_2p^2+a_1p+a_0}$ *having state vector* $\begin{pmatrix} x \; \dot{x} \; \ddot{x} \end{pmatrix}^T$, *we obtain the state–space representation*

$$\dot{X} = \begin{bmatrix} 0 & 1 & 0 \\ 0 & 0 & 1 \\ -a_0 & -a_1 & -a_2 \end{bmatrix} X + \begin{bmatrix} 1 \\ 0 \\ 0 \end{bmatrix} u \quad \text{and} \quad y = \begin{bmatrix} 0 \, 0 \, 1 \end{bmatrix} X \qquad [5.71]$$

*Moreover, if we take as state vector* $\begin{pmatrix} \ddot{x} \; \dot{x} \; x \end{pmatrix}^T$, *the state–space representation is obtained (controllable canonical)*

$$\dot{X} = \begin{bmatrix} -a_2 & -a_1 & -a_0 \\ 1 & 0 & 0 \\ 0 & 1 & 0 \end{bmatrix} X + \begin{bmatrix} 1 \\ 0 \\ 0 \end{bmatrix} u \quad \text{and} \quad y = \begin{bmatrix} 0 \, 0 \, 1 \end{bmatrix} X \qquad [5.72]$$

EXAMPLE 5.8.– *For the transfer function* $H(p) = \frac{b_2p^2+b_1p+b_0}{p^3+a_2p^2+a_1p+a_0}$, *it is assumed*

$$y(p) = (b_2p^2 + b_1p + b_0)\frac{1}{p^3 + a_2p^2 + a_1p + a_0}u(p) = (b_2p^2 + b_1p + b_0)\xi(p) \quad [5.73]$$

*then with as state vector* $\left(\ddot{\xi}\;\dot{\xi}\;\xi\right)^T$, *we get* $y = b_2\ddot{\xi} + b_1\dot{\xi} + b_0\xi$ *wherefrom finally the state representation*

$$\dot{X} = \begin{bmatrix} -a_2 & -a_1 & -a_0 \\ 1 & 0 & 0 \\ 0 & 1 & 0 \end{bmatrix} X + \begin{bmatrix} 1 \\ 0 \\ 0 \end{bmatrix} u \quad \text{and} \quad y = \begin{bmatrix} b_2 \; b_1 \; b_0 \end{bmatrix} X \qquad [5.74]$$

### 5.4.2. *Controllable canonical form*

The transfer function

$$H(p) = \frac{b_2p^2 + b_1p + b_0}{p^3 + a_2p^2 + a_1p + a_0} \qquad [5.75]$$

also corresponds to the following state–space representation

$$\dot{X} = \begin{bmatrix} 0 & 0 & -a_0 \\ 1 & 0 & -a_1 \\ 0 & 1 & -a_2 \end{bmatrix} X + \begin{bmatrix} 1 \\ 0 \\ 0 \end{bmatrix} u \quad \text{and} \quad y = \begin{bmatrix} \beta_1 \; \beta_2 \; \beta_3 \end{bmatrix} X \qquad [5.76]$$

Constants $\beta_i$ are called system Markov parameters, they are such that:

$$H(p) = \frac{b_2p^2 + b_1p + b_0}{p^3 + a_2p^2 + a_1p + a_0} = \frac{\beta_3 + \beta_2(a_2 + p) + \beta_1(p^2 + a_2p + a_1)}{p^3 + a_2p^2 + a_1p + a_0} \qquad [5.77]$$

they are thus solution of the equation

$$\begin{bmatrix} 1 & 0 & 0 \\ a_2 & 1 & 0 \\ a_1 & a_2 & 1 \end{bmatrix} \begin{bmatrix} \beta_1 \\ \beta_2 \\ \beta_3 \end{bmatrix} = \begin{bmatrix} b_2 \\ b_1 \\ b_0 \end{bmatrix} \qquad [5.78]$$

and correspond to the first terms of the expansion $H(p) = \sum_{i=1}^{\infty} \beta_i p^{-i}$.

### 5.4.3. *Observability canonical form*

This representation is the dual form of the previous one.

$$\dot{X} = \begin{bmatrix} 0 & 1 & 0 \\ 0 & 0 & 1 \\ -a_0 & -a_1 & -a_2 \end{bmatrix} X + \begin{bmatrix} \beta_1 \\ \beta_2 \\ \beta_3 \end{bmatrix} u \quad \text{and} \quad y = \begin{bmatrix} 1 \; 0 \; 0 \end{bmatrix} X \qquad [5.79]$$

EXAMPLE 5.9.– $H(p) = \frac{p^2+5p+4}{p^3+3p^2+p-5} = \frac{1}{p-1} + \frac{1}{p^2+4p+5}$, *the second element has two complex conjugated poles. We can deduce a state–space representation of this system by grouping the representations of the simple elements obtained (to do as an exercise).*

EXAMPLE 5.10.– Mass suspended by a spring and a damper

*A mass $M$ is suspended using a spring of stiffness $k$ and a damper with a viscous friction coefficients $b$. When its behavior is dynamic, the gravity is assumed to be compensated by an elongation of the spring, a force $F$ is applied vertically to the mass $M$. The system thus defined is described by the differential equation $m\ddot{x} + b\dot{x} + kx - F = 0$. In the state space, if considering as components of the state vector $x_1 = \dot{x}$ and $x_2 = x$, the system is described by:*

$$\begin{pmatrix} \dot{x}_1 \\ \dot{x}_2 \end{pmatrix} = \begin{pmatrix} -b/m & -k/m \\ 1 & 0 \end{pmatrix} \begin{pmatrix} x_1 \\ x_2 \end{pmatrix} + \begin{pmatrix} 0 \\ 1/m \end{pmatrix} F \text{ and } y = \begin{pmatrix} 0 \; 1 \end{pmatrix} \begin{pmatrix} x_1 \\ x_2 \end{pmatrix} \qquad [5.80]$$

EXAMPLE 5.11.– *One may also consider $x_1 = x$ and $x_2 = \dot{x}$, which gives:*

$$\begin{pmatrix} \dot{x}_1 \\ \dot{x}_2 \end{pmatrix} = \begin{pmatrix} 0 & 1 \\ -k/m & -b/m \end{pmatrix} \begin{pmatrix} x_1 \\ x_2 \end{pmatrix} + \begin{pmatrix} 0 \\ 1/m \end{pmatrix} F \text{ and } y = \begin{pmatrix} 1 \; 0 \end{pmatrix} \begin{pmatrix} x_1 \\ x_2 \end{pmatrix} \qquad [5.81]$$

### 5.4.4. *Observable canonical form*

#### 5.4.4.1. *Companion form (observable canonical form)*

Starting from the same transfer function, we are able to propose another canonical form by writing the differential equation as follows:

$$\begin{aligned} &a_0 y(t) + \frac{d}{dt}(a_1 y(t) + \frac{d}{dt}(a_2 y(t) + \frac{d}{dt}(a_3 y(t) \\ &\quad + \frac{d}{dt}(\cdots + \frac{d}{dt}(a_{n-1} y(t) + \frac{d}{dt} y(t)) \cdots )))) = u(t) \end{aligned} \qquad [5.82]$$

To this end, we choose the following state variables:

$$\begin{aligned} x_1 &= y(t) \\ x_2 &= a_n x_1 + \dot{x}_1 \\ x_3 &= a_{n-1} x_1 + \dot{x}_2 \\ &\vdots \\ x_{n-1} &= a_3 x_1 + \dot{x}_{n-2} \\ x_n &= a_2 x_1 + \dot{x}_{n-1} \\ a_1 x_1 + \dot{x}_n &= u \end{aligned} \qquad [5.83]$$

which can be rewritten as:

$$\begin{aligned}
\dot{x}_1 &= -a_n x_1 + x_2 \\
\dot{x}_2 &= -a_{n-1} x_1 + x_3 \\
&\vdots \\
\dot{x}_{n-2} &= -a_3 x_1 + x_{n-1} \\
\dot{x}_{n-1} &= -a_2 x_1 + x_n \\
\dot{x}_n &= -a_1 x_1 + u \\
y &= x_1
\end{aligned} \qquad [5.84]$$

State–space representation matrices are then expressed as:

$$A = \begin{pmatrix} -a_n & 1\,0 \cdots 0 \\ -a_{n-1} & 0\,1 \cdots 0 \\ \vdots & \vdots\,\vdots\,\vdots\;\;\vdots \\ -a_2 & 0\,0 \cdots 1 \\ -a_1 & 0\,0 \cdots 0 \end{pmatrix}; \quad B = \begin{pmatrix} 0 \\ 0 \\ \vdots \\ 0 \\ 1 \end{pmatrix}; \quad C = \begin{pmatrix} 1\,0 \cdots 0 \end{pmatrix}; \quad D = 0 \qquad [5.85]$$

REMARK 5.8.– *This form of state–space representation is a observable canonical form.*

Consider the case where the numerator of the transfer function has a degree greater than zero (general case). Let:

$$H(p) = \frac{Y(p)}{U(p)} = \frac{b_m p^m + b_{m-1} p^{m-1} + \cdots + b_1 p + b_0}{p^n + a_{n-1} p^{n-1} + a_{n-2} p^{n-2} + \cdots + a_0} \qquad [5.86]$$

Recall that the numerator cannot have a degree greater than the denominator. The numerator and the denominator have the same degree only if outputs are directly connected by means of gains. To address the general case, we introduce the variable $Z(p)$, and we write:

$$\frac{Y(p)}{U(p)} = \frac{Y(p)}{Z(p)} \frac{Z(p)}{U(p)} \qquad [5.87]$$

with

$$\begin{aligned}
\frac{Y(p)}{Z(p)} &= b_m p^m + b_{m-1} p^{m-1} + \cdots + b_1 p + b_0 \qquad [5.88] \\
\frac{Z(p)}{U(p)} &= \frac{1}{p^n + a_{n-1} p^{n-1} + a_{n-2} p^{n-2} + \cdots + a_0}
\end{aligned}$$

We already have the state–space representation of the second equation. With regard to the first and taking the inverse Laplace transform, we get:

$$y(t) = b_m \frac{d^m z(t)}{dt^m} + b_{m-1} \frac{d^{m-1} z(t)}{dt^{m-1}} + b_{m-2} \frac{d^{m-2} z(t)}{dt^{m-2}} + \cdots + b_0 z(t) \quad [5.89]$$

If as above, we take $n-1$ successive derivatives and $z$ as state variable, matrices $A$ and $B$ remain the same and we now get:

$$C = \begin{pmatrix} b_0 & b_1 & b_2 & \cdots & b_{n-1} \end{pmatrix} \quad [5.90]$$

1234

EXAMPLE 5.12.– *For the transfer function*

$$H(p) = \frac{b_2 p^2 + b_1 p + b_0}{p^3 + a_2 p^2 + a_1 p + a_0} \quad [5.91]$$

*it can be written as*

$$(p^3 + a_2 p^2 + a_1 p + a_0) y(p) = (b_2 p^2 + b_1 p + b_0) u(p) \quad [5.92]$$

*By dividing member by member by* $p^3$, *we obtain*

$$y(p) = \frac{1}{p}\left( b_2 u(p) - a_2 y(p) - \frac{1}{p}\left( b_1 u(p) - a_1 y(p) - \frac{1}{p}\left( b_0 u(p) - a_0 y(p) \right) \right) \right) \quad [5.93]$$

*then the integrators outputs are considered as state vector.*

$$\dot{X} = \begin{bmatrix} -a_2 & 1 & 0 \\ -a_1 & 0 & 1 \\ -a_0 & 0 & 0 \end{bmatrix} X + \begin{bmatrix} b_2 \\ b_1 \\ b_0 \end{bmatrix} u \quad \text{and} \quad y = \begin{bmatrix} 1 & 0 & 0 \end{bmatrix} X \quad [5.94]$$

*This gives us the representation above which is the dual representation of the previous: If* $(A, B, C)$ *is a realization of*

$$H(p) = C.(pI - A)^{-1} B = B^T.(pI - A^T)^{-1} C^T \quad [5.95]$$

*then* $(A^T, C^T, B^T)$ *is another realization of* $H(p)$.

### 5.4.5. *Diagonal canonical form*

The diagonal canonical form is obtained using the simple elements decomposition. In the cases in which the system has no multiple poles, it yields

$$H(p) = \frac{b_2 p^2 + b_1 p + b_0}{p^3 + a_2 p^2 + a_1 p + a_0} = \frac{r_1}{p - p_1} + \frac{r_2}{p - p_2} + \frac{r_3}{p - p_3} \qquad [5.96]$$

and thus by system linearity, we can write:

$$\dot{X} = \begin{bmatrix} p_1 & 0 & 0 \\ 0 & p_2 & 0 \\ 0 & 0 & p_3 \end{bmatrix} X + \begin{bmatrix} 1 \\ 1 \\ 1 \end{bmatrix} u \quad \text{and} \quad y = \begin{bmatrix} r_1 & r_2 & r_3 \end{bmatrix} X \qquad [5.97]$$

#### 5.4.5.1. *Jordan form (diagonal or parallel form)*

Suppose that all poles of the transfer function be distinct. The transfer function can therefore be expanded in the form of rational fractions:

$$H(p) = b_o + \frac{r_1}{p - p_1} + \frac{r_2}{p - p_2} + \cdots + \frac{r_n}{p - p_n} = \frac{Y(p)}{U(p)} \qquad [5.98]$$

Carrying out the expansion member by member, we get:

$$Y(s) = b_o U(s) + \frac{r_1}{s - s_1} U(s) + \frac{r_2}{s - s_2} U(s) + \cdots + \frac{r_n}{s - s_n} U(s) \qquad [5.99]$$

State variables are chosen such that they verify:

$$\begin{array}{ll} X_1(p) = \frac{1}{p-p_1} U(p) & \dot{x}_1 = p_1 x_1 + u \\ X_2(p) = \frac{1}{p-p_2} U(p) & \dot{x}_2 = p_2 x_2 + u \\ \vdots & \Longrightarrow \vdots \\ X_n(p) = \frac{1}{p-p_n} U(p) & \dot{x}_n = p_n x_n + u \\ Y(p) = b_o U(p) + r_1 X_1(p) + \cdots + r_n X_n(p) & y = r_1 x_1 + \cdots + r_n x_n + b_o u \end{array} \qquad [5.100]$$

The matrices are the following:

$$A = \begin{pmatrix} p_1 & \cdots & 0 \\ 0 & \ddots & 0 \\ 0 & \cdots & p_n \end{pmatrix} \quad B = \begin{pmatrix} 1 \\ \vdots \\ 1 \end{pmatrix} \quad C = \begin{pmatrix} r_1 & \cdots & r_n \end{pmatrix} \quad D = b_o \qquad [5.101]$$

REMARK 5.9.– *Matrix $A$ is diagonal; its diagonal elements are the poles of $H(p)$. It is the Jordan form of a matrix in which all eigenvalues are distinct. If the system has multiple poles, $H(p)$ assumes the form:*

$$H(p) = b_o + H_1(p) + H_2(p) + \cdots + H_q(p), \qquad (q < n) \tag{5.102}$$

*with*

$$H_j(s) = \frac{r_{1j}}{p - p_j} + \frac{r_{2j}}{(p - p_j)^2} + \cdots \frac{r_{n_j j}}{(p - p_j)^i} \tag{5.103}$$

*and matrices $A$ and $B$ are as such:*

$$A = \begin{pmatrix} A_1 & 0 & \cdots & 0 \\ 0 & A_2 & \cdots & 0 \\ \vdots & \vdots & \vdots & \vdots \\ 0 & 0 & 0 & A_k \end{pmatrix} \quad B = \begin{pmatrix} B_1 \\ B_2 \\ \vdots \\ B_k \end{pmatrix} \quad C = \begin{pmatrix} C_1 & \cdots & C_k \end{pmatrix} \quad D = b_o \tag{5.104}$$

*with*

$$A_j = \begin{pmatrix} p_j & 0 & 0 & \cdots & 0 \\ 1 & p_j & 0 & \cdots & 0 \\ 0 & 1 & p_j & \cdots & 0 \\ \vdots & \vdots & \vdots & \vdots & \vdots \\ 0 & 0 & 0 & \cdots & p_j \end{pmatrix} \quad B_j = \begin{pmatrix} 1 \\ 0 \\ \vdots \\ \vdots \\ 0 \end{pmatrix} \quad C_j = \begin{pmatrix} r_{1j} & \cdots & r_{n_j j} \end{pmatrix} \tag{5.105}$$

### 5.4.6. *Change in state-space representations and change in basis*

A system can be represented according to different state–space representations; it depends on the state variables with which we choose to write the state equations. The interest in changing state–space representation or changing basis can lie either in bringing up, in the state vector, measurable states or having physical meaning or to allow the simplification of the computation of a correction element or an observer (shifting to a canonical form). It should be noted that canonical forms include the minimum number of coefficients with different values of 1 and 0 .

Let $(A, B, C, D)$ be the state–space representation or realization of a system; we can write:

$$\dot{X} = A.X + Bu(t) \tag{5.106}$$

$$y(t) = C.X + Du \tag{5.107}$$

Let a change in variable be defined by:

$$X = TZ$$

Consider a invertible transformation matrix $T$ in $R^{n \times n}$ making it possible to transform the state vector $X$ into vector $Z$ such that $Z = T^{-1}X$ or $X = TZ$. $T$ is a non-singular matrix. The system can be described on the basis of the new state vector by:

$$\dot{Z} = T^{-1}A.TZ + T^{-1}Bu(t) \qquad [5.108]$$

$$y(t) = C.TZ + Du \qquad [5.109]$$

The new state equations are therefore:

$$A' = T^{-1}AT \qquad B' = T^{-1}B \qquad C' = CT \qquad D' = D \qquad [5.110]$$

A change in state variables does not affect the nature of the variables. We now verify that the transfer function of the system, its characteristic equation, its poles and zeros are unchanged.

The impulse response is of course preserved despite the transformation:

$$y(t) = C'e^{A't}B' = CTe^{T^{-1}ATt}T^{-1}B \qquad (e^{T^{-1}ATt} = T^{-1}e^{At}T) \qquad [5.111]$$

We therefore obtain the same impulse response:

$$y(t) = Ce^{At}B \qquad [5.112]$$

The transfer function that is also the Laplace transform of the impulse response is therefore invariant when a change takes place in state.

The transfer function is realizable in different forms and can therefore lead to different choices of state variables; this leads to different state–space representations.

EXAMPLE 5.13.– *Consider the transfer function* $H(p) = \frac{2p+3}{(p+1)(p+2)}$. *This system can also appear as the state representation of the following governable canonical form*

$$\dot{X} = \begin{bmatrix} -3 & -2 \\ 1 & 0 \end{bmatrix} X + \begin{bmatrix} 1 \\ 0 \end{bmatrix} u \quad \text{and} \quad y = \begin{bmatrix} 2 \; 3 \end{bmatrix} X \qquad [5.113]$$

*Choosing the transformation matrix*

$$T = \begin{bmatrix} 2 & -1 \\ -1 & 1 \end{bmatrix} \qquad [5.114]$$

*The state vector $Z$ is defined such that $Z = T^{-1}X$ or $X = TZ$; the diagonal form is thus obtained:*

$$\begin{pmatrix} \dot{z}_1 \\ \dot{z}_2 \end{pmatrix} = \begin{pmatrix} -2 & 0 \\ 0 & -1 \end{pmatrix} \begin{pmatrix} z_1 \\ z_2 \end{pmatrix} + \begin{pmatrix} 1 \\ 1 \end{pmatrix} u \quad \text{and} \quad y = \begin{pmatrix} 1 & 1 \end{pmatrix} \begin{pmatrix} z_1 \\ z_2 \end{pmatrix} \qquad [5.115]$$

This can be verified by dividing the transfer function into simple elements:

$$H(p) = \frac{2p+3}{(p+1)(p+2)} = \frac{1}{p+1} + \frac{1}{p+2} \qquad [5.116]$$

### 5.4.7. *Examples of systems to be modeled: the inverse pendulum*

The cart is modeled as a material point of mass $M$ and the pendulum as a point of mass $m$.

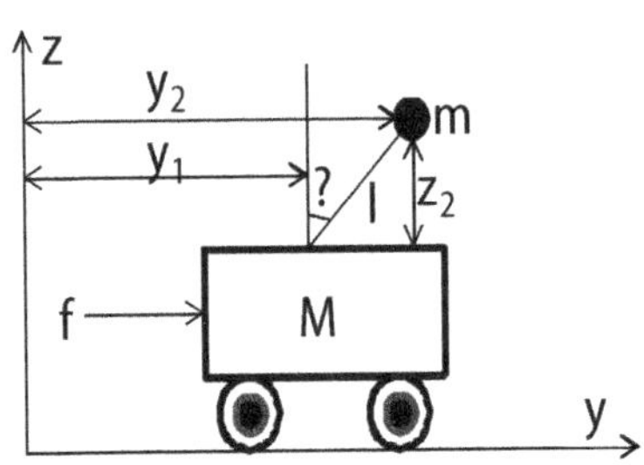


**Figure 5.10.** *Inverse pendulum*

The kinetic energy of the cart is given as:

$$T_1 = \frac{1}{2} M \dot{y}_1^2 \qquad [5.117]$$

The kinetic energy of the pendulum is given as:

$$T_2 = \frac{1}{2} m (\dot{y}_2^2 + \dot{z}_2^2) \qquad [5.118]$$

In addition, we have:

$$\begin{aligned} y_2 &= y_1 + l\sin(\theta) \Longrightarrow \dot{y}_2 = \dot{y}_1 + l\dot{\theta}\cos(\theta) \\ z_2 &= l\cos(\theta) \Longrightarrow \dot{z}_2 = -l\dot{\theta}\sin(\theta) \end{aligned} \quad [5.119]$$

Hence, the total kinetic energy is given by:

$$T = T_1 + T_2 = \frac{1}{2}M\dot{y}_1^2 + \frac{1}{2}m[\dot{y}_1^2 + 2\dot{y}_1 l\dot{\theta}\cos(\theta) + l^2\dot{\theta}^2] \quad [5.120]$$

On the other hand, the potential energy of the pendulum is written as:

$$V = mgz_2 = mgl\cos(\theta) \quad [5.121]$$

The choice of generalized coordinates has to be made: $y_1$ and $\theta$. Therefore, the Lagrange equations are written as:

$$\begin{aligned} \frac{d}{dt}\left(\frac{\partial L}{\partial \dot{y}_1}\right) - \frac{\partial L}{\partial y_1} &= f \\ \frac{d}{dt}\left(\frac{\partial L}{\partial \dot{\theta}}\right) - \frac{\partial L}{\partial \theta} &= 0 \end{aligned} \quad [5.122]$$

All calculations are made, and the result is the following system of equations:

$$\begin{aligned} (M+m)\ddot{y}_1 + ml\cos(\theta)\ddot{\theta} - ml\sin(\theta)\dot{\theta}^2 &= f \\ ml\cos(\theta)\ddot{y}_1 + ml^2\ddot{\theta} - mgl\sin(\theta) &= 0 \end{aligned} \quad [5.123]$$

These equations are nonlinear. In order to be able to control this system, it must be linearized. To do this, it is assumed:

– that $\theta$ is small, henceforth $\cos(\theta) \simeq 1$ and $\sin(\theta) \simeq \theta$;

– that generalized velocities are small enough so that the squared term can be neglected $\dot{\theta}^2$.

There remain therefore two second-order linear equations:

$$\begin{aligned} (M+m)\ddot{y}_1 + ml\ddot{\theta} &= f \\ ml\ddot{y}_1 + ml^2\ddot{\theta} - mgl\theta &= 0 \end{aligned} \quad [5.124]$$

The system is therefore of the fourth order. We take $y_1$ and $\theta$ for state variables and the state vector is completed by $\dot{y}_1$ and $\dot{\theta}$.

#### 5.4.7.1. *System state description*

We have a system with an input $f$ and two outputs $y_1$ and $\theta$. Using the usual notations of state representation, we define the following state vector:

$$x = \begin{pmatrix} y_1 \\ \theta \\ \dot{y}_1 \\ \dot{\theta} \end{pmatrix}, \qquad u = f \qquad [5.125]$$

hence the state equation is

$$\dot{x} = \begin{pmatrix} 0 & 0 & 1 & 0 \\ 0 & 0 & 0 & 1 \\ 0 & -\frac{mg}{M} & 0 & 0 \\ 0 & \frac{(M+m)g}{Ml} & 0 & 0 \end{pmatrix} x + \begin{pmatrix} 0 \\ 0 \\ \frac{1}{M} \\ -\frac{1}{Ml} \end{pmatrix} u \qquad [5.126]$$

and the output equation is:

$$\begin{pmatrix} y_1 \\ \theta \end{pmatrix} = \begin{pmatrix} 1 & 0 & 0 & 0 \\ 0 & 1 & 0 & 0 \end{pmatrix} x \qquad [5.127]$$

#### 5.4.7.2. *External description of the transfer function*

To provide an external description of the system (inverted pendulum), we consider a simplified version of the inversed pendulum to get a SISO-type system. To this end, it is assumed that the axis of the pendulum is fixed (the variable $y_1$ disappears), but there is an engine that performs a torque reaction $\gamma(t)$ to control the position of the terminal point of the pendulum. The equations thus can be expressed as:

$$\begin{aligned} &\dot{\theta} = \omega \\ &\ddot{\theta} = \Omega^2\theta + u \qquad \text{with} \qquad \Omega^2 = \tfrac{g}{l} \text{ and } u = \tfrac{\gamma(t)}{ml^2} \end{aligned} \qquad [5.128]$$

In this case, the state description amounts to:

$$x = \begin{pmatrix} \theta \\ \dot{\theta} \end{pmatrix}, \qquad u = \frac{\gamma(t)}{ml^2} \qquad [5.129]$$

with the state equation:

$$\dot{x} = \begin{pmatrix} 0 & 1 \\ \Omega^2 & 0 \end{pmatrix} x + \begin{pmatrix} 0 \\ 1 \end{pmatrix} u \qquad [5.130]$$

and the output equation:

$$\theta = \begin{pmatrix} 1 \; 0 \end{pmatrix} x \qquad [5.131]$$

The matrices of the new state description are expressed as:

$$A = \begin{pmatrix} 0 & 1 \\ \Omega^2 & 0 \end{pmatrix} \qquad B = \begin{pmatrix} 0 \\ 1 \end{pmatrix} \qquad C = \begin{pmatrix} 1 \; 0 \end{pmatrix} \qquad D = 0 \qquad [5.132]$$

Using two possible methods (computation of $G(s)$ from matrices $A$, $B$, $C$, $D$, or solving the Laplace transform of the differential equation), it shall be verified that the transfer function of the system is:

$$G(s) = \frac{1}{s^2 - \Omega^2} = \frac{1}{(s - \Omega)(s + \Omega)} \qquad [5.133]$$

We note that a pole of $G(s)$ is positive therefore the system is unstable.

### 5.4.8. *System phase–plane representation*

Consider the case of a second-order signal or free system defined by its differential equation:

$$\ddot{y}(t) = f(\dot{y}(t), y(t)) \qquad [5.134]$$

Taking $x_1 = y(t)$ and $x_2 = \dot{y}(t)$, it is possible to associate thereto the following state representation:

$$\dot{X}(t) = \begin{pmatrix} x_2 \\ -f(x_1, x_2) \end{pmatrix} = AX; \quad y(t) = [1; 0].X \qquad [5.135]$$

State variables $x_1$ and $x_2$ are also called phase variables; they determine at every moment the state of the system that can be represented by a point $M$ in the plane $(x_1, x_2)$. This plane is called the phase plan. Point $M$ is characterized by state vector $X(t)$ and its time evolution from an initial point $M_o$ describes a phase curve. Phase curves can be graded in time and depend on the initial point $M_o$ (the initial state). By eliminating the time variable, an equation is obtained parameterized according to the coordinates $(x_1, x_2)$ defining a network of curves. The plot of these curves provides a visualization of the evolution of the system.

A large number of linear or nonlinear systems can be approximated by the second order. This method of representation allows us to easily draw conclusions about the evolution and stability of a system. These disadvantages originate from the fact that a graphical representation is impossible for systems with higher orders than two and is only applicable for autonomous systems (without input control and whose differential equation is not explicitly time dependent) having a unique solution (for the state equation). It should be noted that the principle remains valid for systems of order greater than 2, although the graphical representation be impossible. In the following, we illustrate this method for the case of a second-order linear system.

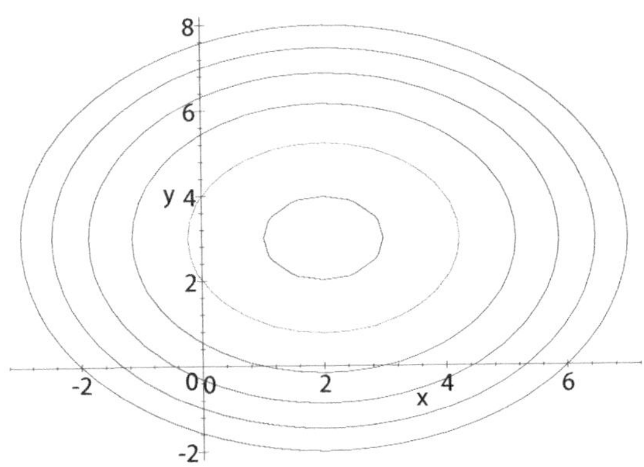


**Figure 5.11.** *Conservative system*

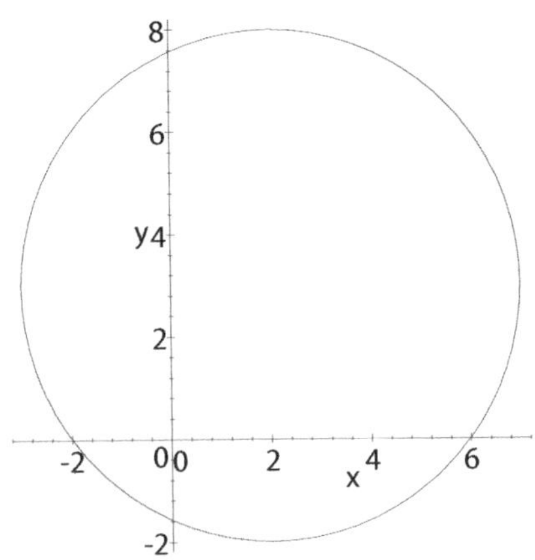


**Figure 5.12.** *Circles*

#### 5.4.8.1. *Case of a second-order linear system*

For the representation of a second-order system in the phase plane, we consider equation:

$$\ddot{y}(t) + 2z\omega_o\dot{y}(t) + \omega_o^2 y(t) = 0 \tag{5.136}$$

$z$ is the damping and $\omega_o$ the angular frequency of the system. Poles $p_1, p_2$ are the roots of the characteristic equation:

$$r + 2z\omega_o r + \omega_o = 0 \tag{5.137}$$

According to damping values $z$, several possible cases can be distinguished.

– *Case 1*: If $p_1$and $p_2$ are two real negative roots of the characteristic equation:

$$y(t) = x_1 = C_1 e^{p_1 t} + C_2 e^{p_2 t} \tag{5.138}$$

$$\dot{y}(t) = x_2 = p_1 C_1 e^{p_1 t} + p_2 C_2 e^{p_2 t} \tag{5.139}$$

where

$$e^{p_1 t} = \frac{[p_2 x_1 - x_2]}{[p_2 - p_1]C_1} \quad \text{and} \quad e^{p_2 t} = \frac{[p_1 x_1 - x_2]}{[p_1 - p_2].C_2} \tag{5.140}$$

where by raising these equations to the power $p_1$ and $p_2$, we get:

$$(p_2 x_1 - x_2)^{\frac{p_2}{p_1}} = C(p_1 x_1 - x_2) \tag{5.141}$$

Constant $C$ depends on constants $C_1$ and $C_2$. This equation represents a family of parabolas tending to zero in $t$, as indicated in Figure 5.13.

– *Case 2*: If $p_1$ and $p_2$ are 2 real positive roots, then it yields, by proceeding as previously, the result, as shownin Figure 5.14.

– *Case 3*: In the case where $p_1$ and $p_2$ are two real roots of opposite signs, equation [5.141] this times belongs to a family of hyperbolas, because $p_2/p_1 < 0$, whose asymptotes have equations:

$$x_2 = p_1.x_1 \quad \text{and} \quad x_2 = p_2.x_1 \tag{5.142}$$

One of the asymptotes (the one corresponding to $p_i$ <0) has its branches directed toward the origin ($x_1$= 0, $x_2$= 0). The representation is shown in Figure 5.15.

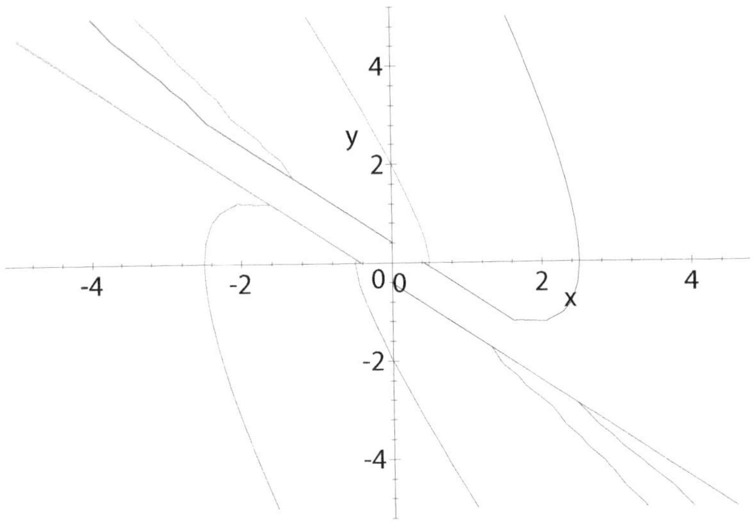


**Figure 5.13.** *Two real negative roots of the characteristic equation*

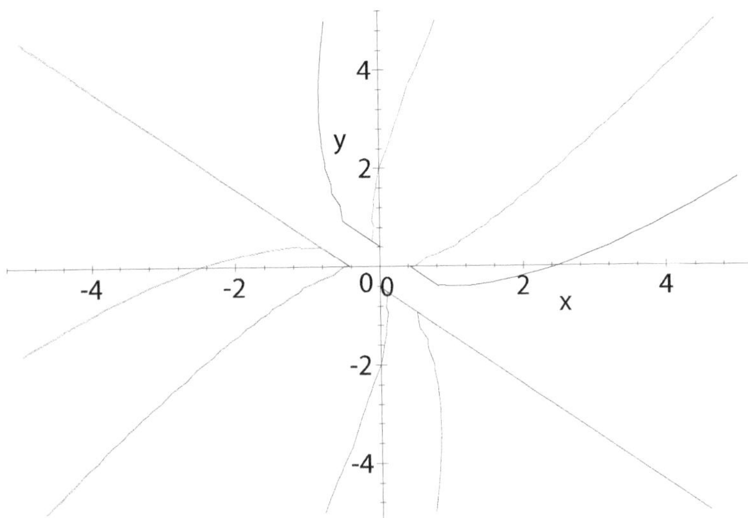


**Figure 5.14.** *Two positive real roots of the characteristic equation*

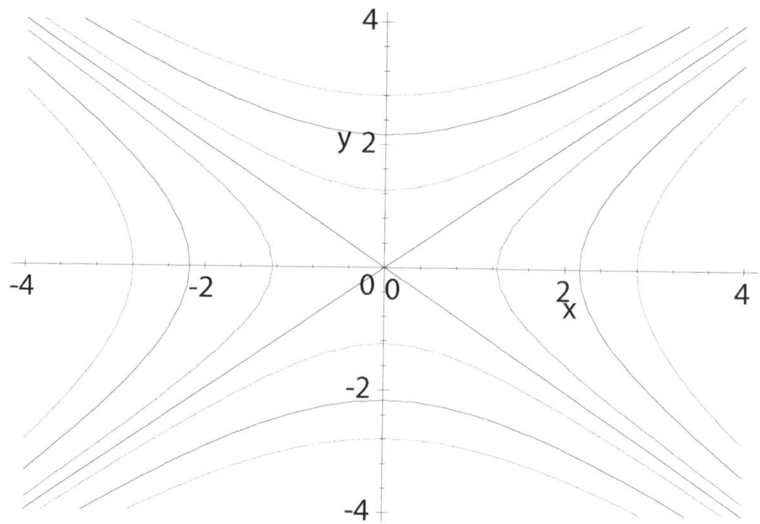


**Figure 5.15.** *Two real roots with opposite signs of the characteristic equation*

– *Case 4*: If $p_1$and $p_2$ are two complex roots with real negative parts, the system is stable and damped oscillating. The solutions are of the form:

$$x_1 = Ae^{p_1}.\sin(\omega t + \varphi) \quad \text{and} \quad x_2 = Be^{p_2}.\sin(\omega t + \psi) \qquad [5.143]$$

$A$, $B$, $\varphi$ and $\psi$ depend on initial conditions. The trajectories have the shape of spirals going toward zero as shown in Figure 5.16.

– *Case 5*: If $p_1$ and $p_2$ are two complex roots with real positive parts, this yields spirals ranging from zero to infinity (see Figure 5.17).

– *Case 6*: If $p_1$and $p_2$ are two pure imaginary roots, we obtain ellipses as presented in Figure 5.18.

$$x_1 = A.\sin(\omega t + \varphi) \text{ and } x_2 = B.\sin(\omega t + \psi) \qquad [5.144]$$

Previously, we have presented the phase–plane representation. This representation is useful for the analysis of the stability of a system but it is limited to the second-order case.

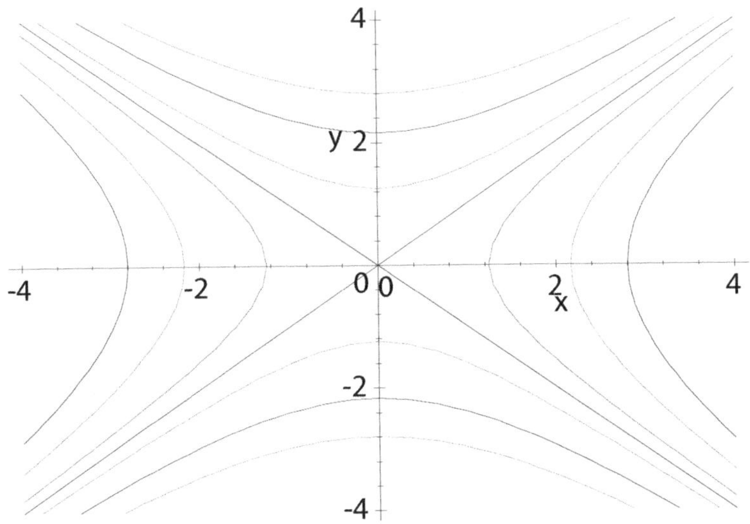


**Figure 5.16.** *Two complex roots of the characteristic equation with real negative parts*

## 5.5. Some application exercises

EXERCISE 1.– Determine three different state–space representations (companion, Jordan, series) of the process and corresponding block diagrams, modeled by the following transfer function:

$$H(s) = \frac{1}{s^3 + 10s^2 + 27s + 18} \qquad [5.145]$$

SOLUTION EXERCISE 1.– The three state–space representations of the process modeled by

$H(s) = \frac{1}{s^3+10s^2+27s+18}$ are given as follows: companion form, written in the temporal domain:

$$\begin{cases} x_1 = y \\ x_2 = \dot{y} = \dot{x}_1 \\ x_3 = \ddot{y} = \dot{x}_2 \\ \dot{x}_3 = -10\ddot{y} - 27\dot{y} - 18y + u = -10x_3 - 27x_2 - 18x_1 + u \end{cases} \qquad [5.146]$$

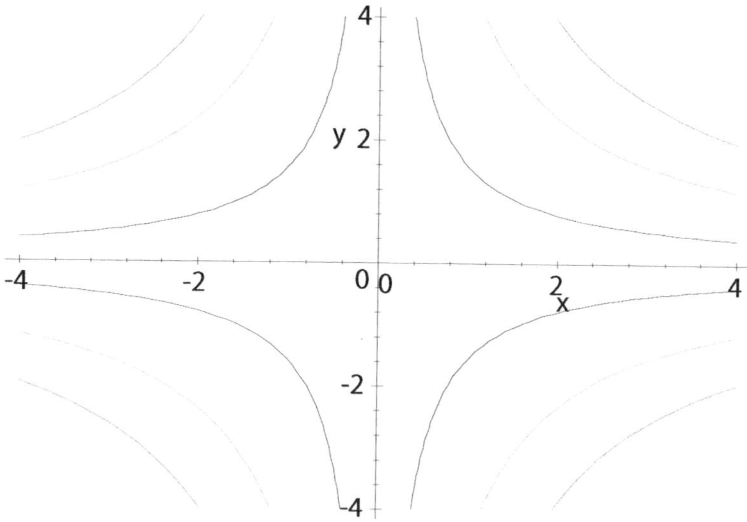


**Figure 5.17.** *Two complex roots of the characteristic equation with real positive parts*

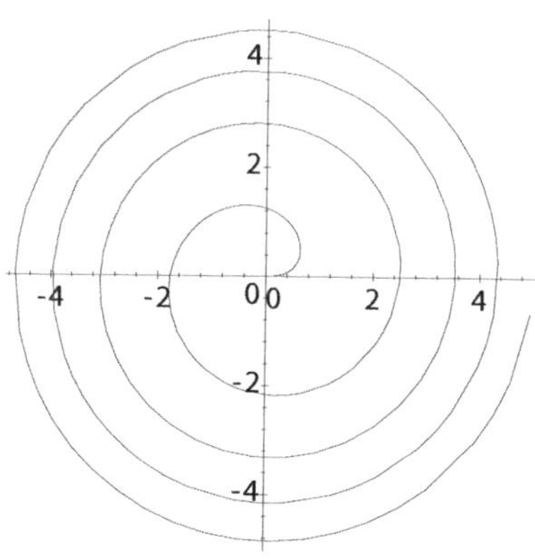


**Figure 5.18.** *Two pure imaginary roots of the characteristic equation (stable system)*

In matrix form, we have:

$$\begin{cases} \begin{bmatrix} \dot{x}_1 \\ \dot{x}_2 \\ \dot{x}_3 \end{bmatrix} = \begin{bmatrix} 0 & 1 & 0 \\ 0 & 0 & 1 \\ -18 & -27 & -10 \end{bmatrix} \begin{bmatrix} x_1 \\ x_2 \\ x_3 \end{bmatrix} + \begin{bmatrix} 0 \\ 0 \\ 1 \end{bmatrix} u \quad \text{(space–state equation)} \\ y = \begin{bmatrix} 1 \; 0 \; 0 \end{bmatrix} \begin{bmatrix} x_1 \\ x_2 \\ x_3 \end{bmatrix} \quad \text{(output equation)} \end{cases} \qquad [5.147]$$

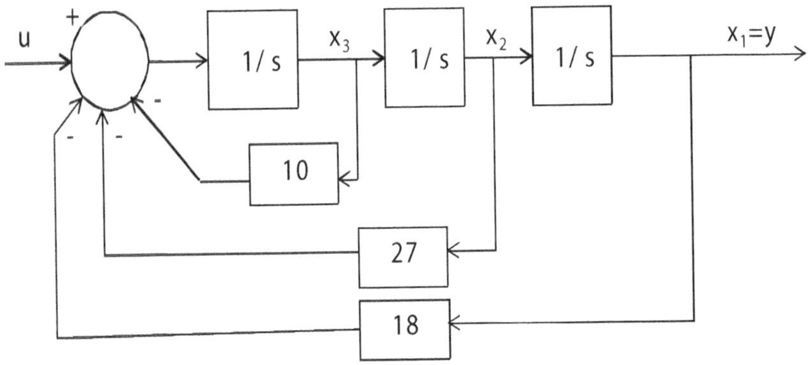


**Figure 5.19.** *State–space representation of the modeled process: companion form*

The modal form (Jordan form) or parallel decomposition is written in the temporal domain as:

$$H(s) = \frac{Y(s)}{U(s)} = \frac{1}{s^3 + 10s^2 + 27s + 18} = \frac{1}{(s+1)(s+3)(s+6)} \qquad [5.148]$$

The transfer function $H(s)$ can be rewritten as simple elements:

$$H(s) = \frac{Y(s)}{U(s)} = \frac{1}{10}\frac{1}{s+1} - \frac{1}{6}\frac{1}{s+3} + \frac{1}{15}\frac{1}{s+6} \qquad [5.149]$$

$$\begin{cases} X_1(s) = \frac{1}{s+1}U(s) & \longrightarrow & \dot{x}_1 = -x_1 + u \\ X_2(s) = \frac{1}{s+3}U(s) & \longrightarrow & \dot{x}_2 = -3x_2 + u \\ X_3(s) = \frac{1}{s+6}U(s) & \longrightarrow & \dot{x}_3 = -6x_3 + u \\ Y(s) = \frac{1}{10}X_1(s) - \frac{1}{6}X_2(s) + \frac{1}{15}X_3(s) & \longrightarrow & y = \frac{1}{10}x_1 - \frac{1}{6}x_2 + \frac{1}{15}x_3 \end{cases} \qquad [5.150]$$

In matrix form:

$$\begin{cases} \begin{bmatrix} \dot{x}_1 \\ \dot{x}_2 \\ \dot{x}_3 \end{bmatrix} = \begin{bmatrix} -1 & 0 & 0 \\ 0 & -3 & 0 \\ 0 & 0 & -6 \end{bmatrix} \begin{bmatrix} x_1 \\ x_2 \\ x_3 \end{bmatrix} + \begin{bmatrix} 1 \\ 1 \\ 1 \end{bmatrix} u \\ y = \begin{bmatrix} \frac{1}{10} & -\frac{1}{6} & \frac{1}{15} \end{bmatrix} \begin{bmatrix} x_1 \\ x_2 \\ x_3 \end{bmatrix} \end{cases}$$

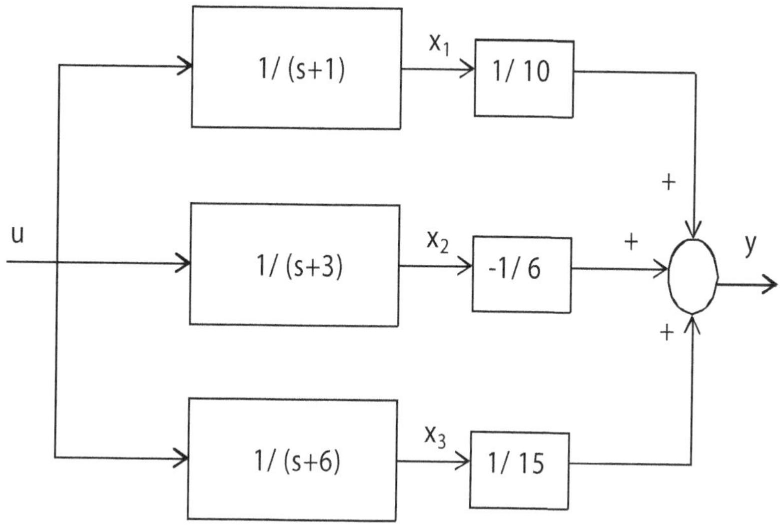


**Figure 5.20.** *State–space representation of the modeled process: Jordan form*

The cascade form or series decomposition is written in the time domain as:

$$H(s) = \frac{1}{s^3 + 10s^2 + 27s + 18} = \frac{1}{(s+1)(s+3)(s+6)} \qquad [5.151]$$

$$\begin{cases} X_1(s) = \frac{1}{s+1} U(s) & \longrightarrow & \dot{x}_1 = -x_1 + u \\ X_2(s) = \frac{1}{s+3} X_1(s) & \longrightarrow & \dot{x}_2 = -3x_2 + x_1 \\ X_3(s) = \frac{1}{s+6} X_2(s) & \longrightarrow & \dot{x}_3 = -6x_3 + x_2 \\ Y(s) = X_3(s) & \longrightarrow & y = x_3 \end{cases} \qquad [5.152]$$

In matrix form:

$$\begin{cases} \begin{bmatrix} \dot{x}_1 \\ \dot{x}_2 \\ \dot{x}_3 \end{bmatrix} = \begin{bmatrix} -1 & 0 & 0 \\ 1 & -3 & 0 \\ 0 & 1 & -6 \end{bmatrix} \begin{bmatrix} x_1 \\ x_2 \\ x_3 \end{bmatrix} + \begin{bmatrix} 1 \\ 0 \\ 0 \end{bmatrix} u \\ y = \begin{bmatrix} 0 \; 0 \; 1 \end{bmatrix} \begin{bmatrix} x_1 \\ x_2 \\ x_3 \end{bmatrix} \end{cases} \quad [5.153]$$

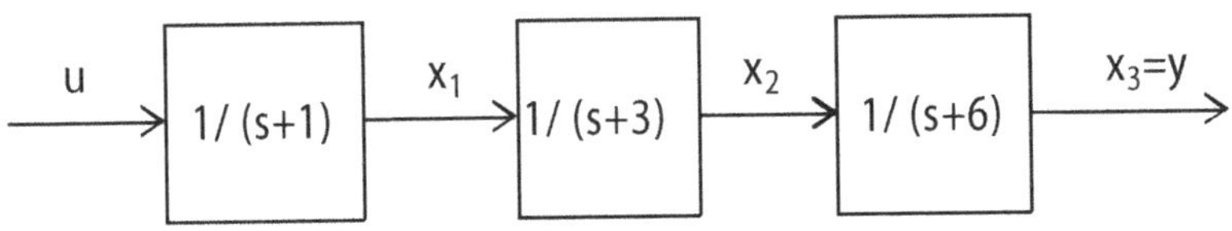


**Figure 5.21.** *State–space representation of the modeled process: cascade form*

EXERCISE 2. – Positioning using a linear motor

A mobile coil is guided in translation on a magnetic circuit. Four permanent magnets ensure the magnetic flux. A reading head (not shown) is driven by the mobile coil. A spring of stiffness $k$ ensures that it returns to position 0.

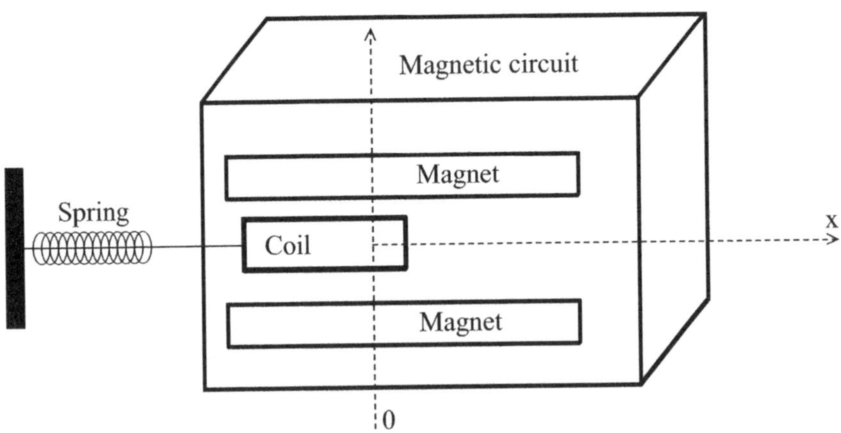


**Figure 5.22.** *Linear motor*

We denote $R$ and $L$ as the resistance and the inductance of the coil; $i$ is the current in the coil; $e$ its counter electromotive force ($e = \alpha \frac{dx}{dt}$); $k$ is the stiffness of the spring;

$f$ is the coefficient of viscous friction; $m$ is the mass of the mobile assembly (coil + head); $\beta$ is such that the motive force exerted on the coil is given by ($F_a = \beta i$); $x$ is the position of the coil with respect to the equilibrium.

– *System modeling:* we apply a voltage at the terminals of the coil.

1) Write the electrical equation governing the system.

2) Write the fundamental equation of the dynamics of the system between the input $u$ and the output $x$.

3) Deduce therefrom the functional diagram of the system in the form of blocks of transfer functions.

4) Give the state–space representation of the system having the state vector:

$$y = [i\ x\ \tfrac{dx}{dt}]^T \qquad [5.154]$$

– *System analysis:*

1) Determine the characteristic polynomial of the system state matrix.

2) Derive therefrom the stability criterion.

3) The following numeric values are provided:

$$\begin{aligned}&R = 8\,\Omega;\ L = 8\,\text{mH};\ k = 50\,\text{N/M};\ M = 200\,\text{g};\ f = 0\,\text{NS/M};\\ &\beta = 5\,\text{N/A};\ \alpha = 5\,\text{VS/M}\end{aligned} \qquad [5.155]$$

i) Numerically represent the state matrix.

ii) Calculate the transfer function:

$$H(s) = \frac{X(s)}{U(s)} \qquad [5.156]$$

By changing representation (transfer function $\rightarrow$ state–space representation), the determinant is given as:

$$\begin{aligned}\Delta(s) &= s^3 + 1,000s^2 + 15,875s + 250,000\\ &= (s + 984.1271)(s^2 + 15.87292s + 254.0322)\end{aligned} \qquad [5.157]$$

by reducing the block diagram (to get a standard functional diagram).

iii) Compute the poles of $H(s)$.

iv) Can a pole be neglected? In this case, what are the characteristic parameters of the system?

SOLUTION EXERCISE 2. –

– Electric equation:

$$u(t) = Ri(t) + L\frac{di(t)}{dt} + e \quad [5.158]$$

– Mechanic equation:

$$m\frac{d^2x(t)}{dt^2} = \beta i(t) - kx(t) - f\frac{dx(t)}{dt} \quad [5.159]$$

– Coupling equation:

$$e = \alpha\frac{dx(t)}{dt} \quad [5.160]$$

By applying the Laplace transform to equations [5.158]–[5.160], we achieve:

$$\begin{cases} U(s) = (R + Ls)I(s) + E(s) \\ F_a(s) = \beta I(s) = (Ms^2 + fs + k)X(s) \\ E(s) = \alpha sX(s) \end{cases} \quad [5.161]$$

By rearranging this equation system, we arrive at:

$$\begin{cases} I(s) = \frac{1}{R+Ls}(U(s) - E(s)) \\ E(s) = \alpha sX(s) \\ sX(s) = \frac{1}{Ms}(F_a(s) - fsX(s) - kX(s)) \end{cases} \quad [5.162]$$

Hence, the following block diagram is obtained (Figure 5.23).

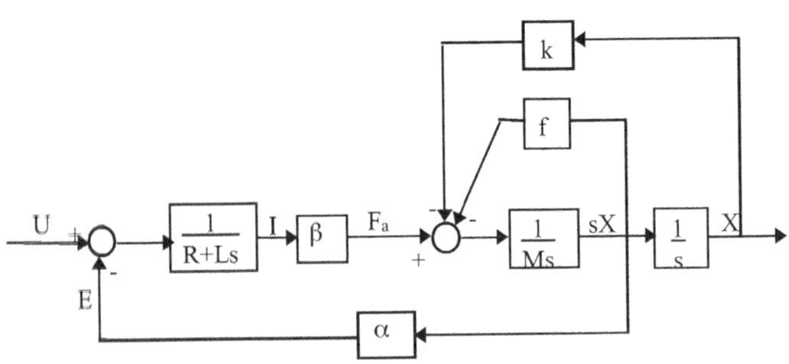


**Figure 5.23.** *Block diagram Exercise 2*

Equations [5.158]–[5.160] give us the following system:

$$\frac{di}{dt} = \frac{u}{L} - \frac{R}{L}i - \alpha\frac{dx}{dt} \qquad [5.163]$$
$$\frac{d^2x}{dt^2} = \frac{\beta}{m}i - \frac{k}{m}x - \frac{f}{m}\frac{dx}{dt}$$

Choosing the next state–space vector:

$$y = [i \;\; x \;\; \frac{dx}{dt}]^T \qquad [5.164]$$

We achieve the following state–space representation:

$$\dot{y} = Ay + Bu \qquad \text{(State–space equation)} \qquad [5.165]$$
$$x = Cy \qquad \text{(Output equation)}$$

with

$$A = \begin{pmatrix} -\frac{R}{L} & 0 & -\frac{\alpha}{L} \\ 0 & 0 & 1 \\ \frac{\beta}{m} & -\frac{k}{m} & -\frac{f}{m} \end{pmatrix}; \quad B = \begin{pmatrix} \frac{1}{L} \\ 0 \\ 0 \end{pmatrix}; \quad C = (0\;1\;0) \qquad [5.166]$$

The characteristic polynomial is:

$$P(\lambda) = \det(\lambda I - A) = \lambda^3 + \lambda(\frac{f}{m} + \frac{k}{L}) + \lambda(\frac{k}{m} + \frac{Rf + \alpha\beta}{Lm}) + \frac{kR}{Lm} \qquad [5.167]$$

Stability using the Routh criterion:

– all terms are present and positive;

– all the words in the first column of the Routh table are positive. System of rank $3 \Rightarrow a_1a_2 - a_3a_0 > 0$. In effect:

$$a_3s^3 + a_2s^2 + a_1s + a_0 = 0 \qquad [5.168]$$

$$\left| \begin{array}{ll} a_3 & a_1 \\ a_2 & a_0 \\ \frac{a_1a_2 - a_0a_3}{a_2} & \\ a_0 & \end{array} \right. \Rightarrow a_1a_2 - a_3a_0 > 0 \qquad [5.169]$$

Let:

$$\frac{f}{m}(\frac{k}{m} + \frac{Rf + \alpha\beta}{Lm}) + \frac{R}{L}\frac{Rf + \alpha\beta}{Lm} > 0 \qquad [5.170]$$

This term is always positive, therefore the system is asymptotically stable.

Numerical application:

$$A = \begin{pmatrix} -1000 & 0 & -625 \\ 0 & 0 & 1 \\ 25 & -250 & 0 \end{pmatrix}; \quad B = \begin{pmatrix} 125 \\ 0 \\ 0 \end{pmatrix}; \quad C = \begin{pmatrix} 0 & 1 & 0 \end{pmatrix} \quad [5.171]$$

The transfer function is given by:

$$H(s) = \frac{X(s)}{U(s)} = C(sI - A)^{-1}B \quad [5.172]$$

Calculate first the term $(sI - A)^{-1}$ :

$$sI - A = \begin{pmatrix} s + 1,000 & 0 & 625 \\ 0 & s & -1 \\ -25 & 250 & s \end{pmatrix} \quad [5.173]$$

hence the determinant:

$$\begin{aligned} \Delta(s) &= (1,000 + s)(s^2 + 250) + 625(25s) \\ &= s^3 + 1,000s^2 + 15,875s + 250,000 \\ &= (s + 984.1271)(s^2 + 15.87292s + 254.0322) \end{aligned} \quad [5.174]$$

where:

$$(sI - A)^{-1} = \frac{1}{\Delta(s)} \begin{pmatrix} s^2 + 250 & 156,250 & -625s \\ 25 & s(s + 1,000) + 15,625 & s + 1,000 \\ 25s & -250(s + 1,000) & s(s + 1,000) \end{pmatrix} \quad [5.175]$$

then the transfer function is given by:

$$\begin{aligned} H(s) &= \frac{1}{\Delta(s)} \begin{pmatrix} 0 & 1 & 0 \end{pmatrix} \begin{pmatrix} \times & \times & \times \\ 25 & \times & \times \\ \times & \times & \times \end{pmatrix} \begin{pmatrix} 125 \\ 0 \\ 0 \end{pmatrix} \\ &= \frac{3,125}{\Delta(s)} = \frac{3,125}{(s + 984.1271)(s^2 + 15.87292s + 254.0322)} \end{aligned} \quad [5.176]$$

The transfer function can also be written in the form :

$$H(s) = \frac{K\omega_n^2}{(1 + T_1 s)(s^2 + 2\xi\omega_n s + \omega_n^2)} \quad [5.177]$$

with $K = 3125$, $T_1 = 10^{-3}$, $\omega_n = 16$ and $\xi = 0.5$

The pole can be neglected, $s = -984.12$, because it is located far off on the left side in the complex plane. Its effect in the temporal domain disappears very quickly. The dynamic is thus described only by the dominant complex conjugate poles located near the imaginary axis.

EXERCISE 3.– Study of car suspension

We propose to analyze a car suspension process through the following simplified study, recalling the study of the movement of the chassis of mass $M$ on a wheel ($m$ is the mass of the wheel).

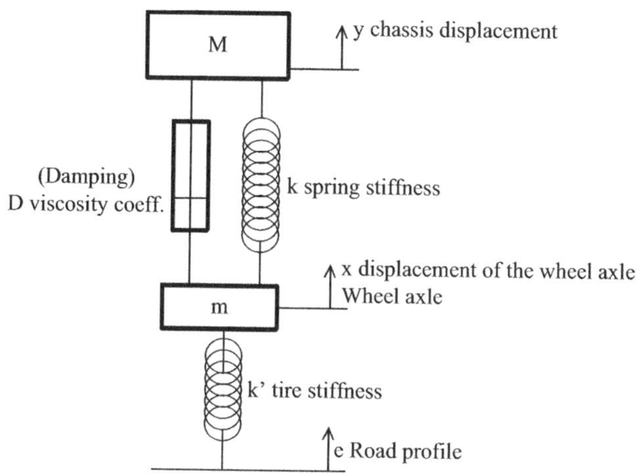


**Figure 5.24.** *Car suspension process*

1) Write the system of differential equations representing the laws of mechanics.

2) We neglect the elasticity of the tire ($x = e$).

3) Determine the transfer function $Y(s)/E(s)$ between input $e$ (profile of the road) and output $y$ (chassis displacement).

4) Give the appearance of the response (chassis motion) to a Heaviside function (sidewalk) according to the value of the parameters.

5) Study the asymptotic Bode diagram based on the relative values of the parameters (optional question). Derive a physical explanation for the vibration phenomenon felt in some cars around 100 km/h.

EXERCISE 4.– Study of an electronic circuit

Determine the transfer function of the following op–amp assembly assumed as perfect. What is this complex gain? Derive its application.

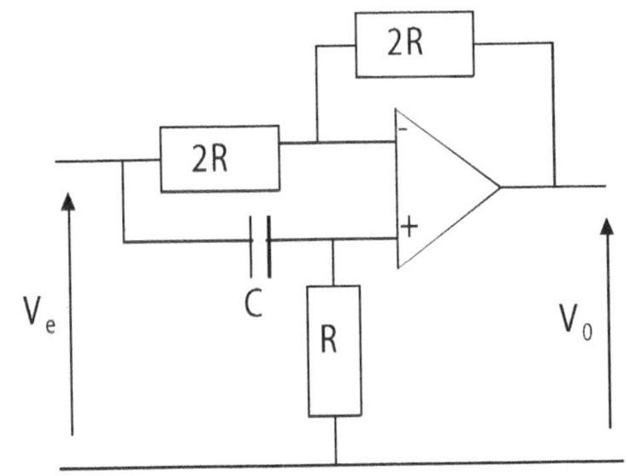


**Figure 5.25.** *Op–amp circuit*

EXERCISE 5.– Electromechanical system

A machine tool is driven by a direct current electric motor controlled by the induced voltage $u(t)$. In the course of the machining work, it is submitted to a resistance torque $C_r(t)$.

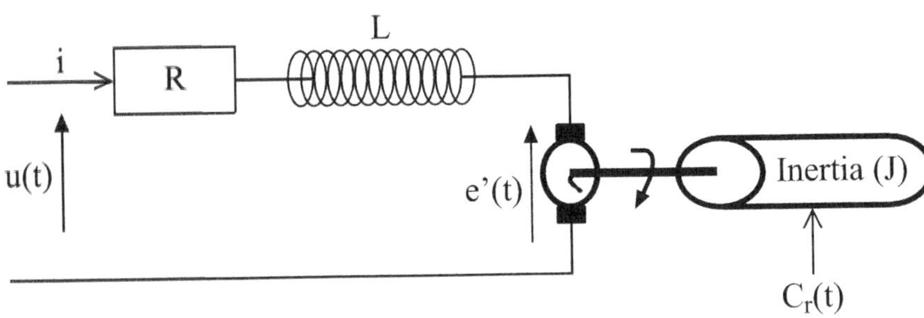


**Figure 5.26.** *Direct current electric motor*

Let $L$ and $R$ be the inductance and the resistance of the armature. $J$ is the inertia time and $f$ is the viscous friction of rotating parts. For such a motor, motor torque is given by $C_m = ki$ and the Counter-electro Motive Force (c.e.m.f.) $e'(t)$ is given by $e'(t) = k\Omega$ ($k$ is the torque constant in $\mathrm{m} \cdot \mathrm{N/A}$).

1) What are the input and output quantities of this process?

2) Calculate the transfer matrix that models it. Derive therefrom the functional diagram of the system.

3) Deduce the shape of responses $\Omega(t)$ and $i(t)$ for a Heaviside function of voltage $u_0$ ($C_r = 0$), then a Heaviside function of resistance torque $C_{r0}$ ($u = u_0 = cte$).

EXERCISE 6.– Economic system

An economic model simplifying relating prices $q(t)$, revenues $r(t)$ and wages $s(t)$ is described by the following equations:

$$\frac{dr}{dt}(t) = -ar(t) + cs(t) \quad [5.178]$$

$$\frac{dq}{dt}(t) = -dq(t) + br(t) + fs(t) \quad [5.179]$$

1) Determine the time constants of the system.

2) Determine the transfer function of the system:

$$H(p) = \frac{Q(p)}{S(p)} \quad [5.180]$$

*Numerical application*: $a = 1/4$, $b = 1/6$, $c = 1/5$, $d = 1/8$, $f = 2/15$, time $t$ is expressed in weeks.

EXERCISE 7.– Control of liquid level

We want to regulate the output flow $Q_s$ of a tank with surface $S$, which contains a liquid whose height is denoted as $H$. This tank is supplied by a valve whose flow is $Q_e$. In addition, the system is subjected to external perturbations that affect the level of liquid in the tank.

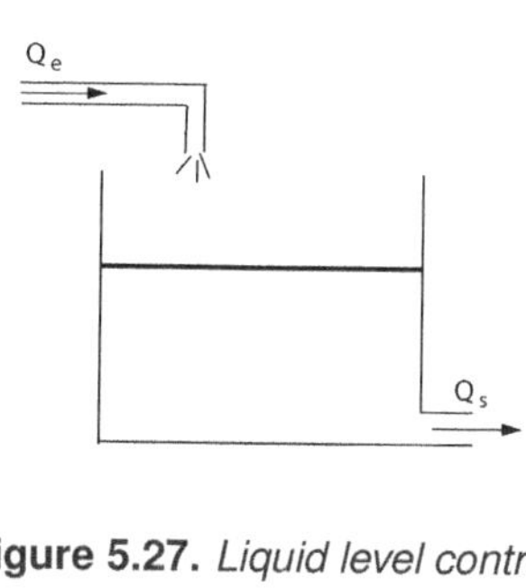


**Figure 5.27.** *Liquid level control*

Determine the differential equation that governs the system. After linearization around the equilibrium height $H$, determine the transfer function of the system and the functional diagram. Verify the validity of the linearization.

# 6

# Dynamic Responses and System Performance

This chapter explores temporal and frequency responses, accuracy and performance.

## 6.1. Introduction to linear time-invariant systems

The purpose of this chapter is the analysis of the dynamic behavior and performance of linear time-invariant (LTI) systems in order to develop control. We have at our disposal a system defined by its state–space representation or its transfer function, and its dynamic behavior is studied in the time domain and then in the frequency domain. Next, the knowledge of time and frequency characteristics of the first- and second-order systems may be applied for the identification of linear systems. This identification can be achieved in several different manners by employing the response of the physical system in test signals. As a first step, we will only discuss conventional methods of identification, then we will introduce parametric methods and recursive methods.

## 6.2. Transition matrix of an LTI system

### 6.2.1. *Transition matrix*

Consider the case of an autonomous linear system ($A$ that may be variable over time): $\dot{x} = A(t)x(t)$, $x(t_o)$ is the initial condition at $t = t_o$ of this system. It can be shown by using the linearity hypothesis that the solution of this differential equation can be written as: $x(t) = \Phi(t, t_o)x(t_o)$, where $\Phi(t, t_o)$ is the state-transition matrix of dimensions $n \times n$ of the system.

$\Phi(t, t_o)$ is the matrix that has to be applied to the initial conditions vector $x(t_o)$ to obtain the system state at time $t$: this is the *state-transition matrix*.

DEFINITION 6.1.– *The system state-transition matrix is defined by* $\Phi(t_1, t_2)$ *such that* $x(t_2) = \Phi(t_2, t_1)x(t_1)$.

PROPOSITION 6.1.– Properties of the state-transition matrix:

– $\Phi(t, t) = I_n \; \forall t \in R$, the transition from a state to itself is done by the identity;

– $\Phi(t_3, t_1) = \Phi(t_3, t_2)\Phi(t_2, t_1)$, this expresses a transition through an intermediate state;

– $\frac{d}{dt}\Phi(t, t_o) = A(t)\Phi(t, t_o)$, the derivative is obtained by multiplying by $A(t)$ matrix;

– $\Phi(t_1, t_2)^{-1} = \Phi(t_2, t_1)$, the transition matrix is invertible.

## 6.3. Evolution equation of an LTI system

### 6.3.1. *State evolution equation*

To evaluate the solution of the system, we will divide the study into two parts: an example involving an LTV system with zero initial conditions and then the general case. After this, we will consider the particular case of LTI systems.

– Example of a system having zero initial conditions: $x(t_o) = 0$, and $\dot{x} = A(t)x(t) + B(t)u(t)$.

The effect of initial conditions on the solution is non-existent because $\Phi(t, t_o)x(t_o) = 0$; suppose zero input at any time except for $t = \tau$, where it is equal to $u(\tau)$. Therefore, only a delta function is considered with height (area under the curve) $u(\tau)\delta(t - \tau)$ to study the effect on the output. For $t < \tau$, we will get a zero state $x(t) = 0$ until the moment just before $\tau$. We denote $dx(\tau) = B(\tau)u(\tau)d\tau$, the influence of the input $u(\tau)$ at time $\tau$, on the system output. After this instant, for $t > \tau$, (that change in state becoming a non-zero initial condition at $t = \tau$), we will obtain the variation caused by this influence:

$$dx(t) = \Phi(t, \tau)dx(\tau) = \Phi(t, \tau)B(\tau)u(\tau)d\tau \quad [6.1]$$

To express the solution of the system, for any input $u(t)$, we integrate the previous equation to take into account all of the input from $t_o$ to $t$:

$$x(t) = \int_{t_o}^{t} \Phi(t, \tau)dx(\tau) = \int_{t_o}^{t} \Phi(t, \tau)B(\tau)u(\tau)d\tau \quad [6.2]$$

– General case: by linearity using the superposition principle, if we take the initial conditions into account, we obtain:

$$x(t) = \Phi(t, t_o)x(t_o) + \int_{t_o}^{t} \Phi(t, \tau)B(\tau)u(\tau)d\tau \qquad [6.3]$$

– Example of an LTI process: $A(t) = A$ and $B(t) = B; \forall t \in \mathbb{R}$

It can be shown that the transition matrix becomes time independent, it depends only on the duration of the time interval $(t - t_o)$:

$$\Phi(t, t_o) = \Phi(t - t_o) \qquad [6.4]$$

To simplify, we assume $t_o = 0$, a Taylor series $x(t)$ can be expanded around $t_o = 0$ :

$$x(t) = x(0) + \dot{x}(0)\frac{t}{1!} + \ddot{x}(0)\frac{t^2}{2!} + ... + x^{(n)}(0)\frac{t^n}{n!} + .. \qquad [6.5]$$

With no external forces, we have:

$$\dot{x}(t) = Ax(t), \ddot{x}(t) = A\dot{x}(t) = A^2x(t), ... , x^{(n)}(t) = A^n x(t), ...$$

where we deduce:

$$x(t) = \left(I + A\frac{t}{1!} + A^2\frac{t^2}{2!} + ... + A^n\frac{t^n}{n!} + ..\right) x(0) = e^{At}x(0) \qquad [6.6]$$

and thus for an LTI system:

$$\Phi(t) = e^{At} \qquad [6.7]$$

In the following section, we present a way to evaluate the transition matrix, in the case of an LTI system.

*Solution of the homogeneous state equation of an LTI system*

The homogeneous equation reflects the free behavior of the system, without external stress:

$$\dot{x}(t) = Ax(t) \qquad [6.8]$$

The solution is given by:

$$x(t) = e^{At}x(0) \qquad [6.9]$$

$$\text{with} \quad e^{At} = I + At + A^2\frac{t^2}{2} + \cdots$$

If $n$ is the dimension of $x(t)$, then $e^{At}$ is a matrix $(n, n)$. Consequently, the solution $x(t)$ can be written as:

$$x(t) = e^{At}x(0) = \Phi(t)x(0) \qquad [6.10]$$

### 6.3.2. *Transition matrix computation*

The transition matrix $\Phi(t) = e^{At}$ can be computed in a number of ways.

1) The transition matrix can be evaluated by diagonalization of matrix $A$, if all eigenvalues are distinct: $e^{At} = Pe^{\Lambda t}P^{-1}$ ($P$ is the matrix of eigenvectors, $\Lambda$ a diagonal matrix).

2) The transition matrix can be calculated using the *Caley Hamilton* theorem that expresses that every matrix satisfies the characteristic equation.

THEOREM 6.1.– If $A$ is a matrix $(n \times n)$ with the characteristic equation:

$$\Delta(\lambda) = \lambda^n + a_{n-1}\lambda^{n-1} + a_{n-2}\lambda^{n-2} + \ldots + a_1\lambda + a_0 = 0 \qquad [6.11]$$

then, we can write:

$$\Delta(A) = A^n + a_{n-1}A^{n-1} + a_{n-2}A^{n-2} + \ldots + a_1A + a_0I = 0 \qquad [6.12]$$

The last equation shows that $\forall N \geq n$, matrix $A^N$can be written as a linear combination of $I, A, A^2, \ldots, A^{n-1}$. The same happens for any power series of $A$. We can thus write:

$$f(A) = \sum_{k=0}^{\infty} b_k A^k = \sum_{k=0}^{n-1} \alpha_k A^k \qquad [6.13]$$

Coefficients $\alpha_i$ $(i = 0, \ldots, n-1)$ can be computed from the equations obtained using the eigenvalues of matrix $A$ ($\lambda_i$ designates the eigenvalues of $A$) because we can write:

$$f(\lambda_i) = \sum_{k=0}^{n-1} \alpha_k \lambda_i^k \qquad [6.14]$$

If $\lambda_i \neq \lambda_j$, then we have $n$ equations with $n$ unknowns and all the coefficients $\alpha_k$ can be determined. In the case of multiple eigenvalues, the method must be amended:

Let $\lambda_i$ equal the eigenvalue of multiplicity $m$, so that the characteristic polynomial is expressed as:

$$\Delta(\lambda) = (\lambda - \lambda_i)^m \Delta_1(\lambda) \quad [6.15]$$

The differentiation of this equation, with respect to $\lambda$ yields:

$$\Delta'(\lambda) = m(\lambda - \lambda_i)^{m-1}\Delta_1(\lambda) - (\lambda - \lambda_i)^m \Delta_1'(\lambda) = (\lambda - \lambda_i)^{m-1}\Delta_2(\lambda) \quad [6.16]$$

Therefore, we see that:

$$\Delta(\lambda_i) = \Delta'(\lambda_i) = \Delta"(\lambda_i) = \cdots = \Delta^{m-1}(\lambda_i) = 0 \quad [6.17]$$

hence the function $f(\lambda_i) = \sum_{k=0}^{n-1} \alpha_k \lambda_i^k$ can be derived to give $(m-1)$ independent equations, which will make it possible to determine coefficients $\alpha_k$.

3) The transition matrix can be computed using the inverse Laplace transformation to the section of the state equation that corresponds to a differential equation without second member, and whose solution is sought for: $\dot{x} - Ax(t) = 0$.

The state transition matrix of a linear system can be defined by: $\Phi(t) = \mathcal{L}^{-1}(pI - A)$.

Consider the series expansion of $(pI - A)^{-1} = \frac{I}{p} + \frac{A}{p^2} + \frac{A^2}{p^3} + \frac{A^3}{p^4} + .....$, from this expansion the inverse Laplace transform is deduced: $\Phi(t) = I + At + \frac{A^2t^2}{2!} + \frac{A^3t^3}{3!} + .... = e^{At}$

$$e^{At} = \mathcal{L}^{-1}[(sI - A)^{-1}] \quad [6.18]$$

## 6.4. Time response to the excitation of continuous linear systems

### 6.4.1. *System response*

The linear system is described in the state space by the following representation:

$$\dot{x} = A.x + Bu(t) \quad [6.19]$$

$$y(t) = C.x + Du(t) \quad [6.20]$$

By applying the Laplace transformation to these equations and considering the initial conditions $x(0)$, we get:

$$X(p) = (pI - A)^{-1}(B.U(p) + x(0)) \quad [6.21]$$

$$Y(p) = (C.(pI - A)^{-1}B + D).U(p) + (C.(pI - A)^{-1}x(0) \quad [6.22]$$

To determine the temporal response of the system, we can evaluate the inverse Laplace transform of the previous equation. If we remember equation [6.21], this allows us to derive the evolution of the system output [1]:

$$y(t) = Cx(t) + Du(t) = Ce^{At}x(0) + C\int_0^t e^{A(t-\tau)}Bu(\tau)d\tau + Du(t) \quad [6.23]$$

or more generally, we can write:

$$y(t - t_o) = Ce^{A(t-t_o)}x(t_o) + C\int_{t_o}^t e^{A(t-\tau)}Bu(\tau)d\tau + Du(t) \quad [6.24]$$

### 6.4.2. *Solution the state equation*

The complete solution of the state equation (external non-zero stresses):

$$\dot{x}(t) = Ax(t) + Bu(t) \quad [6.25]$$

The solution is of the form (variation of the constant):

$$x(t) = e^{At}K(t) \quad [6.26]$$

by deriving, we get: $\dot{x}(t) = Ae^{At}K(t) + e^{At}\dot{K}(t)$

Replacing $x(t)$ by $x(t) = e^{At}K(t)$ in the previous equation yields:

$$\dot{x}(t) = Ae^{At}K(t) + Bu(t) \quad [6.27]$$

hence we have:

$$\dot{K}(t) = e^{-At}Bu(t) \quad [6.28]$$

or still:

$$K(t) = K(t_0) + \int_{t_0}^t e^{-A\tau}Bu(\tau)d\tau \quad [6.29]$$

1 A convolution product of time functions has as Laplace transformation, the product of transforms of functions.

The general solution of a forced system is therefore given by:

$$x(t) = e^{A(t-t_0)}x(t_0) + \int_{t_0}^{t} e^{A(t-\tau)}Bu(\tau)d\tau \qquad (K(t_0) = e^{-At_0}x(t_0)) \qquad [6.30]$$

The first term corresponds to the free behavior and the second to the forced behavior.

If the system output is not directly connected to the input by gains ($D = 0$, which is a frequent case), the following output equation is obtained:

$$y(t) = Cx(t) \qquad [6.31]$$

and consequently the output vector is given by (if $t_0 = 0$):

$$y(t) = Ce^{At}x(0) + \int_{0}^{t} Ce^{A(t-\tau)}Bu(\tau)d\tau \qquad [6.32]$$

#### 6.4.2.1. *Response to impulse excitation*

The entrance is a pulse, therefore consider the case of a pulse at time $\tau'$: $u(t) = \delta(t - \tau')$

The response of the system is:

$$y(t) = Ce^{At}x(0) + \int_{0}^{t} Ce^{A(t-\tau)}B\delta(\tau - \tau')d\tau \qquad [6.33]$$

If the initial conditions are zero, we obtain:

$$h(t - \tau') = y(t) = C\int_{0}^{t} e^{A(t-\tau)}B\delta(\tau - \tau')d\tau = Ce^{A(t-\tau')}B\delta(\tau - \tau') \qquad [6.34]$$

#### 6.4.2.2. *Response to step excitation*

The input is a Heaviside function (with$E > 0$):

$$u(t) = \begin{cases} 0 \ if \ t < 0 \\ E \ if \ t \geq 0 \end{cases} \qquad [6.35]$$

The response of the system is:

$$y(t) = Ce^{At}x(0) + C\int_{0}^{t} e^{A(t-\tau)}BEd\tau \qquad [6.36]$$

by integrating, if $A$ is invertible, it yields:

$$y(t) = Ce^{At}x(0) + CA^{-1}(e^{At} - I)BE \quad [6.37]$$

#### 6.4.2.3. *Response to ramp excitation*

In this case, the input is a ramp (with $a > 0$):

$$u(t) = \begin{cases} 0 \ if \ t < 0 \\ at \ if \ t \geq 0 \end{cases} \quad [6.38]$$

The response of the system is:

$$y(t) = Ce^{At}x(0) + C\int_0^t e^{A(t-\tau)}Ba\tau d\tau \quad [6.39]$$

by integrating, if $A$ is invertible, it yields:

$$y(t) = Ce^{At}x(0) + C\left(A^{-2}(e^{At} - I) - A^{-1}t\right)Ba \quad [6.40]$$

REMARK 6.1.– *The transition matrix $\Phi(t)$ appears in both terms: it therefore plays a fundamental role in the computation of the dynamic response.*

*If the input is a Dirac pulse $\delta(t)$, we see that $h(t) = Ce^{At}B$ is the matrix of impulse responses. It is therefore possible to describe the dynamics of a system in two ways:*

– *with matrices $A, B, C, D$ that provide a parametric representation;*

– *with the measurable impulse response matrix (time functions).*

*The integral that appears in the output expression is an integral of convolution assuming the form:*

$$\int_0^t h(t-\tau)u(\tau)d\tau \quad [6.41]$$

*The response of a time-invariant linear system is thus the convolution product of the input with the impulse response of the system.*

*If matrix $D \neq 0$, the impulse response is equal to $Ce^{At}B + D\delta(t)$.*

### 6.4.3. *Role of eigenvalues of the evolution matrix $A$ within the system dynamics*

To study the role of eigenvalues of $A$, consider a free system that is proportional to $e^{At}$. Suppose that all eigenvalues of matrix $A$ are distinct. Then, there is a set of state variables such that the matrix is diagonal, or there is a transition matrix $P$ that enables the diagonalization of matrix $A$.

$$A = \begin{pmatrix} \lambda_1 & & & \\ & \lambda_2 & & \\ & & \ddots & \\ & & & \lambda_n \end{pmatrix} \quad [6.42]$$

where the $\lambda_i$ are the eigenvalues of $A$. It can be observed that matrix $e^{At}$ is diagonal with diagonal elements $e^{\lambda_i t}$. Each state variable is then of the form

$$x_i(t) = e^{\lambda_i t} x_i(0) \quad [6.43]$$

The free response is therefore entirely determined by the eigenvalues of matrix $A$ and initial conditions. It can immediately be seen that:

– if there is an eigenvalue with a positive real part, a state variable becomes infinite: the *system is unstable*;

– if one or more eigenvalues have a zero real part, one or more state variables do not tend to 0 for all initial conditions, but remain finite: the system is *marginally stable*. In general, a significant difference between a marginally stable system and an unstable system is that any stress makes the output of an unstable system infinite, whereas only some stresses make the output of a marginally stable system infinite;

– if all the eigenvalues of matrix $A$ are real negative, each of the state variables tends toward 0 in a monotonous manner; if there are complex eigenvalues, having negative real parts, some state variables tend to 0 in an oscillatory fashion. Either way, the state vector tends to 0 regardless of initial conditions: the system is *asymptotically stable*.

REMARK 6.2.– *In the case where all eigenvalues are not distinct, the results are the same, with one exception: if there are one or more multiple eigenvalues with zero real part, the system is unstable.*

## 6.5. Sampling and discretization of continuous systems

### 6.5.1. *Choice of the sampling period (Shannon) and integration methods*

For the numerical simulation of signals and systems (calculation of the response to any input) and the integration of differential equations, there are several integration methods. They are identified by their accuracy, computation speed (complexity) and stability. Here, we will present three methods: Euler's method, Runge Kutta's and that using the state transition matrix of the system. Furthermore, the state–space representation corresponding to expressing the system equations in the form of a first-order vector differential equation is very practical for programming a simulator or for the computation of the solution of a system (integration). The differential equations that constitute it are integrated in parallel.

State equation of the system with input $u$ and output $y$

$$\dot{X} = f(X, u(t)) \quad [6.44]$$

$$y(t) = h(X, u(t) \quad [6.45]$$

The solution of this system is expressed through its evolution equation and its transition matrix $\Phi(t)$:

$$X(t) = \Phi(t, t_o)X(t_o) + \int_{t_o}^{t} \Phi(t, \tau)B(\tau)u(\tau)d\tau \quad [6.46]$$

To discretize the system represented by its state–space representation, we consider the presence of a zero-order hold, that is to say that $u(t) = u(k.h)$ for $kh \leq t < (k+1)h$, where $h$ is the sampling period and $k$ the time index in discrete form.

### 6.5.2. *Euler's method*

For the simulation, we must calculate the discrete form of the solution of the equation of the system. Consider the simulation period or integration step $h$ and let $t = k.h$ be the moments at which the solution is evaluated ($k = 0, 1, 2, ...$ is the time index). This method considers the following approximation for the derivative of a time function $\dot{X} \simeq \frac{X_{k+1} - X_k}{h}$. This yields: $X_{k+1} = X_k + h.f(X(k.h), u(k.h)) = X_k + h.f(X_k, u_k)$.

In the linear case

$$\dot{X} = f(X, u(t)) = AX + Bu(t) \quad [6.47]$$

$$y(t) = CX + Du(t) \quad [6.48]$$

we obtain:

$$X_{k+1} = X_k + h.f(X_k, u_k) = X_k + A.h.X_k + B.h.u_k$$
$$= FX_k + B.h.u_k \qquad [6.49]$$
$$y_k = HX_x + Du_k = CX_k + Du_k \qquad [6.50]$$
$$\text{with} \quad F = I + A.h \quad \text{and} \quad G = B.h \qquad [6.51]$$

The choice of the integration step is important for the accuracy of the computation and for numerical stability. On the one hand, it corresponds to the validity and to the quality of the approximation of the derivation and of the approximation of the system by its discrete equivalent $X_{k+1} = FX_k + G.u_k$. The discretization period $h$ must be strictly smaller than half of the smallest time constant of the system $\tau_{\min}$ (Shannon theorem in the linear case). In practice, it requires that it be at least 10 times smaller than $\tau_{\min}$, which explains the rule often empirically introduced that advocates $h \leq \frac{\tau_{\min}}{10}$ for real poles of time constant $\tau \geq \tau_{\min}$ and $h \leq \frac{\pi}{20\omega_{\min}}$ for complex conjugate poles of pulsation $\omega \leq \omega_{\min}$. Despite this rule, in strongly oscillating cases the step integration must be reduced further to avoid numerical instability and improve the accuracy.

REMARK 6.3.– *The equation above shows that matrix $F$ makes use of an approximation of the first-order state transition matrix. This explains the numerical fragility of this integration method in nonlinear and strongly oscillating cases.*

### 6.5.3. *Order n Runge–Kutta method*

The order-4 Runge-Kutta method is the most often used for the simulation of systems, be they linear or not. For the approximation, it uses the successive derivatives of the state equation until the fourth order:

$$X_{k+1} = X_k + \tfrac{h}{6}\left(.f^{(1)}_{k+\frac{1}{2}} + .2f^{(2)}_{k+\frac{1}{2}} + .2f^{(3)}_{k+\frac{1}{2}} + .f^{(4)}_{k+\frac{1}{2}}\right) \qquad [6.52]$$

with $f^{(4)}_{k+\frac{1}{2}} = f^{(n)}(X_{k+\frac{1}{2}}, u_{k+\frac{1}{2}})$ the approximation of the $n$th derivative of $f(X, u)$ computed by extrapolation at times $(k + \frac{1}{2}).h$.

In the case of a linear system

$$\dot{X} = f(X, u(t)) = AX + Bu(t) \qquad [6.53]$$

and

$$y(t) = CX + Du(t) \quad [6.54]$$

we obtain:

$$X_{k+1} = FX_k + G.u_k \quad [6.55]$$

$$y_k = HX_x + Du_k = CX_k + Du_k \quad [6.56]$$

$$\text{with } F = I + Ah + A^2\tfrac{h^2}{2} + A^3\tfrac{h^3}{6} + A^4\tfrac{h^4}{24} \quad [6.57]$$

$$\text{and } G = \left(I + A\tfrac{h}{2} + A^2\tfrac{h^2}{6} + A^3\tfrac{h^3}{24}\right).h \quad [6.58]$$

$$y_k = CX_k + Du_k \quad [6.59]$$

The integration step is simpler to choose because the computational accuracy and the numerical stability of this method are better than for the previous one. The discretization period $h$ must, as in the previous case, verify Shannon's theorem. In practice, it may be 10–20 times larger for a same accuracy as with Euler's method. For strongly oscillating cases (and for cases also called rigid, comprising poles with both very high and low frequencies), the choice of the integration step presents fewer problems by avoiding numerical instability and improve accuracy.

REMARK 6.4.– *The equation above shows for the order-4 Runge–Kutta method (RK34) that the matrix $F$ uses an approximation of the state transition matrix through an expansion in series up to the fourth order, and $G$ to the third order. This explains the difference with Euler's method.*

### 6.5.4. *Method using the state transition matrix with zeroth-order holder*

For the implementation of a simulation and a control in numerical form, we need the discrete equivalent of a continuous system. This method uses the exact solution of the state equation of an LTI system and is therefore applicable for linear systems only.

$$X(t) = \Phi(t - t_o)X(t_o) + \int_{t_o}^{t} \Phi(t - \tau)B(\tau)u(\tau)d\tau \quad [6.60]$$

$$X(t) = e^{A(t-t_0)}X(t_0) + \int_{t_0}^{t} e^{A(t-\tau)}Bu(\tau)d\tau \quad [6.61]$$

$$\Phi(t) = \mathcal{L}^{-1}[(sI - A)^{-1}] = e^{At} = I + At + \frac{A^2t^2}{2!} + \frac{A^3t^3}{3!} + .. + \frac{A^mt^m}{m!} + .. \quad [6.62]$$

We consider the presence of a zero-order holder, that is to say $u(t) = u(k.h)$ for $kh \leq t < (k+1)h$. $h$ is the sampling period and $k$ is the time index in the discrete representation. Considering $t = k.h$ as the initial time, $X(k.h) = X_k$ as the initial state and by applying the evolution equation of the state, we get for $kh \leq t < (k+1)h$:

$$X(t) = e^{A(t-kh)}X(k.h) + \int_{kh}^{t} e^{A(t-s)}Bu(s)ds \quad [6.63]$$

Assuming $s = \tau + k.h$ with $0 \leq \tau < h$ and knowing that $u(s) = u_k$ is constant over a period $h$, we have:

$$X(t) = e^{A(t-kh)}X_k + \int_0^h e^{A(h-\tau)}Bd\tau.u_k \quad [6.64]$$

Finally, for $t = (k+1)h$ we have

$$X_{k+1} = FX_k + G.u_k = e^{A.h}X_k + \int_0^h e^{A(h-\tau)}Bd\tau.u_k \quad [6.65]$$

$$y_k = HX_x + Du_k = CX_k + Du_k \quad [6.66]$$

$$\text{with } F = e^{A.h} = I + Ah + \frac{A^2h^2}{2!} + \frac{A^3h^3}{3!} + .. + \frac{A^mh^m}{m!} + .. \quad [6.67]$$

$$\text{and } G = \int_0^h e^{A(h-\tau)}Bd\tau = \left(I + \frac{Ah}{2!} + \frac{A^2h^2}{3!} + .. + \frac{A^mh^m}{(m+1)!} + ..\right).h \quad [6.68]$$

For the numerical simulation, we obtain a discrete system described by these state equations, which can be represented by a block diagram as shown in Figure 6.1.

### 6.5.5. *Evolution equation for a time-invariant discrete system (DTI)*

For the free discrete system $X_{k+1} = FX_k$, the solution of this system is written as $X_k = A^kX_0$.

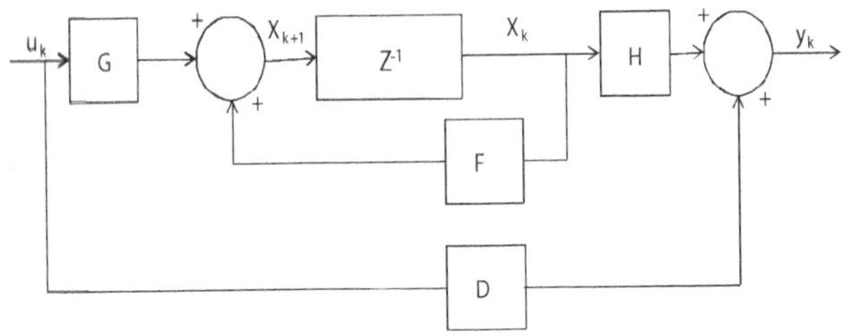


**Figure 6.1.** *Block diagram for a discrete system*

Considering the discrete system $X_{k+1} = FX_k + Gu_k$, we can write the following equations:

$$
\begin{array}{lr}
I & X_k - FX_{k-1} = Gu_k \\
F & X_{k-1} - FX_{k-2} = Gu_{k-1} \\
F^2 & X_{k-2} - FX_{k-3} = Gu_{k-3} \\
 & \ldots\ldots\ldots\ldots\ldots\ldots \\
F^j & X_{k-j} - FX_{k-j-1} = Gu_{k-j-1} \\
F^{k-1} & X_1 - FX_0 = Gu_0
\end{array}
$$

The multiplication (on the left) of each of these equations by the corresponding term of the left column and their summation gives us:

$$X_k - F^k X_0 = \sum_{i=0}^{k-1} F^{k-i-1} Gu_i \qquad [6.69]$$

or still, the discrete convolution equation and the discrete state transition matrix:

$$X_k = F^k X_0 + \sum_{i=0}^{k-1} F^{k-i-1} Gu_i \qquad [6.70]$$

$$\Phi_k = F^k \qquad [6.71]$$

With this equation it is thus possible, as for continuous systems, to calculate the response of a discrete system to any input.

## 6.6. Some temporal responses

The most utilized test signals in automatic control for studying the behavior of a system are:

1) the delta function (impulse response);

2) the Heaviside function (step response);

3) the sinusoidal excitation (harmonic response or frequency response).

The complex gain of a system has one pulse; $\omega_o$ is the value of the transfer function for $p = j\omega_o$ $(\omega = 2j\pi f)$.

### 6.6.1. *Response to an impulse excitation*

The input being applied is the delta function: $u(t) = \delta(t)\ U(p) = 1$. The impulse response is obtained by using the inverse Laplace transformation of the transfer function:

$$Y(p) = H(p) \qquad y(t) = \mathcal{L}^{-1}[H(p)] = h(t) \qquad [6.72]$$

### 6.6.2. *Response to step excitation*

The input being applied is the unit Heaviside function: $u(t) = \Gamma(t)\ U(p) = \frac{1}{p}$. The step response is obtained by inverse Laplace transform of the transfer function multiplied by $U(p)$:

$$Y(p) = \frac{H(p)}{p} \qquad y(t) = \mathcal{L}^{-1}[\frac{H(p)}{p})] = \int_0^t h(s)ds \qquad [6.73]$$

The step response $y(t)$ is the integral of the impulse response $h(t)$.

The order of a system is given by the degree of the denominator of its transfer function. The study of the first- and second-order systems is important because all systems can be represented in the form of a combination or a cascade of first- and/or second-order systems.

EXAMPLE 6.1.– *First-order system: the differential equation linking the input $u(t)$ and the output $y(t)$ of a first-order system is given by*

$$\tau\frac{dy(t)}{dt} + y(t) = Ku(t) \qquad [6.74]$$

*The transfer function is obtained by applying the Laplace transform $(CI = 0)$*

$$H(p) = \frac{Y(p)}{U(p)} = \frac{K}{1 + \tau p} \qquad [6.75]$$

*$K = H(0)$ is the static gain and $\tau$ is the time constant. A first-order system is characterized by its two parameters $K$ and $\tau$.*

*If the stress on input of the first-order system is a Heaviside function, this is referred to as step response and the stationary state is given by*

$$y(\infty) = \lim_{p\to 0} |pY(p)|_{U(p)=1/p} = \lim_{p\to 0} H(p) = H(0) = K \qquad [6.76]$$

*which justifies the static gain term. Its step response is given by*

$$y(t) = K(1 - e^{-t/\tau})U(t) \qquad [6.77]$$

*and is represented by Figure 6.2.*

*It can be observed, for the first-order system, that:*

*1) $y(\tau) = 0.63K$, $y(3\tau) = 0.95K \Rightarrow$ the response time at 5% is thus $t_r = 3\tau$.*

*2) $y(\infty) = K \Rightarrow$ the asymptote $t \to \infty$ corresponds to the static gain of the system.*

*3) The intersection of the tangent $\frac{K}{\tau}t$ and the asymptote $K$ occurs for $t = \tau$ ($y(t) = K$).*

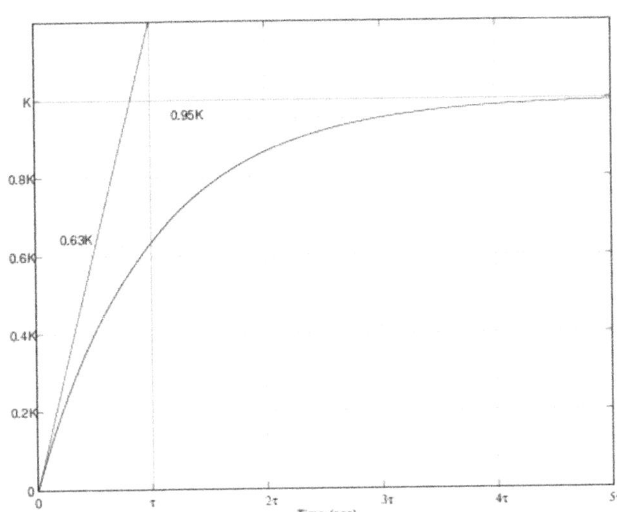


**Figure 6.2.** *Step response*

#### 6.6.2.1. *Response to harmonic excitation or frequency response*

The harmonic response (or frequency response) is the response of a linear system, in steady state, to a sinusoidal excitation: $u(t) = A\sin(\omega t)$. The output of the system is therefore of the form (sinusoid):

$$y(t) = H(\omega)\sin(\omega t + \Phi(\omega)) \quad [6.78]$$

The complex gain is defined by:

$$H(j\omega) = |H(\omega)|\, e^{j\Phi(\omega)} \quad [6.79]$$

The complex gain of a system has one pulse; $\omega_o$ is the value of the transfer function for $p = j\omega_o$ ($\omega = 2j\pi f$).

#### 6.6.2.2. *Time response of a sampled linear system: example of a second-order system*

The differential equation linking input $u(t)$ and output $y(t)$ of a second-order system is given by:

$$\frac{d^2y(t)}{dt^2} + 2\xi\omega_n\frac{dy(t)}{dt} + \omega_n^2 y(t) = K\omega_n^2 u(t) \quad [6.80]$$

Its transfer function is obtained by applying the Laplace transform ($CI\ = 0$)

$$H(p) = \frac{Y(p)}{U(p)} = \frac{K\omega_n^2}{p^2 + 2\xi\omega_n p + \omega_n^2} \quad [6.81]$$

$\omega_n$ is the natural frequency, $\xi$ is the damping coefficient and $K = H(0)$ is the static gain.

REMARK 6.5.– *A second-order system is characterized by its parameters $K$, $\xi$ and $\omega_n$.*

The step response of the second-order system depends on the value of the parameter $\xi$. In effect:

– if $\xi < 0$, the poles of the transfer function have a real positive part, the step response tends exponentially to infinity: the system is unstable;

– if $0 < \xi < 1$, the roots of the characteristic equation are conjugate complex with a negative real part, the step response is therefore of the form

$$y(t) = K\left(1 - \frac{1}{\sqrt{1-\xi^2}} e^{-\xi\omega_n t}\sin(\omega_p t + \phi)\right)U(t)$$

$$\text{with}\ \begin{cases} \sin(\phi) = \sqrt{1-\xi^2} \\ \cos(\phi) = \xi \\ \omega_p = \omega_n\sqrt{1-\xi^2} \end{cases} \qquad [6.82]$$

The response is damped oscillatory with a pseudo-pulse $\omega_p$;

– if $\xi = 1$, the characteristic equation has a double real root, from which the response is of the form:

$$y(t) = K(1 - e^{-\omega_n t} - \omega_n t e^{-\omega_n t})U(t) \qquad [6.83]$$

The answer is damped without oscillations or exceeding the asymptote;

– if $\xi > 1$, the characteristic equation has distinct real roots, the response is thus of the form:

$$y(t) = K\left(1 + \frac{1}{2\sqrt{\xi^2-1}}(\alpha e^{\beta\omega_n t} - \beta e^{\alpha\omega_n t})\right)U(t)$$

$$\text{with}\ \begin{cases} \alpha = -\xi - \sqrt{\xi^2-1} \\ \beta = -\xi + \sqrt{\xi^2-1} \end{cases} \qquad [6.84]$$

The response is more damped than in the case where $\xi = 1$.

Figure 6.3 gives the step responses of the second-order system for values of $\xi$ that correspond to the three cases under study.

REMARK 6.6.– *The response is damped in the case where $\xi \geq 1$. When $0 < \xi < 1$, it is damped oscillatory with a pseudo-pulse $\omega_p$ and a pseudo-period $T_p = \frac{2\pi}{\omega_p}$. This pseudo-period can be measured from the moments corresponding to two successive maxima, as it can be seen in Figure 6.4.*

*The first overshoot $D_1$ occurs at time $t_1$, in which it can be written as:*

$$D_1 = \exp(-\xi\omega_n t_1) \qquad [6.85]$$

*the second overshoot $D_2$ occurs at instant $t_2$, wherefrom it can be written as:*

$$D_2 = \exp(-\xi\omega_n t_2) \qquad [6.86]$$

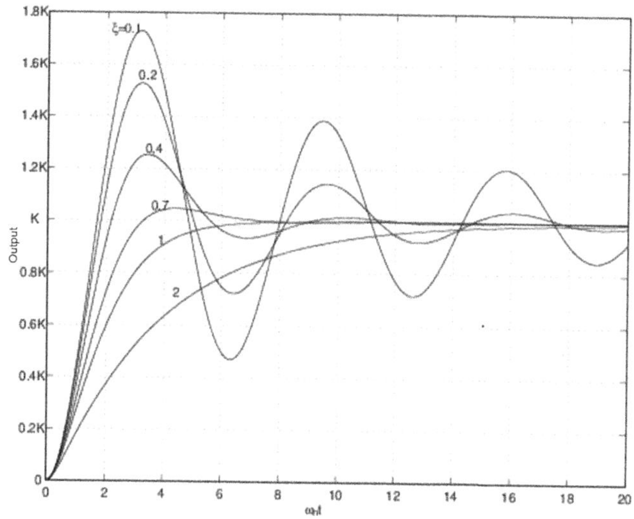


**Figure 6.3.** *Form of the step response with respect to the damping coefficient*

*By performing the ratio of amplitudes $D_1/D_2$, we achieve:*

$$\frac{D_1}{D_2} = \exp(\xi\omega_n(t_2 - t_1)) \qquad [6.87]$$

*with $t_2 - t_1 = T_p = \frac{2\pi}{\omega_p}$ as the pseudo-period, we can therefore write:*

$$\frac{D_2}{D_1} = \exp(\xi\omega_n T_p) = \exp\left(\frac{2\pi\xi}{\sqrt{1-\xi^2}}\right) \qquad [6.88]$$

*From the ratio $\frac{D_1}{D_2}$, we can extract the value of the damping coefficient $\xi$, which yields:*

$$\xi = \frac{\ln(\frac{D_1}{D_2})}{\sqrt{\left(\ln(\frac{D_1}{D_2})\right)^2 + 4\pi^2}} \qquad [6.89]$$

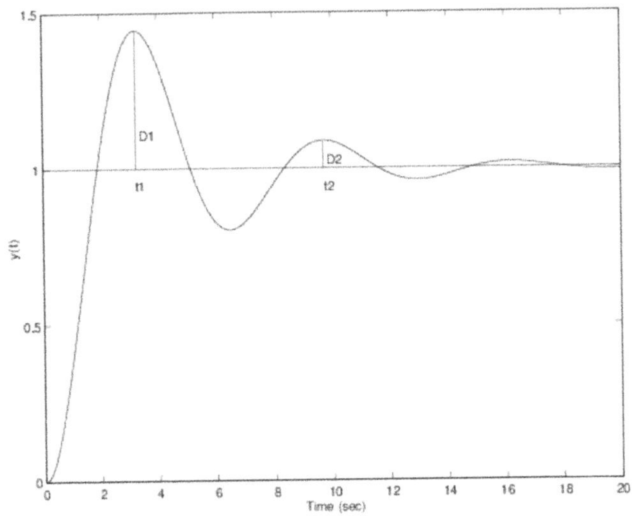


**Figure 6.4.** *Damped response*

The transitional behavior of the step response is characterized by the rise time and the settling time. The rise time is the time the response takes to reach the asymptote for the first time. It is uniquely defined for $\xi < 1$ and is given by:

$$t_m = \frac{1}{\omega_n\sqrt{1-\xi^2}}(\pi - Arc\cos(\xi)) \quad [6.90]$$

Furthermore, the rise time is defined as being between 10% and 90% of the final value, which is given by:

$$t_{10-90\%} = \frac{1.8}{\omega_n} \quad \text{if} \quad \xi = 0.5 \quad [6.91]$$

The settling time of $n\%$ is defined by the time that the response takes to approach a range around the final value which is of $\pm n$. It is given by the following formula:

$$t_{e(n\%)} = \frac{1}{\xi\omega_n}\ln(\frac{100}{n}) \quad [6.92]$$

For example, for $n = 2\%$, we have:

$$t_{e(2\%)} = \frac{4}{\xi\omega_n} \quad [6.93]$$

## 6.7. Transfer function frequency responses

The frequency response of a system $H(f)$ is a complex function of the frequency and can be represented by Bode (two modulus and argument curves are plotted with respect to frequency), Black (a curve graduate in pulse representing the modulus according to the argument) and Nyquist plots (a curve graded in pulse representing the imaginary part according to the real part).

### 6.7.1. *Bode plot*

The gain $H(j\omega)$ ($\omega = 0 \to \infty$) describes the harmonic response of the system. It can be represented in a:

Bode plot (modulus according to $\omega$ and phase according to $\omega$):

– Modulus in $db = |H(j\omega)|_{db} = 20\log_{10}|H(j\omega)|$ with respect to $\log_{10}\omega$.

– Phase in degrees $Arg(H(j\omega)) = \Phi(\omega)$ according to $\log_{10}\omega$.

#### 6.7.1.1. *Example of a first-order system*

To this end, we calculate the modulus and the phase of the transfer function for $p = j\omega$:

$$H(j\omega) = \frac{K}{1+j\omega T} = \frac{K}{1j\frac{\omega}{1/T}} = \frac{K}{1+j\frac{\omega}{\omega_c}} \quad (\omega_c \text{ is the cut-off pulse}) \qquad [6.94]$$

The modulus and the phase $G(j\omega)$ for $K = 1$ are given by:

$$|H(j\omega)|_{db} = 20\log_{10}\left(\left|\frac{1}{1+j\frac{\omega}{\omega_c}}\right|\right) = -20\log_{10}\left(\sqrt{1+(\frac{\omega}{\omega_c})^2}\right) \qquad [6.95]$$

$$\Phi(\omega) = Arg(H(j\omega)) = -arc\tan(\frac{\omega}{\omega_c})$$

– Bode plot: the Bode plot and the plot of the modulus in dB according to the pulse $\omega$ and the phase plot with respect to $\omega$.

REMARK 6.7.– *If the transfer function takes the form* $H(p) = \frac{K}{p+as}$, *we should write it*

$$H(p) = \frac{K}{a}\frac{1}{1+\frac{p}{a}} = K'\frac{1}{1+\frac{p}{a}} \qquad [6.96]$$

*and plot* $H(p)/K'$.

REMARK 6.8.– *Before tracing the actual Bode plot, we begin with an asymptotic plot. For this, the cutoff pulse $\omega_c$ is identified, then we proceed as follows:*

$$\omega \ll \omega_c \longrightarrow \begin{array}{l} |G(j\omega)|_{dB} = 0dB \\ \Phi(\omega) = 0^\circ \end{array} \qquad [6.97]$$

$$\omega \gg \omega_c \longrightarrow \begin{array}{l} |G(j\omega)|_{dB} = -20\log_{10}(\frac{\omega}{\omega_c}) \\ \Phi(\omega) = -90^\circ \end{array}$$ *(slope -6dB/octave or -20dB/decade)*

*For $\omega = \omega_c$, we have:*

$$\begin{aligned} |G(j\omega_c)|_{dB} &= -3dB \\ \Phi(\omega_c) &= -45^\circ \end{aligned} \qquad [6.98]$$

– The Bode plot of the system is given by Figure 6.5.

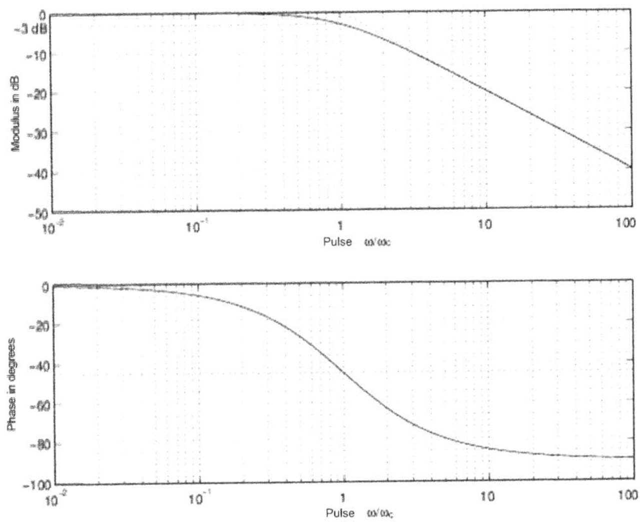


**Figure 6.5.** *Modulus and phase Bode plot of a first-order system*

### 6.7.1.2. *Example of a second-order system*

The frequency response is partly derived from the transfer function assuming $p = j\omega$:

$$H(j\omega) = \frac{K}{1 - (\frac{\omega}{\omega_n})^2 + 2j\xi\frac{\omega}{\omega_n}} \qquad [6.99]$$

The modulus and the phase $H(j\omega)$ are given by:

$$|H(j\omega)| = \frac{K}{\sqrt{\left(1-(\frac{\omega}{\omega_n})^2\right)^2 + 4\xi^2(\frac{\omega}{\omega_n})^2}} \qquad [6.100]$$

$$\Phi(\omega) = Arg(H(j\omega)) = -arc\tan\left(\frac{2\xi\frac{\omega}{\omega_n}}{1-(\frac{\omega}{\omega_n})^2}\right)$$

– Bode plot: before actually tracing the Bode plot, we first start by making an asymptotic plot. To do this, the natural frequency $\omega_n$ is identified and we then proceed in the following manner:

$$\omega \ll \omega_n \longrightarrow \begin{array}{l} |H(j\omega)|_{dB} = 0dB \\ \Phi(\omega) = 0° \end{array} \qquad [6.101]$$

$$\omega \gg \omega_n \longrightarrow \begin{array}{l} |H(j\omega)|_{dB} = -40\log_{10}(\frac{\omega}{\omega_c}) \quad \text{(slope -12dB/octave or -40dB/decade)} \\ \Phi(\omega) = -180° \end{array}$$

For $\omega = \omega_n$, we have:

$$|H(j\omega_n)|_{dB} = 20\log_{10}(\frac{K}{2\xi}) \qquad [6.102]$$

$$\Phi(\omega_n) = -90°$$

The modulus curve shows a maximum for values of $\xi < \frac{\sqrt{2}}{2}$. This maximum is:

$$M_r = \frac{K}{2\xi\sqrt{1-\xi^2}} \qquad [6.103]$$

and it occurs for:

$$\omega = \omega_r = \omega_n\sqrt{1-2\xi^2} \qquad [6.104]$$

The system Bode plot for $K = 1$ is shown in Figures 6.6 and 6.7.

### 6.7.2. *Nyquist plot*

*Nyquist plot*: the Nyquist plot is done by plotting the imaginary part and the real part with respect to $\omega$ or by making a plot in polar coordinates using the Bode plot (modulus and phase).

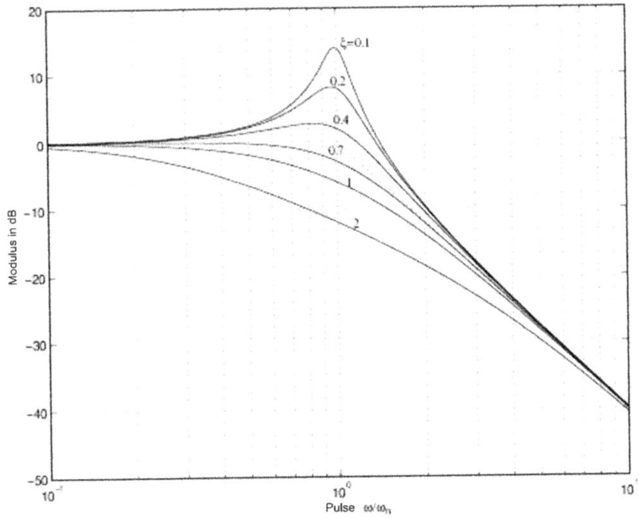


**Figure 6.6.** *Modulus Bode plot of a second-order system*

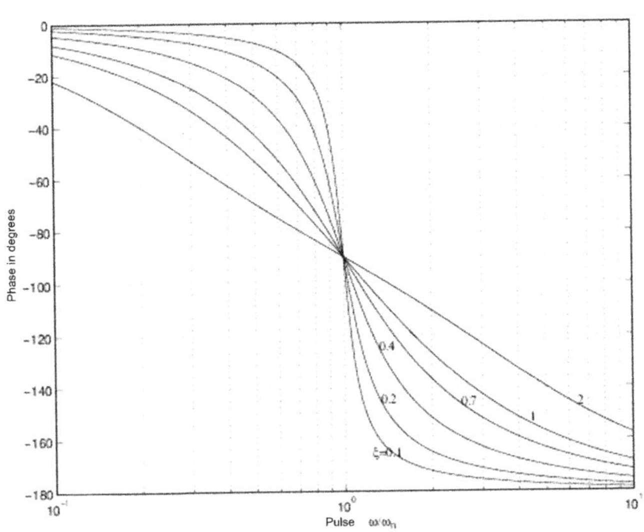


**Figure 6.7.** *Phase Bode plot of a second-order system*

*Nyquist plot*: the plot of the Nyquist locus can be achieved by plotting the real part and the imaginary part of $H(j\omega)$ according to $\omega$, with $H(j\omega)$ written as (for $K = 1$):

$$H(j\omega) = \frac{1-(\frac{\omega}{\omega_n})^2}{\left(1-(\frac{\omega}{\omega_n})^2\right)^2 + 4\xi^2(\frac{\omega}{\omega_n})^2} - j\frac{2\xi(\frac{\omega}{\omega_n})}{\left(1-(\frac{\omega}{\omega_n})^2\right)^2 + 4\xi^2(\frac{\omega}{\omega_n})^2} \qquad [6.105]$$

– It can also be achieved from the Bode plot; for this, a plot is done in polar coordinates.

The plot of the Nyquist locus for a first-order system is shown in Figure 6.8.

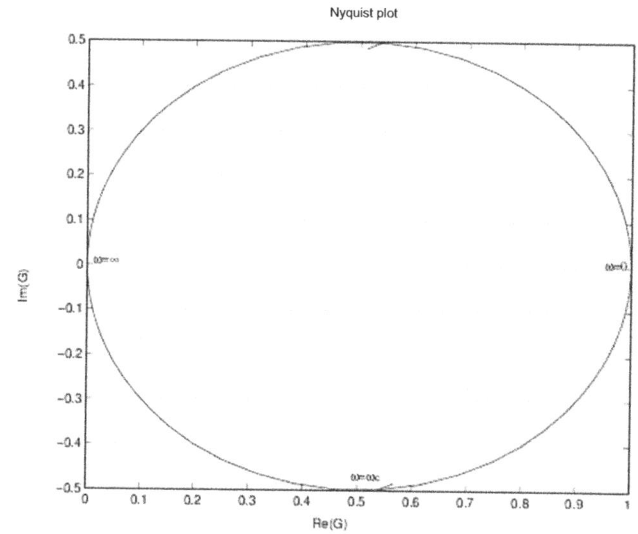


**Figure 6.8.** *Nyquist plot of a first-order system*

– The plot obtained for $K = 1$ is a circle with center $(0.5, 0)$ and radius equal to $0.5$. If we choose $K \neq 1$, we will have a circle of center $(\frac{K}{2}, 0)$ and with radius $\frac{K}{2}$.

The plot for a second-order system is shown in Figure 6.9.

### 6.7.3. *Black–Nichols plot*

Black–Nichols (the argument on the *x*-axis and the modulus in ordinate, graduated in $\omega$).

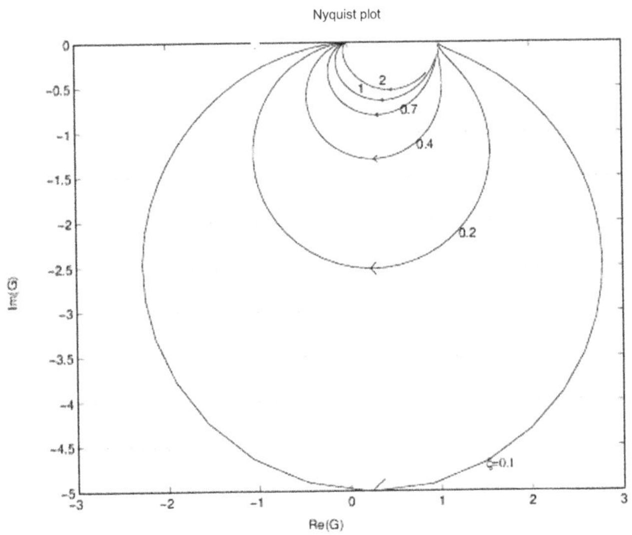


**Figure 6.9.** *Nyquist plot of a second-order system*

The static gain of a system if it exists is given by $H(0)$. It corresponds to the steady state when the input is a Heaviside function (Laplace's final value theorem).

Black–Nichols plot: The Black–Nichols plot is the plot of the modulus with respect to the phase for any $\omega$. It is obtained by using the Bode plot.

– The Black–Nichols plot is shown in Figure 6.10.

REMARK 6.9.– *We note the presence of a nomograph on the plot called Black nomograph, which represents isogain and isophase curves for the closed-loop transfer function. We will see further how to use the nomograph.*

Black–Nichols plot: The Black–Nichols plot for a second-order system is shown in Figure 6.11.

## 6.8. Parametric identification

In this section, we present some conventional methods for the identification of linear and time-invariant continuous systems. These methods are based on the temporal responses of the system to identify, especially the step response. The identification is a representation of an unknown system by a minimal transfer function equivalent to the exact transfer function of the system. It is therefore achieved by making an analogy with the minimal order system that gives a step response as close as possible to the step response of the actual system. For example, if we have a system that has dominant poles (close to the second order), we can propose as a minimal transfer function to be identified a second-order function whose characteristic parameters are the static gain $K$, the natural frequency $\omega_n$ and the damping coefficient $\xi$.

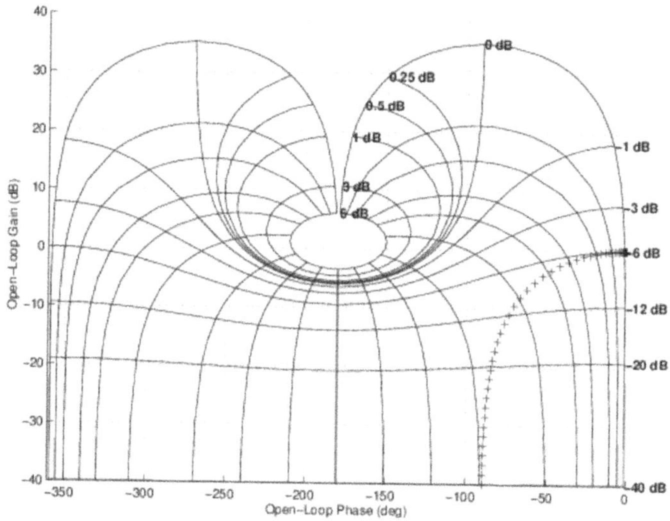


**Figure 6.10.** *Black–Nichols plot of a first-order system*

### 6.8.1. *Identification by analogy*

The general structure of a transfer function is given by

$$G(s) = \frac{b_m s^m + b_{m-1} s^{m-1} + \cdots + b_1 s + b_0}{s^n + a_n s^{n-1} + a_{n-1} s^{n-1} + \cdots + a_2 s + a_1} \qquad [6.106]$$

for a Heaviside input, the output of the system is given by

$$Y(s) = G(s)U(s) = G(s) \cdot \frac{1}{s} \qquad [6.107]$$

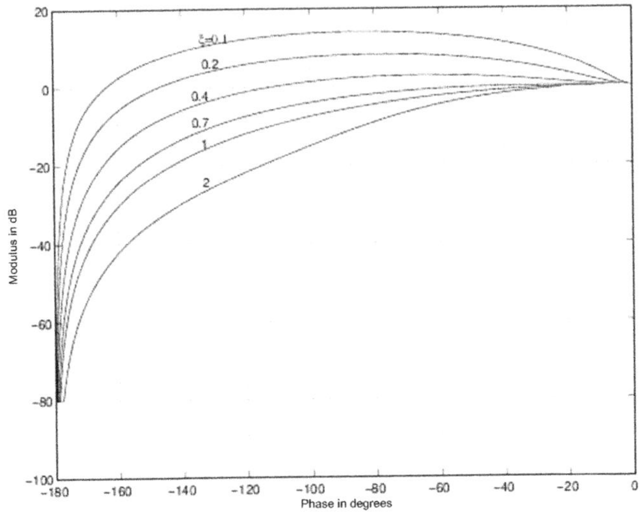


**Figure 6.11.** *Black–Nichols plot of a second-order system*

This allows us to compute, by using the properties of the Laplace transformation, the output at very specific times, for example:

$$\begin{aligned} y(0^+) &= \lim_{t\to 0} y(t) = \lim_{s\to\infty} sY(s) = \lim_{s\to\infty} G(s) \\ y(\infty) &= \lim_{t\to\infty} y(t) = \lim_{s\to 0} sY(s) = \lim_{s\to 0} G(s) \\ y'(0^+) &= \lim_{t\to 0} y'(t) = \lim_{s\to\infty} s[sY(s) - y(0^+)] = \lim_{s\to\infty} s[G(s) - y(0^+)] \end{aligned} \quad [6.108]$$

#### 6.8.1.1. *Identification based on a temporal response*

There are several scenarios available to us to use the result of the response at the origin:

– when the order of the numerator is equal to the order of the denominator ($m = n$):

$$\begin{aligned} y(0^+) &= b_n \\ y'(0^+) &= b_{n-1} - b_n a_n \end{aligned} \quad [6.109]$$

– when there is a difference of a degree between the denominator and the numerator ($m = n - 1$):

$$y(0^+) = 0 \qquad [6.110]$$
$$y'(0^+) = b_{n-1} \quad \text{(non-zero tangent at the origin)}$$

– when there is a difference of at least two degrees between the denominator and the numerator ($m \leq n - 2$):

$$y(0^+) = 0 \qquad [6.111]$$
$$y'(0^+) = 0 \quad \text{(horizontal tangent at the origin)}$$

#### 6.8.1.2. *Response at infinity*

There are also several scenarios to make use of the result of the response at $\infty$:

– when ($b_0 = 0$, $a_1 \neq 0$):

$$y(\infty) = 0 \quad \text{(series derivation)} \qquad [6.112]$$

– when ($b_0 \neq 0$, $a_1 = 0$):

$$y(\infty) = 0 \quad \text{(series integrator)} \qquad [6.113]$$

– when ($b_0 \neq 0$, $a_1 \neq 0$):

$$y(\infty) = \frac{b_0}{a_1} \quad \text{(static gain)} \qquad [6.114]$$

REMARK 6.10.– *An association of the results of the response and its differential at the original and of the response at $\infty$, we can propose a structure for a transfer function. Once the structure is chosen, one simply has to identify the parameters characterizing this transfer function.*

### 6.8.2. Parameters identification: examples of systems

EXAMPLE 6.2.– *When considering a first-order system, we simply have to determine these parameters in order to obtain the transfer function that represents the system. We consider a transfer function of a first-order system:*

$$G(s) = \frac{K}{1 + \tau s} \qquad [6.115]$$

*The parameters characterizing this transfer function are $K$ and $\tau$. We use the responses at origin and at $\infty$ to determine them:*

$$y(0^+) = \lim_{s\to\infty} G(s) = 0 \qquad [6.116]$$

$$y'(0^+) = \lim_{s\to\infty} s[G(s) - y(0^+)] = \lim_{s\to\infty} sG(s) = \frac{K}{\tau}$$

$$y(\infty) = K$$

*The tangent at the origin cuts the asymptote $K$ at time $t = \tau$. As a result, the value of the static gain $K$ can be determined by measuring the value at $\infty$ (final value) and the time constant $\tau$ is determined by measuring the time corresponding to the intersection of the tangent at the origin and the asymptote.*

EXAMPLE 6.3.– *When the system has a step response corresponding to a second-order system in the case of complex poles, we are faced with a system known as a system with dominant poles. In this case, We propose a transfer function of the form:*

$$G(s) = \frac{K\omega_n^2}{s^2 + 2\xi\omega_n s + \omega_n^2} \qquad [6.117]$$

*where the characteristic parameters are the static gain $K$, the natural frequency $\omega_n$ and the damping coefficient $\xi$. The responses at the origin and at $\infty$ allow us to write:*

$$y(0^+) = 0 \qquad [6.118]$$

$$y'(0^+) = 0$$

$$y(\infty) = K$$

*The value of gain $K$ is thus directly read on the plot; it corresponds to the asymptote to the curve for a very large $t$ ($t \to \infty$). To determine the natural frequency $\omega_n$ and the damping coefficient $\xi$, we will use the measures of the pseudo-period and the first two maxima of the response. The latter will be used by the logarithmic decrement method to determine the damping coefficient $\xi$. The natural frequency is calculated based on the value $\xi$ and on the measurement of the pseudo-period:*

$$\xi = \frac{\ln(\frac{D_1}{D_2})}{\sqrt{\left(\ln(\frac{D_1}{D_2})\right)^2 + 4\pi^2}} \quad \textit{(logarithmic decrement)} \qquad [6.119]$$

$$\omega_p = \omega_n\sqrt{1-\xi^2} = \frac{2\pi}{T_p} = \frac{2\pi}{t_2 - t_1} \Longrightarrow \omega_n = \frac{2\pi}{(t_2 - t_1)\sqrt{1-\xi^2}}$$

When the response is not oscillating and is random, there is a method that is used to approximate systems by a multiple real-pole transfer function. This method is discussed in the following section.

### 6.8.3. *Strejc method (minimal dephasing)*

This method is applicable to minimal dephasing systems. It also makes it possible to identify delays. The transfer function to identify is of the form:

$$G(s) = \frac{Ke^{-\tau s}}{(1+Ts)^n} \tag{6.120}$$

with $\tau$ the delay, $n$ the multiplicity of the pole that is also the order of the transfer function, $K$ the static gain and $T$ the multiple time constant.

The step response is shown in Figure 6.12.

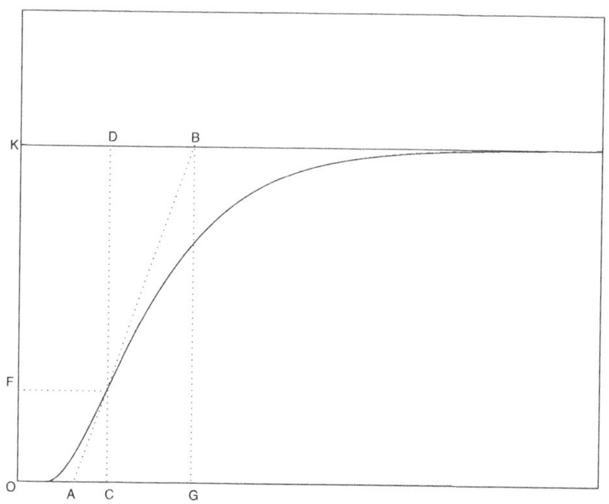


**Figure 6.12.** *Step response Strejc method*

The Strejc table is given as follows.

The values of $n$, $T$ and $\tau$ are determined using the Strejc table:

– $n$ is determined from columns $OF/K$ and $CG/AG$;

– $T$ is determined from columns $AG/T$ and $CG/T$.

– $\tau$ is determined from columns $OA/AG$ and $OA/T$, writing:

$$\frac{\tau}{T} = \frac{OA}{T}(\text{measured}) - \frac{OA}{T}(\text{Table (6.1)}) \qquad [6.121]$$

| $n$ | $OF/K$ | $OA/AG$ | $CG/AG$ | $AG/T$ | $OA/T$ | $CG/T$ |
|---|---|---|---|---|---|---|
| 1 | 0 | 0 | 1 | 1 | 0 | 1 |
| 2 | 0.26 | 0.104 | 0.74 | 2.7 | 0.28 | 2 |
| 3 | 0.32 | 0.22 | 0.68 | 3.7 | 0.8 | 2.5 |
| 4 | 0.35 | 0.32 | 0.65 | 4.46 | 1.42 | 2.9 |
| 5 | 0.37 | 0.41 | 0.63 | 5.12 | 2.1 | 3.22 |
| 6 | 0.38 | 0.49 | 0.62 | 5.7 | 2.8 | 3.5 |
| 7 | 0.39 | 0.57 | 0.61 | 6.2 | 3.55 | 3.77 |
| 8 | 0.40 | 0.64 | 0.60 | 6.7 | 4.3 | 4.02 |

**Table 6.1.** *Strejc table*

## 6.9. Dynamics of linear systems

### 6.9.1. *Link between frequency domain and time domain*

We will see that we can perform a temporal study of the control system from frequency analysis. To do this, we shall establish the links that exist between the temporal and the frequency domains. To illustrate these links, we will start from the transfer function of a second-order system

$$G(s) = \frac{\omega_n^2}{s^2 + 2\xi\omega_n + \omega_n^2} \qquad [6.122]$$

and we can write that this transfer function corresponds to an open-loop transfer function, closed by a unit feedback, of the form:

$$H(s) = \frac{\omega_n^2}{s(s + 2\xi\omega_n)} \Longrightarrow G(s) = \frac{H(s)}{1 + H(s)} = \frac{\omega_n^2}{s^2 + 2\xi\omega_n + \omega_n^2} \qquad [6.123]$$

Knowing the temporal behavior of a second-order system described by $G(s)$, we will show the relationship between the frequency characteristics of the function $H(s)$ and the temporal characteristics of this closed-loop transfer function $G(s)$. We will then generalize this result for higher order systems.

## 6.10. System performance and accuracy

### 6.10.1. *Damping factor of a system*

A close relationship can be found between the damping coefficient $\xi$ of the transfer function $G(s)$ and the phase gap defined for the transfer function in BO $H(s)$. It is given by:

$$\xi \simeq \frac{\Delta\Phi}{100} \qquad [6.124]$$

for phase gaps ranging up to $70°$.

REMARK 6.11.– *The phase margin is often used to specify the desired performance of the controlled system. In addition, it is often considered that the margin of the particular system above is a good approximation for other higher order systems.*

### 6.10.2. *System speed and transient*

We have seen that rise time or settling time characterizes the transitional system and consequently its speed at adopting to sudden variations. We have a relationship between this transient and bandwidth. The larger the bandwidth of the system is, the faster the system is and vice versa.

### 6.10.3. *System static error, speed, sensitivity to noise and accuracy*

Consider the control system in Figure 6.13:

where $U(p)$ is the control input; $B(p)$ is the disturbance input; $Y(p)$ is the system output; $\varepsilon(p)$ is the error.

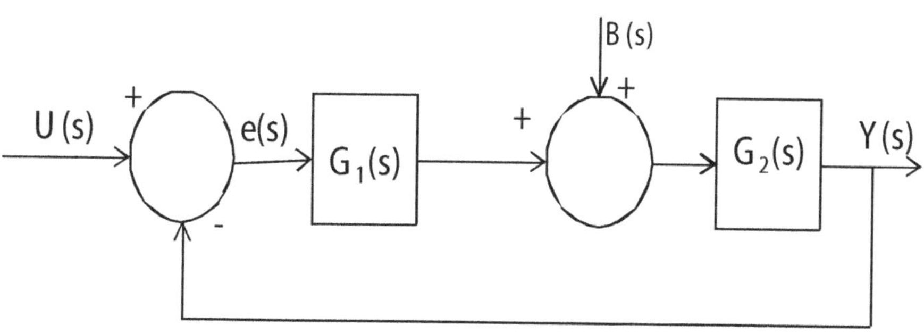


**Figure 6.13.** *Noisy controlled system*

The purpose of control is to forward to the output a law determined in general by the input. Among the means of measuring the quality of a servo, we find the accuracy with which the law is followed. The aim will therefore be to seek to minimize the error $\varepsilon$.

Being a linear system, the superposition principle applies and the total error can be written as:

$$\varepsilon(p) = \varepsilon_u(p) + \varepsilon_b(p) \quad [6.125]$$

where

$$\varepsilon_u(p) = \frac{1}{1 + G_1(p)G_2(p)} U(p) \quad \text{(error associated with the input } U(s)\text{)} \quad [6.126]$$

and

$$\varepsilon_b(p) = \frac{-G_2(p)}{1 + G_1(p)G_2(p)} B(p) \quad \text{(error associated with the input } B(s)\text{)} \quad [6.127]$$

Errors $\varepsilon_u(p)$ and $\varepsilon_b(p)$ comprise a transient part and a permanent part. The transient part characterizes the dynamic accuracy where we measure the error in response to random variations of $U(p)$ and $B(p)$. The permanent part characterizes the static accuracy and corresponds to the error when $t \rightarrow \infty$ in response to canonical signals (Heaviside, ramp, etc.). Static accuracy is studied as follows:

#### 6.10.3.1. *B(s)=0 and U(s)≠0*

In this case:

$$\varepsilon(p) = \varepsilon_u(p) = \frac{1}{1 + G_1(p)G_2(p)} U(p) \quad [6.128]$$

The product $G_1(p)G_2(p)$, which is the open-loop transfer function of the system, is written in the form

$$G_1(p)G_2(p) = \frac{KN(p)}{p^v D(p)} \quad \text{with} \quad \frac{N(0)}{D(0)} = 1 \quad [6.129]$$

Final value theorem:

$$\lim_{t\rightarrow\infty} \varepsilon_u(t) = \lim_{p\rightarrow 0} p\varepsilon_u(p) = \lim_{p\rightarrow 0} pU(p) \cdot \frac{1}{1 + \frac{KN(p)}{p^v D(p)}} \quad [6.130]$$

The error is a function of the input and of $v$ (integration number).

1) *Heaviside function input* (static error or position error):

$$U(p) = \frac{E_0}{p} \quad \Rightarrow \quad \lim_{t\to\infty} \varepsilon_1(t) = \lim_{p\to 0} p\varepsilon_u(p) = \lim_{p\to 0} \frac{E_0}{1 + \frac{KN(p)}{p^v D(p)}} \qquad [6.131]$$

This final error is zero if $v \geq 1$ (the system presents at least one integration). If $v = 0$, the static error is equal to $\frac{E_0}{1+K}$.

2) *Ramp input* (speed error):

$$U(p) = \frac{V_0}{p^2} \quad \Rightarrow \quad \lim_{t\to\infty} \varepsilon_2(t) = \lim_{p\to 0} p\varepsilon_u(p) = \lim_{p\to 0} \frac{V_0}{p} \cdot \frac{1}{1 + \frac{KN(p)}{p^v D(p)}} \qquad [6.132]$$

This final error is zero if $v \geq 2$ (the system presents at least a double integration). If $v = 1$, it is equal to $\frac{V_0}{K}$. If $v = 0$, it tends toward infinity.

| $v$ | $\varepsilon_1(U(p) = 1/p)$ | $\varepsilon_2(U(p) = 1/p^2)$ |
|---|---|---|
| 0 | $\frac{1}{1+K}$ | $\infty$ |
| 1 | 0 | $\frac{1}{K}$ |
| 2 | 0 | 0 |

**Table 6.2.** *Values of the speed errors for a ramp input*

3) *Sine input* ($u(t) = E_0 pin(\omega_0 t)$): the error tends toward a harmonic signal of amplitude:

$$\mid \varepsilon_u \mid = \frac{E_0}{\mid 1 + G_1(p)G_2(p) \mid_{p=j\omega_0}} \qquad [6.133]$$

#### 6.10.3.2. $B(s) \neq 0$ *and* $U(s) = 0$

In this case:

$$\varepsilon(p) = \varepsilon_b(p) = \frac{-G_2(p)}{1 + G_1(p)G_2(p)} B(p) \qquad [6.134]$$

*Heaviside disturbance*: $(B(p) = B_0/p)$ : Final value theorem:

$$\lim_{t\to\infty} \varepsilon(t) = \lim_{p\to 0} \frac{-G_2(p)}{1 + G_1(p)G_2(p)} B_0 = \lim_{p\to 0} \frac{-1}{\frac{1}{G_2(p)} + G_1(p)} B_0 \qquad [6.135]$$

It is easily shown that this error is zero when $p \to 0$ if:

$$\lim_{p\to 0} G_2(p) = 0 \qquad [6.136]$$

$$G_1(p) \simeq \frac{K}{p^v} \quad \text{with } v \geq 1 \qquad [6.137]$$

The condition $G_2(p) = 0$ when $p \to 0$ is rarely satisfied (derivative system). Therefore, a system constituting of at least an integration upstream of the application point of the disturbance (in $G_1(p)$) presents a zero static error in response to a disturbance step response.

### 6.10.4. *Conclusion*

In order to have small or zero errors, it is necessary to:

– have one or more integrations;

– have a high gain.

However, these conditions lead to a low stability margin or even to unstable systems, so there is a need for correction.

From this step of the analysis, we can conclude that the requirements for a good accuracy can cause instability of the looped system, and satisfying stability does not mean good performance for the controlled system. In Chapter 7, we will cover the notion of specifications that reflects the performance of the controlled system (stability, accuracy, speed, etc.) and the notion of control to satisfy the specifications.

## 6.11. Some application exercises

EXERCISE 1.– For the system described below, determine the state vector for a unit step function on input (with $x_1(0) = x_2(0) = 0$).

$$\begin{bmatrix} \dot{x}_1 \\ \dot{x}_2 \end{bmatrix} = \begin{bmatrix} 0 & 1 \\ -2 & -3 \end{bmatrix} \begin{bmatrix} x_1 \\ x_2 \end{bmatrix} + \begin{bmatrix} 0 \\ 1 \end{bmatrix} U \qquad [6.138]$$

EXERCISE 2.– The state vector is given by

$$x(t) = e^{At}x(0^+) + \int_0^t e^{A(t-\tau)}Bu(\tau)d\tau \qquad [6.139]$$

where

$$x(t) = L^{-1}\{[sI - A]^{-1}\}x(0^+) + L^{-1}\{[sI - A]^{-1}Bu(s)\} \;\; (L^{-1} = (T.L)^{-1}) [6.140]$$

There are several methods for calculating $e^{At}$:

1) Laplace transformation:

$$e^{At} = L^{-1}\{[sI - A]^{-1}\} \tag{6.141}$$

2) diagonalization of matrix A (if $\lambda_i \neq \lambda_j$):

$$e^{At} = Pe^{\Lambda t}P^{-1} \qquad P \text{ is the eigenvectors matrix} \tag{6.142}$$

3) Cayley–Hamilton's theorem: every matrix satisfies its own characteristic equation. If matrix A is $n \times n$ and admits the characteristic equation:

$$\Delta(\lambda) = \lambda^n + a_{n-1}\lambda^{n-1} + a_{n-2}\lambda^{n-2} + \ldots + a_1\lambda + a_0 = 0 \tag{6.143}$$

then, we can write:

$$\Delta(A) = A^n + a_{n-1}A^{n-1} + a_{n-2}A^{n-2} + \ldots + a_1A + a_0I = 0 \tag{6.144}$$

The last equation shows that any matrix $A^N$, $N \geq n$, can be written as a linear combination of $I, A, A^2, \ldots, A^{n-1}$. The same happens for any power series of $A$. As a result, we can write:

$$f(A) = \sum_{k=0}^{\infty} b_k A^k = \sum_{k=0}^{n-1} \alpha_k A^k \tag{6.145}$$

Coefficients $\alpha_i$ ($i = 0, \ldots, n-1$) can be computed from the equations obtained using the eigenvalues of matrix $A$ ($\lambda_i$ designates the eigenvalues of $A$), because we can write:

$$f(\lambda_i) = \sum_{k=0}^{n-1} \alpha_k \lambda_i^k \qquad (\lambda \text{ are the eigenvalues of } A) \tag{6.146}$$

If $\lambda_i \neq \lambda_j$ then we have $n$ equations with $n$ unknowns and all the coefficients $\alpha_k$ can be determined. In the case of multiple eigenvalues, the method must be amended: let $\lambda_i$ be the eigenvalue of multiplicity $m$.

$$\Delta(\lambda) = (\lambda - \lambda_i)^m \Delta_1(\lambda) \tag{6.147}$$

This equation is differentiated, that is:

$$\begin{aligned}\Delta'(\lambda) &= m(\lambda - \lambda_i)^{m-1}\Delta_1(\lambda) - (\lambda - \lambda_i)^m \Delta_1'(\lambda) \\ &= (\lambda - \lambda_i)^{m-1}\Delta_2(\lambda)\end{aligned} \tag{6.148}$$

Therefore, it can be seen that:

$$\Delta(\lambda_i) = \Delta'(\lambda_i) = \Delta"(\lambda_i) = \cdots = \Delta^{m-1}(\lambda_i) = 0 \qquad [6.149]$$

hence the function $f(\lambda_i) = \sum_{k=0}^{n-1} \alpha_k \lambda_i^k$ can be differentiated to give $(m-1)$ independent equations, which can be used to determine coefficients $\alpha_k$. We apply the Cayley-Hamilton theorem, to calculate $e^{At}$. Matrix $A$ has two distinct eigenvalues ($\lambda_0 = -1$, $\lambda_1 = -2$). It then can be expressed that:

$$e^{At} = C_0(t)I + C_1(t)A \qquad [6.150]$$

This yields two equations with two unknowns:

$$\begin{cases} e^{-t} = C_0(t) - C_1(t) \\ e^{-2t} = C_0(t) - 2C_1(t) \end{cases} \Longrightarrow \begin{cases} C_0(t) = 2e^{-t} - e^{-2t} \\ C_1(t) = e^{-t} - e^{-2t} \end{cases} \qquad [6.151]$$

wherefrom:

$$e^{At} = \begin{bmatrix} 2e^{-t} - e^{-2t} & e^{-t} - e^{-2t} \\ -2e^{-t} + 2e^{-2t} & -e^{-t} + 2e^{-2t} \end{bmatrix} \qquad [6.152]$$

and

$$\int_0^t e^{A(t-\tau)} Bu(\tau) d\tau = \begin{bmatrix} \frac{1}{2} - e^{-t} + \frac{1}{2}e^{-2t} \\ e^{-t} - e^{-2t} \end{bmatrix} \qquad [6.153]$$

EXERCISE 3.– Given the following system:

$$y((k+2)T) + 0.2y((k+1)T) - 0.63y(kT) = u(kT) \qquad [6.154]$$

Study the stability and the unit step response.

SOLUTION EXERCISE 3.– The state-space representation of the process modeled by the recurrence equation.

$$y((k+2)T) + 0.2y((k+1)T) - 0.63y(kT) = u(kT) \qquad [6.155]$$

is given as follows:

$$y_k = y(kT) = x_{1k} = x_1(kT) \qquad [6.156]$$

$$y_{k+1} = x_{1(k+1)} = x_{2k}$$

$$y_{k+2} = x_{2(k+1)} = -0.2x_{2k} + 0.63x_{1k} + u_k$$

1) In matrix form, we have

state equation: $$\begin{bmatrix} x_{1(k+1)} \\ x_{2(k+1)} \end{bmatrix} = \begin{bmatrix} 0 & 1 \\ 0.63 & -0.2 \end{bmatrix} \begin{bmatrix} x_{1k} \\ x_{2k} \end{bmatrix} + \begin{bmatrix} 0 \\ 1 \end{bmatrix} u_k \qquad [6.157]$$

output equation: $$y = \begin{bmatrix} 1 & 0 \end{bmatrix} \begin{bmatrix} x_{1k} \\ x_{2k} \end{bmatrix} \qquad [6.158]$$

2) The stability of the system is ensured if the eigenvalues of the evolution matrix $A$ are inside the unit circle.

$$det(\lambda I - A) = 0 \quad \Longrightarrow \quad \begin{cases} \lambda_1 = 0.7 \\ \lambda_2 = -0.9 \end{cases} \qquad [6.159]$$

The system is stable because:

$$\mid \lambda_1 \mid < 1 \quad \text{and} \quad \mid \lambda_2 \mid < 1 \qquad [6.160]$$

3) The unit step response: based on the recurrence equation, calculate the $Z$-transform.

$$(z^2 + 0.2z - 0.63)Y(z) = U(z) \qquad [6.161]$$

The transfer function $H(z)$ is given by:

$$H(z) = \frac{Y(z)}{U(z)} = \frac{1}{z^2 + 0.2z - 0.63} = \frac{1}{(z+0.9)(z-0.7)} \qquad [6.162]$$

the step response:

$$Y(z) = H(z)U(z) \quad \text{with} \quad U(z) = \frac{z}{z-1} \qquad [6.163]$$

wherefrom

$$Y(z) = \frac{z}{(z+0.9)(z-0.7)(z-1)} \qquad [6.164]$$

The simple elements decomposition of the following expression is carried out:

$$\begin{aligned} \frac{Y(z)}{z} &= \frac{1}{(z+0.9)(z-0.7)(z-1)} \\ &= \frac{25}{4}\frac{1}{z+0.9} + \frac{50}{3}\frac{1}{z-0.7} + \frac{100}{57}\frac{1}{z-1} \end{aligned} \qquad [6.165]$$

hence

$$Y(z) = \frac{25}{4}\frac{z}{z+0.9} + \frac{50}{3}\frac{z}{z-0.7} + \frac{100}{57}\frac{z}{z-1} \qquad [6.166]$$

The response is given by:

$$y(nT) = (\frac{25}{4}(-0.9)^n + \frac{50}{3}(0.7)^n + \frac{100}{57})u(nT) \qquad [6.167]$$

EXERCISE 4.– Identification with the semilogarithmic plane method. In this problem, we consider a mechanical system whose step response is represented in Figure 6.14.

*Modeling:*

The system is modeled by:

$$H(s) = \frac{Y(s)}{U(s)} = \frac{K(1 - T_1 s)}{(1 + T_2 s)(1 + T_3 s)} \quad (T_3 \text{ is very large compared to } T_1 \text{ and } T_2) \qquad [6.168]$$

1) Justify the choice of this model.

2) Represent the shape of the impulse response (without making any calculations).

3) Give a state–space representation (state equation and output equation) of the system using two different methods:

4) using the modes method (Jordan's form);

5) from the differential equation (companion form).

6) Give the expression of the step response $y(t)$ of the model according to parameters $K$, $T_1$, $T_2$ and $T_3$, and specify the value of $y'(0^+)$.

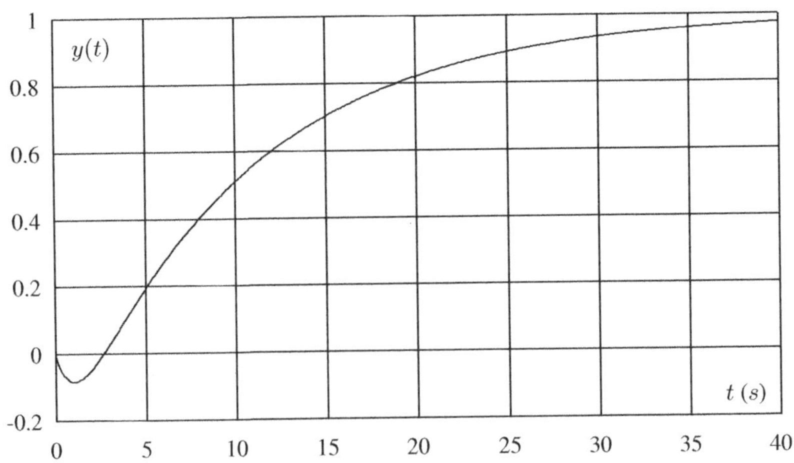


**Figure 6.14.** *Step response of a mechanical system*

*Identification*:

1) What does the expression of $y(t)$ become when $t \gg \max(T_1, T_2)$. It is shown that the plot of:

$$\alpha(t) = Log\frac{K - y(t)}{K} \quad \text{(comply with the hypothesis } t \gg \max(T_1, T_2)\text{)} \qquad [6.169]$$

allows us to identify $T_3$ and provides a relation between $T_2$ and $T_1$. Then, give an estimation of the parameters of the model.

2) The system is associated according to the plot shown in Figure 6.15, $C(s) = k$.

3) What is the shape of the step response of the servo when $k$ is big enough?

4) Show that the experimental survey of this response would identify the complete system (corrected system).

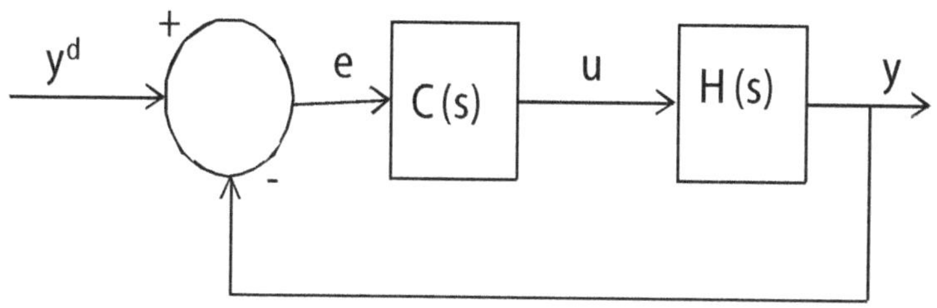


**Figure 6.15.** *Controlled system with no feedback*

SOLUTION EXERCISE 4.– Modeling

Based on the step response given in Figure 6.16, we are able to propose a model structure of the second order.

The derivative at the origin is finite (a degree of difference between the numerator and the denominator) and there is an inflection point, we thus propose a second order. The zero has a real positive part because the derivative at the origin is negative, hence the following transfer function:

$$H(s) = \frac{K(1 - T_1 s)}{(1 + T_2 s)(1 + T_3 s)} \qquad [6.170]$$

The impulse response (see Figure 6.17) is given by the derivative of the step response.

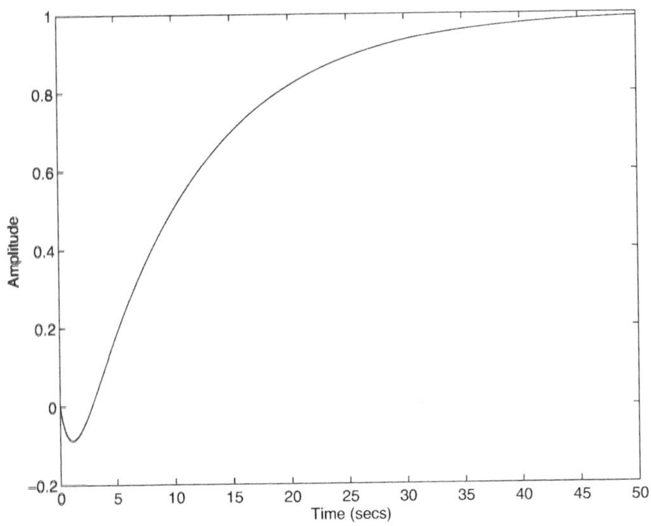


**Figure 6.16.** *Step response*

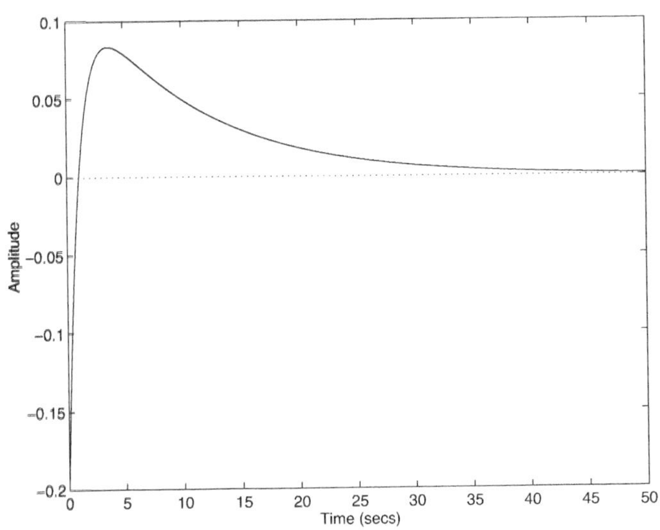


**Figure 6.17.** *Impulse response*

State–space representation of the system *using the modes method (Jordan's form)*:

$$H(s) = \frac{Y(s)}{U(s)} = \frac{K(1-T_1 s)}{(1+T_2 s)(1+T_3 s)}$$

$$\Rightarrow Y(s) = \frac{K}{T_2 - T3}[\frac{T_2+T_1}{1+T_2 s} - \frac{T_1+T_3}{1+T_3 s}]U(s) \quad [6.171]$$

Taking as state variables,

$$X_1(s) = \frac{U(s)}{1+T_2 s} \quad \text{and} \quad X_2(s) = \frac{U(s)}{1+T_3 s} \quad [6.172]$$

it follows that:

$$\begin{cases} \dot{x}_1 = -\frac{1}{T_2}x_1 + \frac{1}{T_2}u \\ \dot{x}_2 = -\frac{1}{T_3}x_2 + \frac{1}{T_3}u \\ y = \frac{K}{T_2 - T3}[(T_1+T_2)x_1 \\ \quad -(T_1+T_3)x_2] \end{cases} \Rightarrow \begin{cases} A = \begin{bmatrix} -\frac{1}{T_2} & 0 \\ 0 & -\frac{1}{T_3} \end{bmatrix}; \; B = \begin{bmatrix} \frac{1}{T_2} \\ \frac{1}{T_3} \end{bmatrix} \\ C = K \begin{bmatrix} \frac{T_1+T_2}{T_2-T_3} & -\frac{T_1+T_3}{T_2-T_3} \end{bmatrix} \end{cases}$$

[6.173]

*Companion form*:

$$H(s) = \frac{Y(s)}{U(s)} = \frac{K(1-T_1 s)}{(1+T_2 s)(1+T_3 s)}$$

$$\Rightarrow Y(s)(1+(T_2+T_3)s+T_2T_3s^2) = K(1-T_1 s)U(s) \quad [6.174]$$

Reordering the equation yields:

$$Y(s) + (T_2+T_3)sY(s) + T_2T_3s(\frac{KT_1}{T_2T_3}U(s) + sY(s)) = KU(s) \quad [6.175]$$

We choose as state vector:

$$X_1(s) = Y(s) \quad \text{and} \quad X_2(s) = \frac{KT_1}{T_2T_3}U(s) + sY(s) \quad [6.176]$$

Moving into the temporal domain:

$$x_1(t) + (T_2+T_3)\dot{x}_1 + (T_2T_3)\dot{x}_2(t) = Ku(t); \quad x_2 = \frac{KT_1}{T_2T_3}u(t) + \dot{x}_1 \quad [6.177]$$

hence the following state equation:

$$\begin{cases} \dot{x}_1 = x_2 - \frac{KT_1}{T_2T_3}u(t) \\ \dot{x}_2 = -\frac{1}{T_2T_3}x_1(t) - \frac{T_2+T_3}{T_2T_3}x_2 + \frac{K}{T_2T_3}[1 + T_1\frac{T_2+T_3}{T_2T_3}]u(t) \\ y = x_1 \end{cases} \quad [6.178]$$

The state–space representation matrices are given as:

$$\begin{cases} A = \begin{bmatrix} 0 & 1 \\ -\frac{1}{T_2T_3} & -\frac{T_2+T_3}{T_2T_3} \end{bmatrix}; \quad B = \begin{bmatrix} -\frac{KT_1}{T_2T_3} \\ \frac{K}{T_2T_3}[1 + T_1\frac{T_2+T_3}{T_2T_3}] \end{bmatrix} \\ C = \begin{bmatrix} 1 \; 0 \end{bmatrix} \end{cases} \quad [6.179]$$

The expression of the step response $y(t)$:

$$\begin{aligned} Y(s) &= H(s)U(s)\mid_{U(s)=\frac{1}{s}} = \frac{K(1 - T_1 s)}{(1 + T_2 s)(1 + T_3 s)} \cdot \frac{1}{s} \\ &= \frac{K}{T_3 - T2}[\frac{T_3 - T2}{s} + \frac{T_1 + T_2}{s + \frac{1}{T_2}} - \frac{T_1 + T_3}{s + \frac{1}{T_3}}] \end{aligned} \quad [6.180]$$

By calculating the inverse Laplace transform, we arrive at:

$$\begin{aligned} & y(t) = K[1 + \frac{T_1 + T_2}{T_3 - T2}e^{-\frac{t}{T_2}} - \frac{T_1 + T_3}{T_3 - T2}e^{-\frac{t}{T_3}}] \\ \text{and} \quad & y'(0) = \lim_{s\to\infty} s^2 Y(s) = -\frac{KT_1}{T_3 T2} \end{aligned} \quad [6.181]$$

*Identification*

When $t \gg \max(T_1, T_2)$, the expression of $y(t)$ becomes:

$$y(t) \simeq K[1 - \frac{T_1 + T_3}{T_3 - T2}e^{-\frac{t}{T_3}}] \quad [6.182]$$

hence the expression of $\alpha(t)$ is given by:

$$\alpha(t) = \text{Log}\frac{K - y(t)}{K} = \text{Log}\frac{T_1 + T_3}{T_3 - T2} - \frac{t}{T_3} \quad [6.183]$$

Taking different values of $\alpha$ at different moments:

$$\left.\begin{matrix} t = 30\,\text{s}\ \alpha = -2.70 \\ t = 40\,\text{s}\ \alpha = -3.7 \end{matrix}\right\} \Rightarrow \left.\begin{matrix} T_3 = 10\,\text{s} \\ t = 40\,\text{s}\ \ \alpha = -3.70 \end{matrix}\right\} \Rightarrow 1.34 = \frac{T_1 + 10}{10 - T_2} \quad [6.184]$$

The derivative at the origin is given by:

$$y'(0) = \lim_{s \to \infty} s^2 Y(s) = -\frac{KT_1}{T_2 T_3} = 0.2 \quad \Rightarrow \quad T_1 = 2T_2 \qquad [6.185]$$

Therefore, the values of $T_1$ and $T_2$ can be extracted as:

$$T_2 \simeq 1.05\,\text{s} \quad \text{and} \quad T_1 \simeq 2.1\,\text{s} \qquad [6.186]$$

The system is associated according to the plot of the following figure, $C(s) = k$.

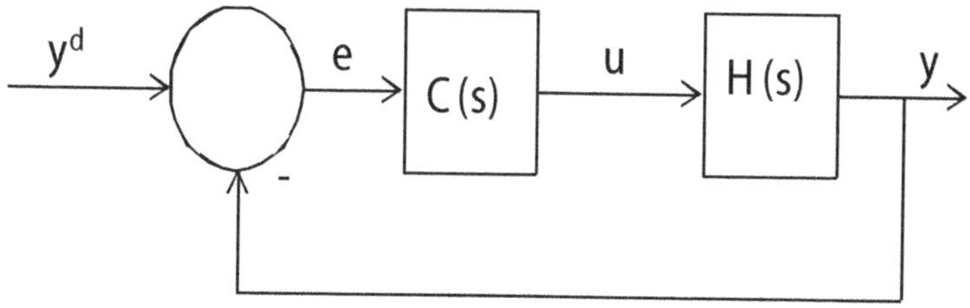


**Figure 6.18.** *Closed-loop controlled system*

The transfer function of the closed-loop system is given by:

$$\begin{aligned} F(s) &= \frac{kK(1 - T_1 s)}{(1 + T_2 s)(1 + T_3 s) + kK(1 - T_1 s)} \\ &= \frac{kK(1 - T_1 s)}{T_2 T_3 s^2 + (T_2 + T_3 - kKT_1)s + 1 + kK} \end{aligned} \qquad [6.187]$$

If $k$ is large $\Rightarrow$ complex poles $\Rightarrow$ oscillating system.

The transfer function can be written as:

$$F(s) = K_1 \cdot \frac{1 - T_1 s}{\frac{1}{\omega_0^2} s^2 + \frac{2\xi}{\omega_0} s + 1} \quad \text{with} \quad \omega_0^2 = \frac{1 + kK}{T_3 T2};$$

$$\xi = \frac{T_2 + T_3 - kKT_1}{1 + kK} \quad \text{and} \quad K_1 = \frac{kK}{1 + kK} \qquad [6.188]$$

The shape of the step response of the control when $k$ is large enough ($k = 4$) is given by the following figure:

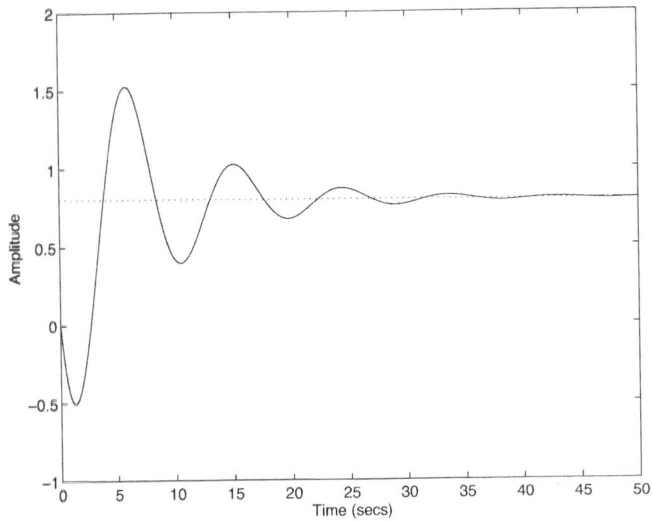


**Figure 6.19.** *Step response of the system in CL*

Based on the experimental measurements of this response, we can derive the characteristic parameters of the system ($K_1, T_1, \omega_0$ and $\xi$) from:

– the pseudo-period $T_p \Rightarrow \omega_p = \omega_0\sqrt{1-\xi^2}$.

– $\xi$ is calculated using the logarithmic decrement.

The final value is given by:

$$y(\infty) = K_1 \quad [6.189]$$

The derivative at the origin is expressed as:

$$y'(0) = -\frac{kKT_1}{T_3T2} \quad [6.190]$$

A system of four equations with four unknowns.

EXERCISE 5.– Identification by analogy

The joined curves represent the unit step response of unknown systems. Determine for each case an adequate structure for the transfer function and identify the parameters of this function.

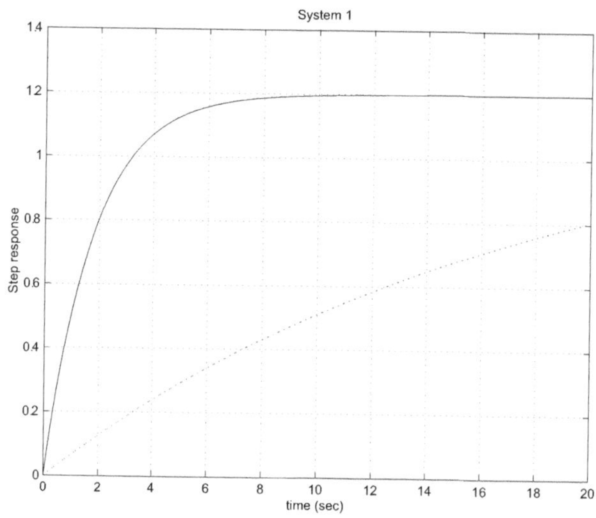


**Figure 6.20.** *Unit step response of an unknown system*

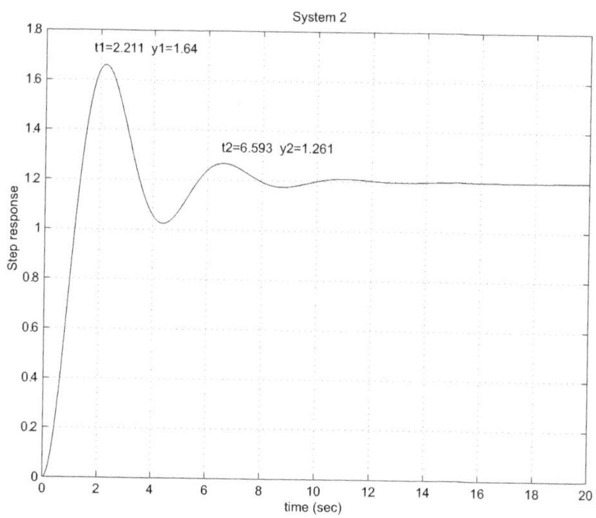


**Figure 6.21.** *Unit step response of an unknown system 2*

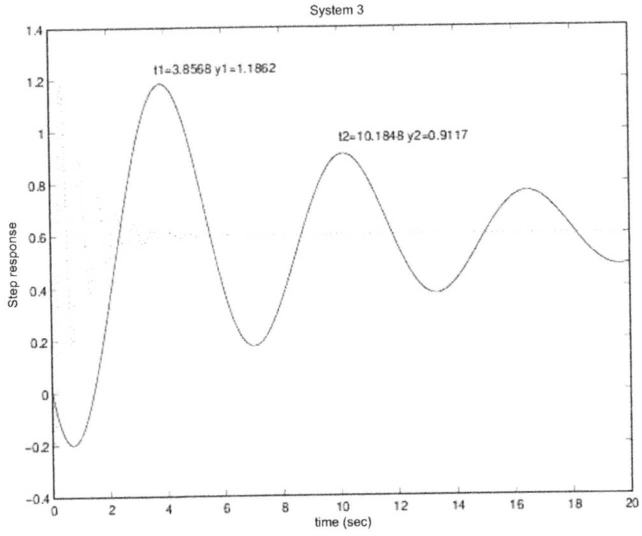


**Figure 6.22.** *Unit step response of an unknown system 3*

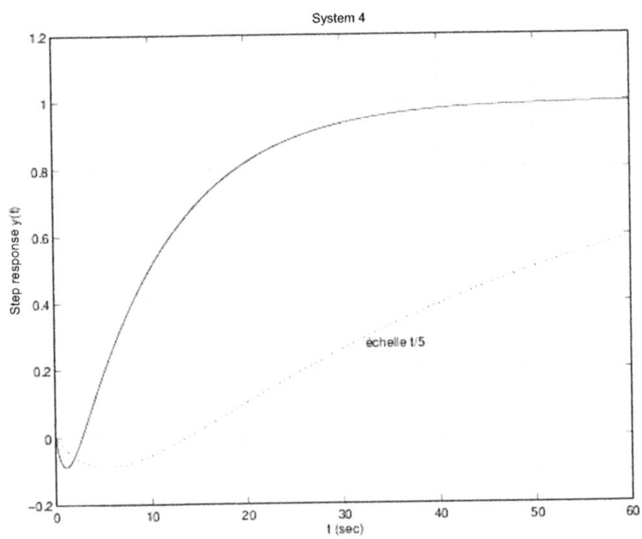


**Figure 6.23.** *Unit step response of an unknown system 4*

EXERCISE 6.– Given the discrete system described by the following recurrence equation:

$$Y((k+2)T) + 0.2Y((k+1)T) - 0.63Y(kT) = U(kT) \qquad [6.191]$$

$$Y(0) = 0, \quad Y(1) = 0 \qquad [6.192]$$

1) Give a state–space representation (companion form).

2) Is this system stable?

3) Calculate the unit step response.

SOLUTION EXERCISE 6.– The state–space representation of the process modeled by the recurrence equation.

$$y((k+2)T) + 0.2y((k+1)T) - 0.63y(kT) = u(kT) \qquad [6.193]$$

is given as follows:

$$\begin{aligned} y_k &= y(kT) = x_{1k} = x_1(kT) \\ y_{k+1} &= x_{1(k+1)} = x_{2k} \\ y_{k+2} &= x_{2(k+1)} = -0.2x_{2k} + 0.63x_{1k} + u_k \end{aligned} \qquad [6.194]$$

In matrix form, we have:

state equation: $$\begin{bmatrix} x_{1(k+1)} \\ x_{2(k+1)} \end{bmatrix} = \begin{bmatrix} 0 & 1 \\ 0.63 & -0.2 \end{bmatrix} \begin{bmatrix} x_{1k} \\ x_{2k} \end{bmatrix} + \begin{bmatrix} 0 \\ 1 \end{bmatrix} u_k \qquad [6.195]$$

output equation: $$y = \begin{bmatrix} 1 & 0 \end{bmatrix} \begin{bmatrix} x_{1k} \\ x_{2k} \end{bmatrix} \qquad [6.196]$$

The stability of the system is ensured if the eigenvalues of the evolution matrix $A$ are inside the unit circle.

$$det(\lambda I - A) = 0 \Longrightarrow \begin{cases} \lambda_1 = 0.7 \\ \lambda_2 = -0.9 \end{cases} \qquad [6.197]$$

The system is stable because:

$$| \lambda_1 | < 1 \quad \text{and} \quad | \lambda_2 | < 1 \qquad [6.198]$$

The unit step response:

From the recurrence equation, calculate the Z-transform.

$$(z^2+0.2z-0.63)Y(z)=U(z) \qquad [6.199]$$

The transfer function $H(z)$ is given by:

$$H(z)=\frac{Y(z)}{U(z)}=\frac{1}{z^2+0.2z-0.63}=\frac{1}{(z+0.9)(z-0.7)} \qquad [6.200]$$

The step response:

$$Y(z)=H(z)U(z) \quad \text{with} \quad U(z)=\frac{z}{z-1} \qquad [6.201]$$

hence

$$Y(z)=\frac{z}{(z+0.9)(z-0.7)(z-1)} \qquad [6.202]$$

The simple elements decomposition of the following expression is carried out:

$$\frac{Y(z)}{z}=\frac{1}{(z+0.9)(z-0.7)(z-1)}=\frac{25}{4}\frac{1}{z+0.9}+\frac{50}{3}\frac{1}{z-0.7}+\frac{100}{57}\frac{1}{z-1} \qquad [6.203]$$

hence

$$Y(z)=\frac{25}{4}\frac{z}{z+0.9}+\frac{50}{3}\frac{z}{z-0.7}+\frac{100}{57}\frac{z}{z-1} \qquad [6.204]$$

The response is given by:

$$y(nT)=(\frac{25}{4}(-0.9)^n+\frac{50}{3}(0.7)^n+\frac{100}{57})u(nT) \qquad [6.205]$$

EXERCISE 7.– A system is described by the following state–space representation:

$$\begin{cases} X_{k+1}=\begin{bmatrix}0.5 & -1\\ 0 & 0.5\end{bmatrix}X_k+\begin{bmatrix}0\\1\end{bmatrix}U_k \\ Y_k=\begin{bmatrix}1 & -1\end{bmatrix}X_k \end{cases} \qquad [6.206]$$

1) Is this system stable?

2) What its transmittance in Z?

3) Derive therefrom a scalar recurrence equation between the input and the output.

SOLUTION EXERCISE 7.– The stability of the system is inferred from the eigenvalues of $A$. The latter are given as:

$$\lambda_1 = \lambda_2 = 0.5 \qquad [6.207]$$

Since $|\lambda_1| < 1$ and $|\lambda_2| < 1$, the system is then stable. The transmittance in $Z$ is given by:

$$H(z) = C[zI - A]^{-1}B \qquad \text{with} \qquad [zI - A]^{-1} = \begin{bmatrix} \frac{1}{z-0.5} & \frac{-1}{(z-0.5)^2} \\ 0 & \frac{1}{z-0.5} \end{bmatrix} \qquad [6.208]$$

hence

$$H(z) = -\frac{z + 0.5}{(z - 0.5)^2} \qquad [6.209]$$

the recurrence equation:

$$H(z) = \frac{Y(z)}{U(z)} \quad \Longrightarrow \quad Y(z)(z^2 - z + 0.5) = -U(z)(z + 0.5) \qquad [6.210]$$

which yields:

$$\begin{aligned} y_{k+2} - y_{k+1} + 0.25y_k &= -u_{k+1} - 0.5u_k \quad \Longrightarrow \\ y_{k+2} &= y_{k+1} - 0.25y_k - u_{k+1} - 0.5u_k \end{aligned} \qquad [6.211]$$

EXERCISE 8.– A device has a transfer function: $H(s) = \frac{1}{s(s+1)}$

1) What is its unit step response?

2) This system is controlled according to the diagram below.

3) Let $C(s) = K$, what is the transfer function of the closed-loop system? Express its characteristic parameters.

4) For $K = 1$, what can be said:

a) of its step response $y(t)$?

b) of permanent position $\varepsilon_p$ and velocity $\varepsilon_v$ errors?

5) For $K = 9$, what can be said of $y(t)$, $\varepsilon_p$ and $\varepsilon_v$?

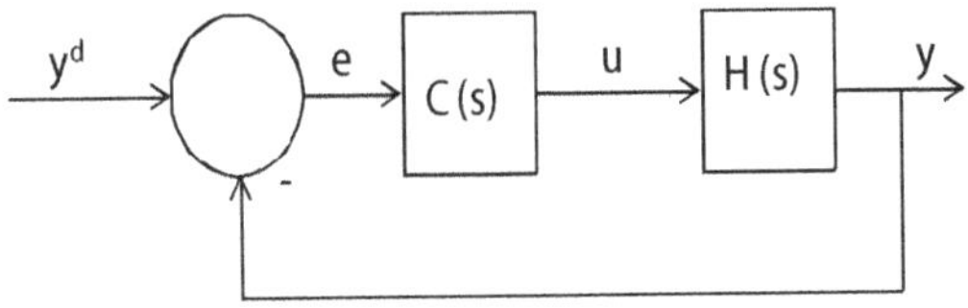


**Figure 6.24.** *Controlled system with correction*

SOLUTION EXERCISE 8. – The unit step response is given by:

$$Y(s) = H(s) \cdot \frac{1}{s} = \frac{1}{s^2(s+1)} = -\frac{1}{s} + \frac{1}{s^2} + \frac{1}{s+1} \qquad [6.212]$$

whence by calculating the inverse Laplace transformation, it follows that:

$$y(t) = (t - 1 + e^{-t})\Gamma(t) \qquad (\Gamma(t) \text{ is the Heaviside function}) \qquad [6.213]$$

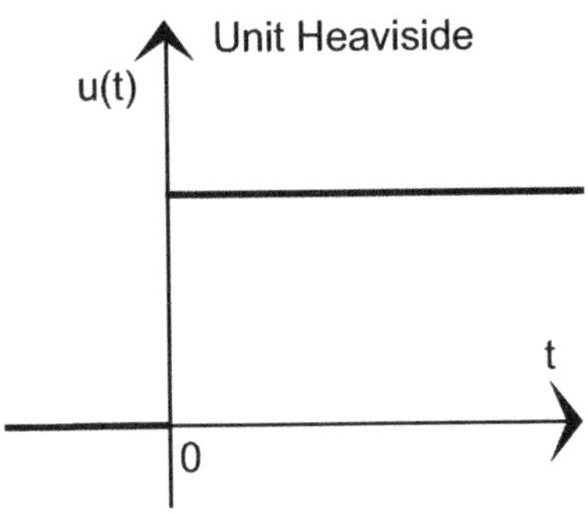


**Figure 6.25.** *Unit step response of* $H(s) = \frac{1}{s(s+1)}$

This system is controlled according to the diagram:

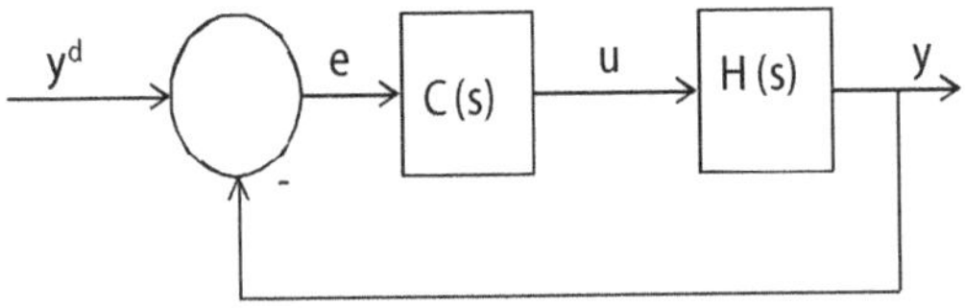


**Figure 6.26.** *Controlled system*

Let $C(s) = K$, the transfer function of the closed-loop system is given as follows:

$$F(s) = \frac{KH(s)}{1 + KH(s)} = \frac{K}{s^2 + s + K} = \frac{\omega_0^2}{s^2 + 2\xi\omega_0 s + \omega_0^2} \qquad [6.214]$$

with the following characteristic parameters

$$\omega_0 = \sqrt{K} \quad \text{and} \quad \xi = \frac{1}{2\sqrt{K}} \qquad [6.215]$$

It can be observed that if $K \nearrow \Longrightarrow \omega_0 \nearrow$ and $\xi \searrow$.

The permanent position $\varepsilon_p$ and velocity $\varepsilon_v$ errors are given as follows:

$$\begin{aligned}\varepsilon_p &= \lim_{s \to 0} s\varepsilon(s) \mid_{u(s)=1/s} \\ &= \lim_{s \to 0} s \cdot \frac{1}{1 + C(s)H(s)} \cdot \frac{1}{s} \quad \Rightarrow \quad \varepsilon_p = \frac{1}{1 + C(0)H(0)} = 0\end{aligned} \qquad [6.216]$$

The system shows an integration, therefore the position error $\varepsilon_p = 0$.

$$\begin{aligned}\varepsilon_v &= \lim_{s \to 0} s\varepsilon(s) \mid_{u(s)=1/s^2} \\ &= \lim_{s \to 0} s \cdot \frac{1}{1 + C(s)H(s)} \cdot \frac{1}{s^2} \quad \Rightarrow \quad \varepsilon_v = \lim_{s \to 0} \frac{1}{sC(s)H(s)} = 1/K\end{aligned} \qquad [6.217]$$

The system shows a single integration, therefore the velocity error $\varepsilon_v = cte$. Its step response of the system thus looped is given by ($\xi < 1$):

$$\begin{aligned}y(t) &= (1 - \frac{e^{-\xi\omega_0 t}}{\sqrt{1 - \xi^2}} \cdot \sin(\omega_0\sqrt{1 - \xi^2}t + \phi))\Gamma(t) \\ &\text{with} \begin{cases} \sin(\phi) = \sqrt{1 - \xi^2} \\ \cos(\phi) = \xi \end{cases}\end{aligned} \qquad [6.218]$$

In conclusion, $K \nearrow \Longrightarrow \varepsilon_v \searrow$ but $y(t)$ is badly damped (compromise between accuracy and stability).

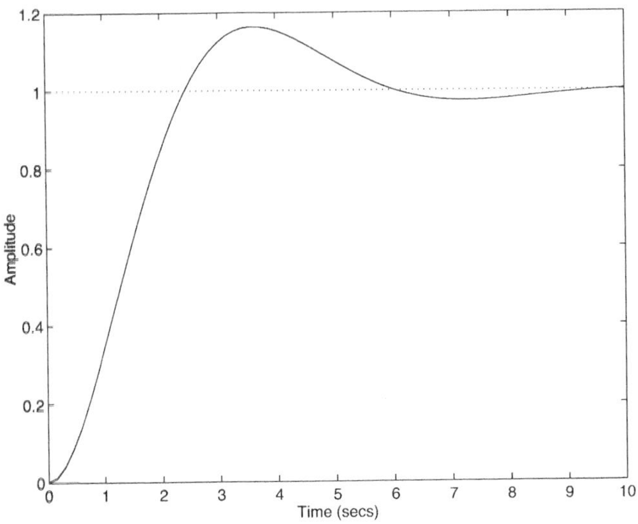


**Figure 6.27.** *Step response* $K = 1 \Rightarrow (\xi = 0.5, \omega_0 = 1, \varepsilon_p = 0, \varepsilon_v = 1)$

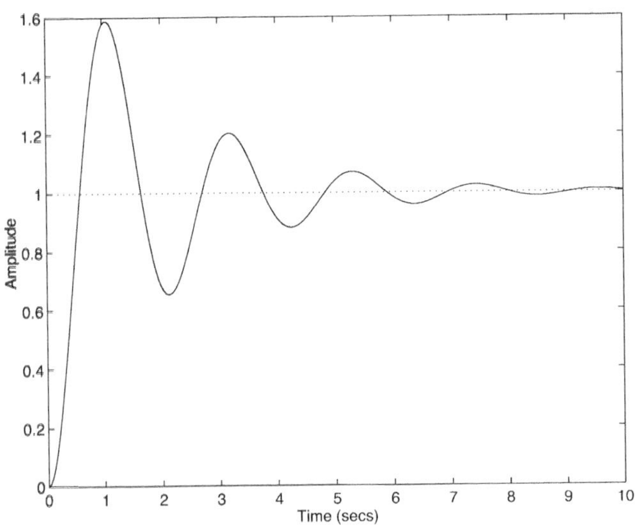


**Figure 6.28.** *Step response* $K = 9 \Rightarrow$
$(\xi = 1/6, \omega_0 = 3, \varepsilon_p = 0, \varepsilon_v = 1/9)$

# 7

# System Stability and Robustness Analysis Methods

This chapter explores the definition and analyses of dynamic systems, stability and robustness criteria.

## 7.1. Introduction

In this chapter, we focus our interest on analysis methods of the stability of linear systems and especially on those concerning looped systems. For this purpose, we will study the case of looped systems with unity feedback because we can always reduce the diagram of a looped system to a functional block diagram with a unity feedback loop, whatever the complexity of its functional diagram.

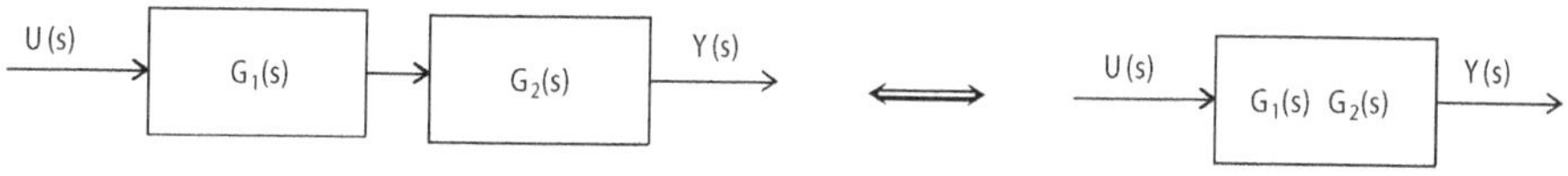


**Figure 7.1.** *Cascade diagrams*

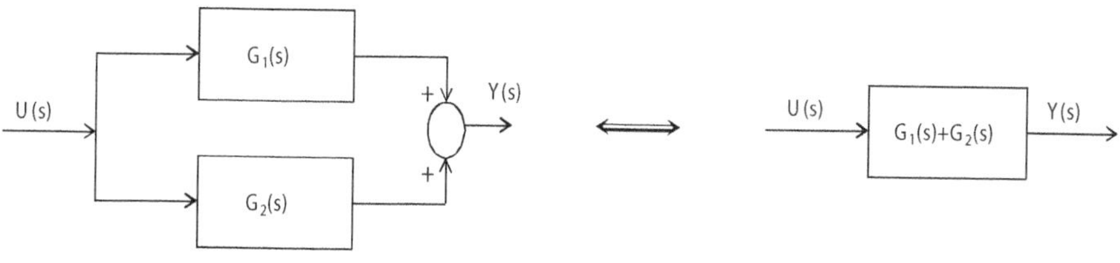


**Figure 7.2.** *Parallel diagrams*

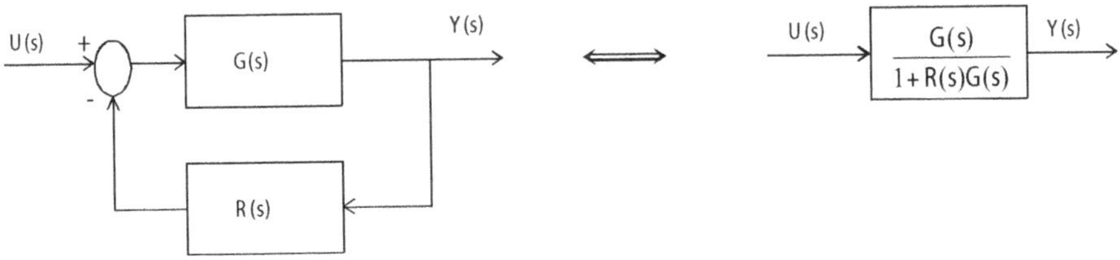


**Figure 7.3.** *Negative feedback parallel diagrams 1*

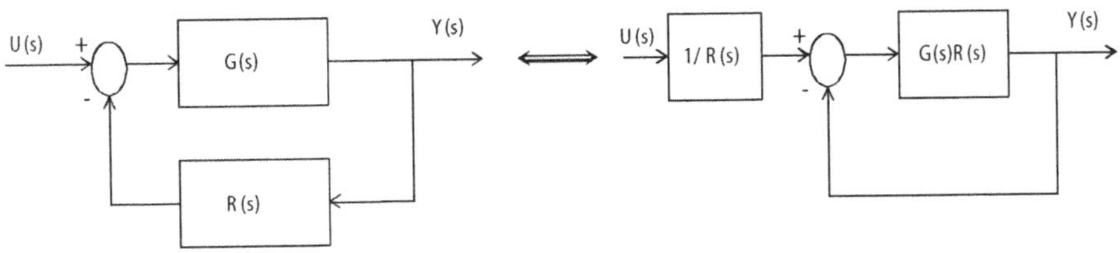


**Figure 7.4.** *Negative feedback parallel diagrams 2*

## 7.2. Definitions related with the stability of a dynamic system

Consider a free dynamic system, described by the following differential vector equation:

$$\dot{x} = f(x,t) \qquad \forall t \in R \tag{7.1}$$

where $x(t)$ is the state vector of dimension $n$ of the system, defining a point in the phase plane, and $f(x,t)$ is a vector function in $R$ that may be nonlinear. We consider here the case of free signals or systems. It should be observed that this system is free but it is not generally autonomous since its equation explicitly depends on time. The equation of an autonomous system can be written in the form $\dot{x} = f(x)$ without explicit time dependence. Thus, a free time-variant system is not autonomous.

Let $\Phi(t_o, x_o, t)$, the vector function, non-disturbed solution of equation [7.1], differentiable with respect to time and such that for any $x \in \mathbb{R}^n$ , $t \in \mathbb{R}$ $x(t_o) = x_o$ initial state and $t_o$ initial time, we have:

$$\Phi(t_o, x_o, t) = x = x(t) \tag{7.2}$$

$$\frac{d(\Phi(t_o, x_o, t))}{dt} = f[\Phi(t_o, x_o, t), t] \tag{7.3}$$

$\Phi(t_o, x_o, t)$ defines a curve or a trajectory of the system in the phase space for an initial state $x_o \in \mathbb{R}^n$ and $t \in \mathbb{R}$.

### 7.2.1. *Equilibrium state of a system*

DEFINITION 7.1.– *A state $x_e$ of the free system is called an equilibrium point if:*

$$f(x_e, t) = 0 \quad \forall t \in R, t > to \qquad [7.4]$$

This expresses that if $x_e$ is an equilibrium point then, in the case without disturbance, the system tends to remain in a nearby neighborhood because at this point velocity $\dot{x}_e = f(x_e, t)$ is zero. When the equilibrium point is $x_e \neq 0$, the equilibrium point can be brought by transformation to the origin of the phase space. For a linear time-invariant system, $\dot{x} = f(x, t) = A.x$, if $A$ is not singular then $x_e = 0$ is the point of equilibrium. If $A$ is singular, there is an infinite number of equilibrium points. The search for equilibrium points does not require solving differential equations ($\dot{x} = f(x, t) = 0$).

DEFINITION 7.2 (*Stable Equilibrium State*).– *A state of equilibrium is stable if after a deviation from this equilibrium, the system remains in the vicinity or tends to return to this equilibrium point. An equilibrium point will be known as unstable equilibrium if after deviating from this point the system tends to move away.*

For example, a pendulum has a stable equilibrium (low vertical position) and an unstable equilibrium (high vertical position, modulus $2k$).

### 7.2.2. *Stable system: bounded input bounded output*

DEFINITION 7.3.– *A system is said to be stable if for any bounded input $u(t)$ ($u(t) < M$) it establishes a correspondence with a bounded output $y(t)$. This notion of stability is known as "BIBO stability" (bounded input, bounded output).*

If $h(t)$ is the impulse response of the system, the following can be stated:

$$y(t) = \int_{-\infty}^{+\infty} u(\tau).h(t - \tau)d\tau = u(t) * h(t) \qquad [7.5]$$

Knowing that $e(t)$ is bounded, $|e(t)| < M \in \mathbb{R}$ we then get:

$$|y(t)| \leq \int_{-\infty}^{+\infty} |u(\tau)| \, . \, |h(t - \tau)| \, d\tau \leq M. \int_{-\infty}^{+\infty} |h(\tau)| \, d\tau \qquad [7.6]$$

hence the system is stable if its impulse response $h(t)$ is an absolutely summable function. A sufficient condition for stability is that $h(t)$ be summable ($\int_{-\infty}^{+\infty} |h(\tau)|\, d\tau$ is finite).

For the definition of the stability of a linear system $H_o(p)$ closed by a feedback loop $R(p)$, we will use the following proposition.

PROPOSITION 7.1.– The following definitions are equivalent:

– a system is stable if its closed-loop (CL) impulse response is absolutely summable ($\int_{-\infty}^{+\infty} |h(\tau)|\, d\tau$ is finite);

– a loop system is stable if transmittance poles in CL have strictly negative real parts. Transmittance in CL is written as:

$$G(p) = \frac{H_o(p)}{1 + H_o(p)R(p)} \qquad [7.7]$$

This definition can be stated as: a loop system is stable if the roots of the characteristic equation (zeros) $1 + H_o(p)R(p) = 1 + H(p) = 0$ do not have a positive or zero real part;

– a loop system is stable for any bounded input if its output is bounded.

## 7.3. Stability criteria

### 7.3.1. *Routh criterion and stability algebraic criterion*

The characteristic equation is given by:

$$1 + H(p) = 0 \quad \text{or} \quad 1 + \frac{N(p)}{D(p)} = 0 \quad \Rightarrow \quad D(p) + N(p) = 0 \qquad [7.8]$$

The last equation or characteristic equation of the loop system $G(p)$ is written as:

$$a_n p^n + a_{n-1} p^{n-1} + \ldots + a_1 p + a_0 = 0 \qquad [7.9]$$

The Routh criterion gives a necessary and sufficient condition for a polynomial to have its roots with strictly negative real parts. From the above equation [7.9] that allows us to define the poles of the CL system, a Routh table is created (Table 7.1).

The first line, called line $p^n$, has as a first coefficient $a_n$ and the following are obtained by reducing the power by decrements of 2 ($a_{n-2}; a_{n-4}; \cdots$). The first coefficient of line $p^{n-1}$ is $a_{n-1}$ and the following are obtained by reducing the power by decrements of 2 ($a_{n-3}; a_{n-5}; \cdots$). The following lines are calculated based on these two lines, according to the indications presented in Table 7.1. Each line of this table is decreased by a column by obtaining a zero at the end of the line. Therefore, the table is reduced to a single non-zero coefficient for line $p^0$. The first column comprises ($n$ + 1) elements. In this sequence of operations, we can multiply or divide a line by a strictly positive term to simplify the calculations.

| | | | | | | |
|---|---|---|---|---|---|---|
| line $p^n$ | $a_n$ | $a_{n-2}$ | $a_{n-4}$ | ... | | |
| line $p^{n-1}$ | $a_{n-1}$ | $a_{n-3}$ | $a_{n-5}$ | ... | | |
| line $p^{n-2}$ | $b_1 = \frac{a_{n-1}a_{n-2}-a_n a_{n-3}}{a_{n-1}}$ | $b_2 = \frac{a_{n-1}a_{n-4}-a_n a_{n-5}}{a_{n-1}}$ | $b_3 = \frac{a_{n-1}a_{n-6}-a_n a_{n-7}}{a_{n-1}}$ | ... | ... ... | ... ... |
| line $p^{n-3}$ | $c_1 = \frac{b_1 a_{n-3}-b_2 a_{n-1}}{b_1}$ | $c_2 = \frac{b_1 a_{n-5}-b_3 a_{n-1}}{b_1}$ | $c_3 = \frac{b_1 a_{n-7}-b_4 a_{n-1}}{b_1}$ | ... | ... 0 | 0 0 |
| $\vdots$ | $\vdots$ | $\vdots$ | $\vdots$ | 0 | 0 0 | 0 0 |
| line $p^1$ | $\vdots$ | $\vdots$ | 0 | 0 | 0 | 0 |
| line $p^0$ | $\vdots$ | 0 | 0 | 0 | | |

**Table 7.1.** *Routh table*

THEOREM 7.1 (Routh criterion).– The roots of the characteristic equation

$$a_n p^n + a_{n-1} p^{n-1} + \ldots + a_1 p + a_0 = 0 \qquad [7.10]$$

have strictly negative real part if and only if all terms $a_i$ are not null of the same sign and if the terms in the first column of the Routh table have the same sign. The number of unstable roots is given by the number of sign changes in the first column.

*Advantages*: As a simple criterion, it does not require the factorization of transmittances or the calculation of the roots.

*Disadvantages*: It does not provide any indication about the degree or margin of stability and does not apply to systems with delay in the direct chain of the unit feedback loop.

EXAMPLE 7.1.– *Let the open-loop (OL) transfer function given by:*

$$H(p) = \frac{1}{p(1+p)(0.5+p)} \qquad [7.11]$$

*The characteristic polynomial of the CL system is obtained by writing the denominator of* ($G(p) = \frac{H(p)}{1+H(p)}$)*:*

$$p(0.5 + 1.5p + p^2) + 1 = 0 \qquad [7.12]$$

*Hence, the characteristic polynomial is expressed as:*

$$p^3 + 1.5p^2 + 0.5p + 1 = 0 \qquad [7.13]$$

*The Routh array is the following:*

| | | | |
|---|---|---|---|
| $p^3$ | $1$ | $0.5$ | $0$ |
| $p^2$ | $1.5$ | $1$ | $0$ |
| $p^1$ | $\frac{1.5\times0.5-1}{1.5} = -\frac{1}{6}$ | $0$ | $0$ |
| $p^0$ | $1$ | $0$ | $0$ |

**Table 7.2.** *Routh table result*

*There are two changes in sign in the first column, therefore the system has two poles with positive real terms. The system is thus unstable.*

#### 7.3.1.1. *Zeroes in the first column of the Routh table*

In the presence of zeroes in the first column of the table with the other terms of the line non-zero, we replace them by $\varepsilon$ small positive and carry on with calculations. When the table is filled, we make $\varepsilon$ tend to zero and observe the sign of the terms of the first column to study the stability.

EXAMPLE 7.2.– *Let the characteristic polynomial of the CL system be:*

$$p^4 + 2p^3 + 4p^2 + 8p + \alpha = 0 \qquad [7.14]$$

*The Routh array is the following:*

| | | | |
|---|---|---|---|
| $p^4$ | $1$ | $4$ | $\alpha$ |
| $p^3$ | $2$ | $8$ | $0$ |
| $p^2$ | $4 - \frac{8}{2} = 0 \rightarrow \varepsilon$ | $\alpha$ | $0$ |
| $p^1$ | $8 - \frac{2\alpha}{\varepsilon} \rightarrow -\frac{2\alpha}{\varepsilon}$ | $0$ | $0$ |
| $p^0$ | $\alpha$ | $0$ | $0$ |

**Table 7.3.** *Routh table results*

*Considering $\varepsilon$ positive, $8-\frac{2\alpha}{\varepsilon}$ is of the sign of $-\frac{2\alpha}{\varepsilon}$ when $\varepsilon$ tends to zero. Therefore, if $\alpha < 0$ there is only a single change in sign, which implies that there is a root with positive real part; as a result, the system is unstable. However, if $\alpha > 0$ there are two changes in sign, which shows that we have two roots with positive real part; in this case, the system is unstable.*

### 7.3.1.2. *A whole line of zeroes in the Routh table*

When filling in the Routh table and in the particular case where a whole line of zeroes appears, this is then a system having poles on the imaginary axis. In this case, we will use the auxiliary equation (obtained using the line just above the line of zeroes) to continue with the Routh table.

| ligne $p^n$ | $a_n$ | $a_{n-2}$ | $a_{n-4}$ | $\cdots$ | $\cdots$ | $\cdots$ |
|---|---|---|---|---|---|---|
| line $p^{n-1}$ | $a_{n-1}$ | $a_{n-3}$ | $a_{n-5}$ | $\cdots$ | $\cdots$ | $\cdots$ |
| line $p^{n-2}$ | $\alpha_1$ | $\alpha_2$ | $\alpha_3$ | $\alpha_4$ | $\cdots$ | $\cdots$ |
| line $p^{n-3}$ | 0 | 0 | 0 | 0 | 0 | 0 |
| line $p^{n-3}$ | $b_1$ | $b_2$ | $b_3$ | $b_4$ | $\cdots$ | $\cdots$ |
| line $p^{n-2}$ | $\vdots$ | $\vdots$ | | | $\cdots$ | $\cdots$ |

**Table 7.4.** *Routh table*

In order to determine new coefficients to replace this line of zeroes, the following operations are carried out:

1) An auxiliary polynomial $P(p)$ is restored by means of the auxiliary line, preceding the zeroes line: $P(p) = \alpha_1 p^{n-2} + \alpha_2 p^{n-4} + \alpha_3 p^{n-6} + \alpha_4 p^{n-8} + \cdots$

2) This polynomial is differentiated with respect to the Laplace variable $p$: $Q(p) = \frac{d}{dp}P(p) = b_1 p^{n-3} + b_2 p^{n-5} + b_3 p^{n-7} + b_4 p^{n-9} + \cdots$

3) The zeroes line is restored or replaced by a line having for coefficients those of the polynomial $Q(p)$, as shown in Table 7.4, and then we continue with the construction of the Routh table.

4) To determine the pure complex conjugate imaginary poles, the roots of the auxiliary equation $P(p) = 0$ are computed. It is also possible to replace the variable $p$ by $j\omega$ and the real and imaginary parts of the auxiliary equation will be canceled out.

EXAMPLE 7.3.– *Let the characteristic polynomial of the CL system be:*

$$p^3 + 5p^2 + 4p + 20 = 0 \qquad [7.15]$$

*The Routh table is the following:*

1) an auxiliary polynomial $P(p)$ is restored by means of the auxiliary line: $P(p) = 5p^2 + 20$;

2) this polynomial is differentiated with respect to the Laplace variable $p$: $Q(p) = \frac{d}{dp}P(p) = 10p + 0$;

3) the line is recovered with the coefficients of polynomial $Q(p)$, line $p^1$ replaced$|10|0|0$. We then continue with the construction of the Routh table;

4) to determine the pure conjugate complex imaginary poles, we compute the roots of the auxiliary equation $P(p) = 5p^2 + 20 = 0$. We have pure imaginary roots $\pm j\omega = \pm 2j$.

| | | | |
|---|---|---|---|
| line $p^3$ | 1 | 4 | 0 |
| line $p^2$ | 5 | 20 | 0 |
| line $p^1$ | 0 | 0 | 0 |
| line $p^1$ Replaced | 10 | 0 | 0 |
| line $p^0$ | 20 | 0 | 0 |

**Table 7.5.** *Routh table results*

EXAMPLE 7.4.– *Let the OL transfer function given by:*

$$H(p) = \frac{K(1+5p)}{p^2(1+0.8p+4p^2)} \quad [7.16]$$

*The characteristic polynomial of the CL system is obtained by writing:*

$$denominator\,of(G(p) = \frac{H(p)}{1+H(p)}) = 0 \Longleftrightarrow p^2(1+0.8p+4p^2) + K(1+5p) = 0 \quad [7.17]$$

*Hence, the characteristic polynomial:*

$$4p^4 + 0.8p^3 + p^2 + 5Kp + K = 0 \quad [7.18]$$

*The Routh table is the following:*

| | | | |
|---|---|---|---|
| $p^4$ | $4$ | $1$ | $K$ |
| $p^3$ | $0.8$ | $5K$ | $0$ |
| $p^2$ | $1-25K$ | $K$ | |
| $p^1$ | $K\frac{4.2-125K}{1-25K}$ | | |
| $p^0$ | $K$ | | |

**Table 7.6.** *Routh table results*

*We study the stability of the loop system according to the parameter $K$. The CL system is thus stable if:*

$$\left.\begin{array}{ll} 1-25K>0 & \Rightarrow K<0.04 \\ 4.2-125K>0 \Rightarrow K<0.0336 \\ \quad K>0 \end{array}\right\} \Longrightarrow 0<K<K_c=0.0336 \qquad [7.19]$$

### 7.3.2. *Jury criterion and discrete system example*

The criterion can be used to determine the number of roots of the characteristic polynomial which are outside of the unit circle (unstable). Given the following characteristic equation:

$$A(z)=a_n z^n+a_{n-1}z^{n-1}+\cdots+a_1 z+a_0=0 \qquad [7.20]$$

To determine the number of roots with a modulus greater than the unit, the Jury table is built as follows.

The first double line is composed of the polynomial coefficient in decreasing order and below the increasing order of the indices. The index placed in superscript represents the line number. The coefficients of a double line are obtained from those of the previous line. To this end, the second coefficient line is multiplied by $\alpha_n=\frac{a_n}{a_0}$ (for the double line $n$ ) and it is subtracted from the first. This gives us $a_0^{n-1}$; $a_1^{n-1}$;… ;$a_{n-2}^{n-1}$; $a_{n-1}^{n-1}$; 0; 0. The second line (of the double line $n-1$) is obtained by reversing the order of the coefficients. The other coefficients can be calculated in the same way.

| | |
|---|---|
| line $z^n$ | $a_0 \quad a_1 \quad \ldots \; a_{n-2} \; a_{n-1} \; a_n \; 0$<br>$a_n \quad a_{n-1} \; \ldots \; a_2 \quad a_1 \quad a_0 \; 0$ |
| line $z^{n-1}$ | $a_0^{n-1} \; a_1^{n-1} \; \ldots \; a_{n-2}^{n-1} \; a_{n-1}^{n-1} \; 0 \; 0$<br>$a_{n-1}^{n-1} \; a_{n-2}^{n-1} \; \ldots \; a_1^{n-1} \; a_0^{n-1} \; 0 \; 0$ |
| line $z^{n-2}$ | $a_0^{n-2} \; a_1^{n-2} \; \ldots \; a_{n-2}^{n-2} \; 0 \; 0 \; 0$<br>$a_{n-2}^{n-2} \; a_{n-3}^{n-2} \; \ldots \; a_0^{n-2} \; 0 \; 0 \; 0$ |
| | ……………………… |
| line $z^0$ | $a_0^0 \quad 0 \; \ldots\ldots\ldots\ldots 0$ |

**Table 7.7.** *Jury table*

Furthermore, the coefficients of the line of rank $k$ are obtained by the following formula:

$$a_i^{k-1}=a_i^k-\alpha_k.a_{k-i}^k=a_i^k-\frac{a_k^k}{a_0^k}.a_{k-i}^k \;\; with \;\; \alpha_k=\frac{a_k^k}{a_0^k} \qquad [7.21]$$

THEOREM 7.2.– If we have $a_0 > 1$, then $A(z) = 0$ has all its coefficients smaller than unity if and only if all coefficients $a_0^k$ of the first column are positive. If none of the $a_0^k$ is zero, then the number of negative $a_0^k$ equals the number of roots with a modulus greater than unity (unstable).

If the $a_0^k$ are positive, then it can be shown that $a_0^0$ is equivalent to the conditions

$$A(1) > 0, \text{ and } \quad (-1)^n A(-1) > 0 \qquad [7.22]$$

These two conditions are necessary for the stability, and they must therefore be tested before the construction of the table.

EXAMPLE 7.5.– *Example of a second-order system:* $A(z) = z^2 + a_1 z + a_0 = 0$.

*We get* $\alpha_n = \alpha_2 = \frac{a_2}{1}$ *and* $\alpha_n = \frac{a_1}{1+a_2} = \frac{a_1(1-a_2)}{1-a_2^2}$

| | | | | | |
|---|---|---|---|---|---|
| line $z^2$ | $1$ | $a_1$ | $a_2$ | $0$ | |
| | $a_2$ | $a_1$ | $1$ | $0$ | |
| line $z$ | $1-a_2^2$ | $a_1(1-a_2)$ | $0$ | $0$ | |
| | $a_1(1-a_2)$ | $1-a_2^2$ | $0$ | $0$ | |
| line $z^0$ | $1-a_2^2-\frac{a_1^2(1-a_2)^2}{1+a_2}$ | $0$ | $0$ | $0$ | |

**Table 7.8.** *Jury table results*

EXAMPLE 7.6.– *The roots of* $A(z) = 0$ *are thus inside the unit circle if:* $1 - a_2^2 > 0$ $\Rightarrow a_2 < 1$ *and* $(1 - a_2^2 - \frac{a_1^2(1-a_2)^2}{1+a_2}) > 0 \Rightarrow (1+a_2)^2 > a_1^2$, *which gives us:*

$$\begin{cases} |a_2| < 1 \\ a_2 > -1 + a_1 \\ a_2 > -1 - a_1 \end{cases} \qquad [7.23]$$

THEOREM 7.3 (Cauchy).– Let $V$ be a function of a complex variable $p$ written as:

$$V(p) = \frac{\prod_{i=1}^{m} (p - z_i)}{\prod_{j=1}^{n} (p - p_j)} \qquad [7.24]$$

Given a closed contour $\Gamma$ in the complex plane formed by the variable $p$ that contains $P$ poles and $Z$ zeroes of $V(p)$. When the complex variable $p$ describes once the closed contour $\Gamma$ in one direction, the variable $V(p)$ describes a closed contour $\Omega$ in the same path direction as variable $p$. This contour $\Omega$ completes $N$ revolutions around the point $(0,0)$, with

$$\overset{\frown}{N} = Z - P \qquad [7.25]$$

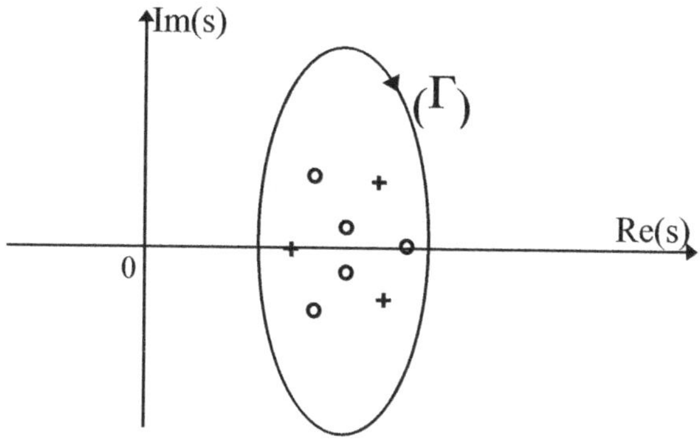


**Figure 7.5.** *Closed Gamma contour*

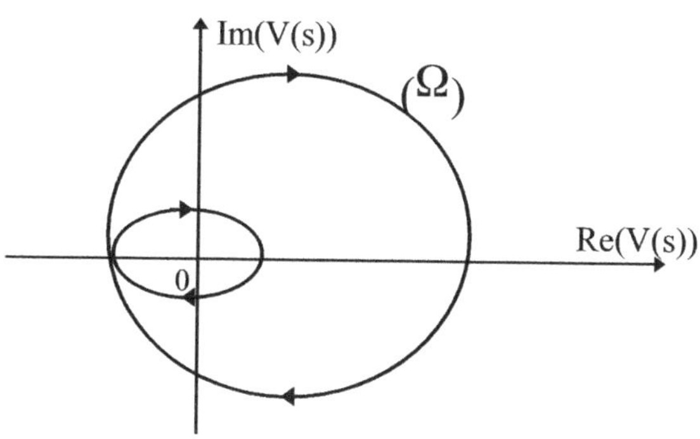


**Figure 7.6.** *Closed Omega contour*

We apply this Cauchy theorem to study the stability of CL linear systems. To do so, we use $\Gamma$ as a closed contour described by the variable $p$, the half right plane of the complex plane. We consider a contour formed by half a circle of infinite radius. The selected function $V(p)$ is the denominator of the CL transfer function:

$$V(p) = 1 + H(p) = 1 + \frac{N(p)}{D(p)} = \frac{D(p) + N(p)}{D(p)} \qquad [7.26]$$

The poles $V(p)$ are the poles of the transfer function of the OL system, and the zeroes of $V(p)$ are the poles of the CL transfer function. As a result, $P$ is the number of poles of the transfer function in OL that is located in the right half plane of the complex plane (poles with a positive real part), and $Z$ is the number of poles of the transfer function CL in the right half plane of the complex plane (poles with a positive real part). For a CL system to be stable (with no poles having a positive real part), it is necessary that $Z = 0$.

REMARK 7.1.– *The Nyquist plot represents the plot of $H(p)$ in the complex plane and not the plot of $V(p) = 1 + H(p)$. The number of revolutions by the path of $H(p)$ would thus be considered around the point $(-1, 0)$ and not around $(0, 0)$. Therefore, $N$ will represent the number of revolutions carried out by $\Omega$ around the point $(-1, 0)$.*

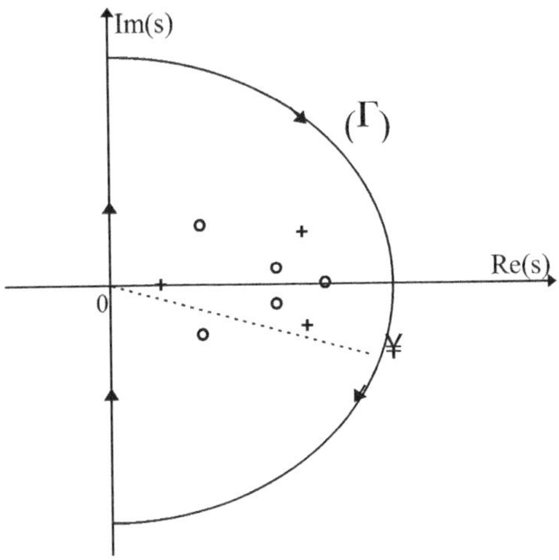


**Figure 7.7.** *Nyquist plot, contour of* $s$

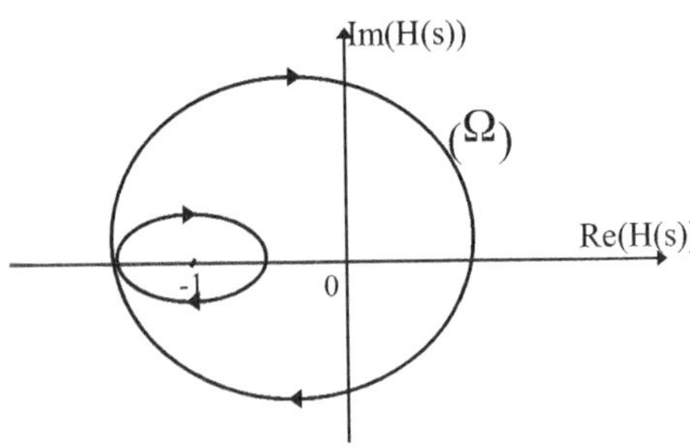


**Figure 7.8.** *Nyquist plot, contour of* $H(s)$

REMARK 7.2.– *The Nyquist locus is symmetrical with respect to the real axis. It is therefore sufficient to draw the locus for $\omega > 0$ and to close it by symmetry with respect to the real axis.*

If the transfer function $H(p)$ has poles at the boundary of the contour $\Gamma$ (for example pole at zero), we write

$$H(p) = \frac{K}{p^n} H_1(p) \approx \frac{K}{p^n} \text{ (when } p \to 0) \quad [7.27]$$

The Cauchy contour excludes this pole:

$$p = \rho e^{j\theta} \text{ with } \rho \to 0 \text{ and } \theta \in [-\pi/2 \quad \pi/2] \quad [7.28]$$

and the Nyquist locus is closed by a circle arc at $\infty$ in the clockwise direction, whose number of half-revolutions is equal to the order of the pole at zero:

$$\frac{K}{p^n} = \frac{K}{\rho^n} e^{-jn\theta} = R \cdot e^{j\Theta} \text{ with } R \to \infty \text{ and } \Theta \in [n\pi/2 \quad -n\pi/2] \quad [7.29]$$

REMARK 7.3.– *If the OL transfer function $H(p)$ can be expressed in the form of $KH_1(p)$, where $K$ is a constant, it is possible to study the stability of the system in CL according to $K$ by plotting $H_1(p)$ with the critical point that becomes $(-\frac{1}{K}, 0)$.*

#### 7.3.2.1. *Inverse criterion (cases in which $P = 0$)*

This criterion is applicable to systems whose OL transfer function has a positive gain and which do not present any poles with positive real part ($P = 0$). It is stated as follows:

– the CL system is stable if proceeding in the increasing direction the Nyquist locus for $\omega > 0$ leaves the critical point $(-1, 0)$ on the left-hand side. It is unstable if the point $(-1, 0)$ is left on the right-hand side.

#### 7.3.2.2. *Stability criterion of systems with minimum phase*

In the case of systems with minimum phase (all poles and zeroes with a negative real part), stability is defined from the Bode plot of the OL system. To this end, stability margins are defined on the gain and the phase:

– gain margin is defined by:

$$\Delta G = 0_{db} - G_{\text{read}}|_{\Phi=-180^\circ} \quad [7.30]$$

– phase margin is defined by:

$$\Delta\Phi = 180^\circ + \Phi_{\text{read}}|_{G=0db} \quad [7.31]$$

REMARK 7.4.– *For a stable CL system with unity feedback, the gain margin or phase margin is positive ($\Delta G > 0$ where $\Delta\Phi > 0$).*

*The multiplication of the transfer function by a gain $K$ causes the gain margin and the phase margin of the system to vary.*

### 7.3.2.3. *Application example*

The transfer function of the OL system $H(s)$ is of the form:

$$H(p) = \frac{K(1+5p)}{p^2(1+2\frac{\xi}{\omega_0}p+\frac{1}{\omega_0^2}p^2)} \quad \text{with} \begin{cases} \omega_0^2 = \frac{1}{4} \Rightarrow \omega_0 = 0.5 \\ 2\frac{\xi}{\omega_0} = 0.8 \Rightarrow \xi = 0.2 \end{cases} \qquad [7.32]$$

The following Bode plot is given as follows.

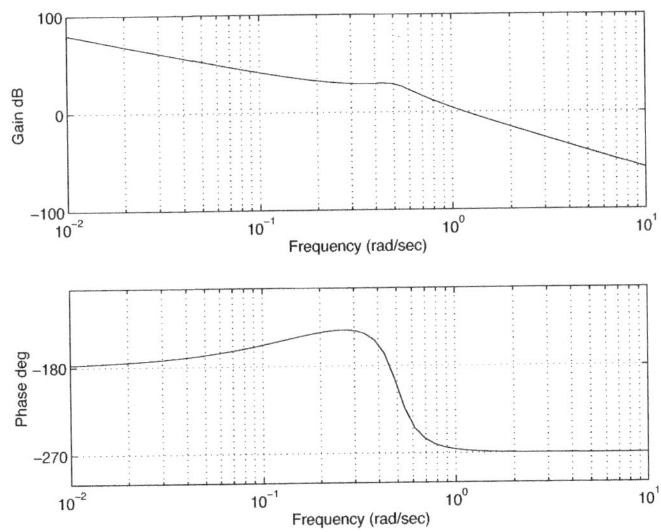


**Figure 7.9.** *Bode plot of* $\frac{H(p)}{K} = \frac{1+5p}{p^2(1+0.8p+4p^2)}$

The Nyquist plot is thus derived based on the Bode plot.

We have $P = 0$, therefore for the system to be stable it is necessary that $N$ be equal to zero. For this purpose, it is necessary that:

$$-\frac{1}{K} < A \qquad [7.33]$$

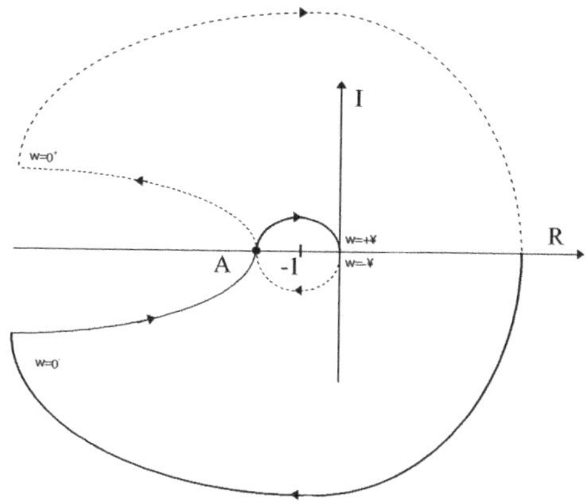


**Figure 7.10.** *Nyquist plot of* $\frac{H(p)}{K} = \frac{1+5p}{p^2(1+0.8p+4p^2)}$

The critical point $A$ is such that $\phi(\omega = \omega_c) = -180°$. We are looking for the $\omega_c$ that corresponds to this:

$$\phi(\omega_c) = -180° \Longrightarrow -180° + arctg(5\omega_c) - arctg(\frac{0.8\omega_c}{1-4\omega_c^2}) = -180° \quad [7.34]$$

which yields

$$arctg(5\omega_c) = arctg(\frac{0.8\omega_c}{1-4\omega_c^2}) \Longrightarrow 20\omega_c^2 = 4.2 \quad [7.35]$$

given

$$\omega_c = 0.458 \text{ rad/s} \quad [7.36]$$

the system is stable if

$$\frac{1}{K} > \frac{1}{\omega_c^2}\sqrt{\frac{1+25\omega_c^2}{(1-4\omega_c^2)^2+0.64\omega_c^2}} = 29.76 \Longrightarrow K < K_c = 0.0336 \quad [7.37]$$

The result can be verified using the auxiliary equation

$$(1-25K_c)p^2 + K_c = 0 \Longleftrightarrow 0.16p^2 + 0.0336 = 0 \Rightarrow p = \pm j\omega_c = \pm j0.458 \quad [7.38]$$

#### 7.3.2.4. *Conclusion*

Stability alone is not sufficient to obtain a good performance of the loop system. We will see later how to design a corrector element in order to satisfy the specification requirements defined *a priori* for the system.

## 7.4. Some application exercises

### 7.4.1. *Exercises: circle criterion, causes of instability and practical cases*

EXERCISE 1.– A discrete system is described by the state–space representation that follows. Is this system stable? What is its transmittance in Z? Derive thereof its recurrence equation.

$$X_{k+1} = \begin{bmatrix} 0.5 & -1 \\ 0 & 0.5 \end{bmatrix} X_k + \begin{bmatrix} 0 \\ 1 \end{bmatrix} u_k \qquad [7.39]$$

$$y_k = \begin{bmatrix} 1 & -1 \end{bmatrix} X_k \qquad [7.40]$$

SOLUTION EXERCISE 1.– The stability of the system is inferred from the eigenvalues of $A$. The latter are given:

$$\lambda_1 = \lambda_2 = 0.5 \qquad [7.41]$$

Since $\mid \lambda_1 \mid < 1$ and $\mid \lambda_2 \mid < 1$, the system is then stable. The transmittance in $Z$ is given by:

$$H(z) = C[zI - A]^{-1}B \qquad \text{with} \qquad [zI - A]^{-1} = \begin{bmatrix} \frac{1}{z-0.5} & \frac{-1}{(z-0.5)^2} \\ 0 & \frac{1}{z-0.5} \end{bmatrix} \qquad [7.42]$$

hence

$$H(z) = -\frac{z + 0.5}{(z - 0.5)^2} \qquad [7.43]$$

The recurrence equation:

$$H(z) = \frac{Y(z)}{U(z)} \quad \Longrightarrow \quad Y(z)(z^2 - z + 0.5) = -U(z)(z + 0.5) \qquad [7.44]$$

which yields:

$$\begin{aligned} y_{k+2} - y_{k+1} + 0.25y_k &= -u_{k+1} - 0.5u_k \\ &\Longrightarrow \quad y_{k+2} = y_{k+1} - 0.25y_k - u_{k+1} - 0.5u_k \end{aligned} \qquad [7.45]$$

EXERCISE 2.– Plot Bode and Nyquist diagrams of processes having as OL transfer function ($T > \tau$):

$$H_1(p) = \frac{K(1+\tau p)}{1-Tp}; \quad H_2(p) = \frac{K(1+\tau p)}{p^2(1+Tp)}; \quad H_3(p) = \frac{K}{p(p-1)}; \qquad [7.46]$$

$$H_4(p) = \frac{K(1+0.2p)}{(1+0.05p)(1+0.1p)(-1+p)};$$

$$H_5(p) = \frac{Kp(1-0.1p)^2}{(1+2p)(1+0.5p)(1+0.02p)} \qquad [7.47]$$

Study the stability of the CL systems above using the Nyquist method. Verify the results obtained by applying Routh's criterion.

SOLUTION EXERCISE 2. – The OL transfer functions $H_i(p)$ of the processes below are all functions of the parameter $K$. As a result, Bode and Nyquist plots are given for $H_i(p)/K$, and then the stability is studied according to this parameter $K$.

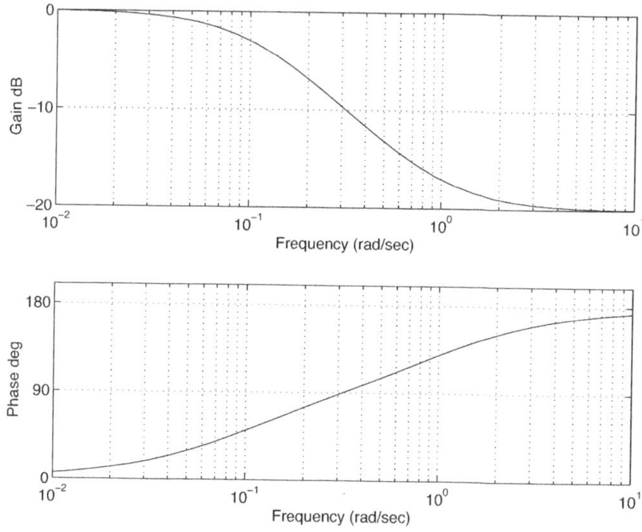


**Figure 7.11.** *Bode plot of* $\frac{H_1(p)}{K} = \frac{1+\tau p}{1-Tp}$

Knowing that $P = 1$, then:

i) $\overset{\curvearrowright}{N} = 0 \Longrightarrow$ stable system in CL, because $Z = 1$.

ii) $\overset{\curvearrowright}{N} = -1 \Longrightarrow$ stable system in CL, since $Z = 0$.

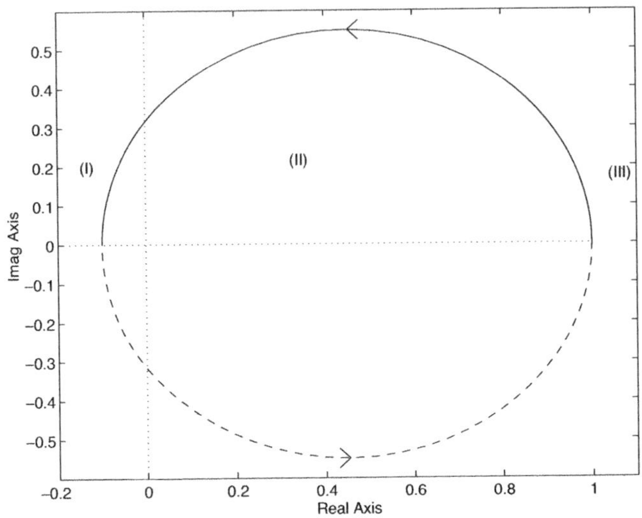


**Figure 7.12.** *Nyquist plot of* $\frac{H_1(p)}{K} = \frac{1+\tau p}{1-Tp}$

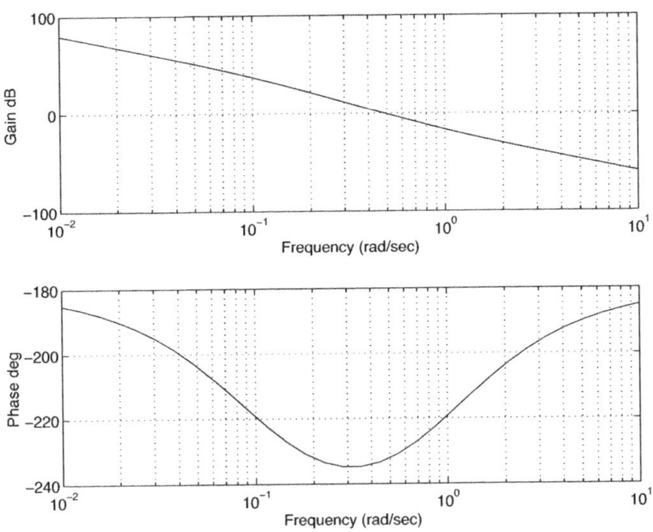


**Figure 7.13.** *Bode plot of* $\frac{H_2(p)}{K} = \frac{1+\tau p}{p^2(1+Tp)}$

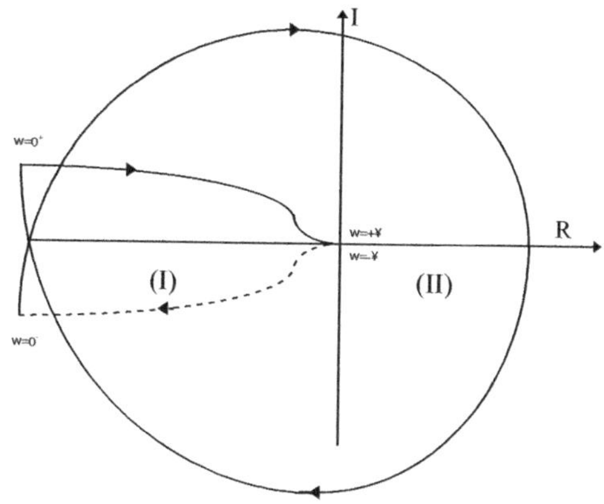


**Figure 7.14.** *Nyquist plot of* $\frac{H_2(p)}{K} = \frac{1+\tau p}{p^2(1+Tp)}$

Knowing that $P = 0$, then:

i) $\overset{\curvearrowright}{N} = 2 \Longrightarrow$ unstable system in CL, because $Z = 2$;

ii) $\overset{\curvearrowright}{N} = 1 \Longrightarrow$ unstable system in CL, because $Z = 1$.

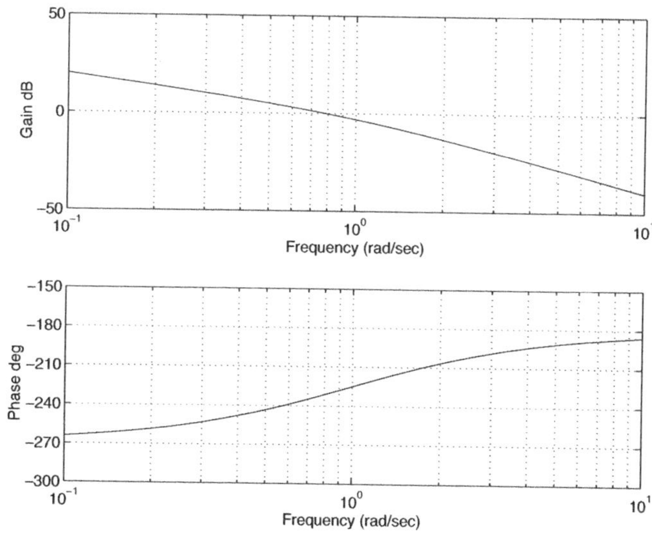


**Figure 7.15.** *Bode plot of* $\frac{H_3(p)}{K} = \frac{1}{p(p-1)} = \frac{1}{-p(1-p)}$

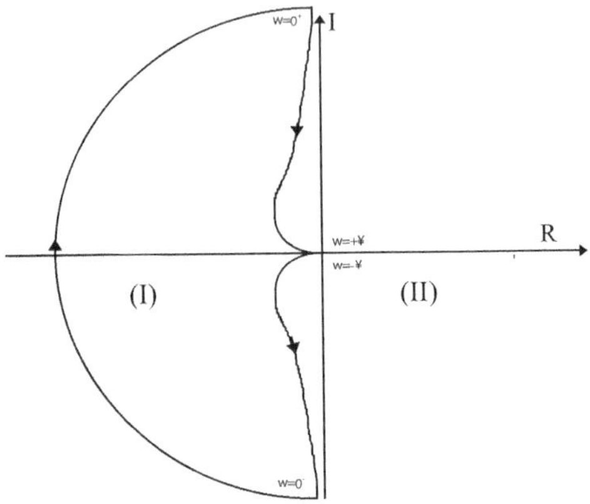


**Figure 7.16.** *Nyquist plot of* $\frac{H_3(p)}{K} = \frac{1}{p(p-1)} = \frac{1}{-p(1-p)}$

Knowing that $P = 1$, then:

i) $\overset{\curvearrowright}{N} = 1 \Longrightarrow$ unstable system in CL, because $Z = 2$;

ii) $\overset{\curvearrowright}{N} = 0 \Longrightarrow$ unstable system in CL, because $Z = 1$.

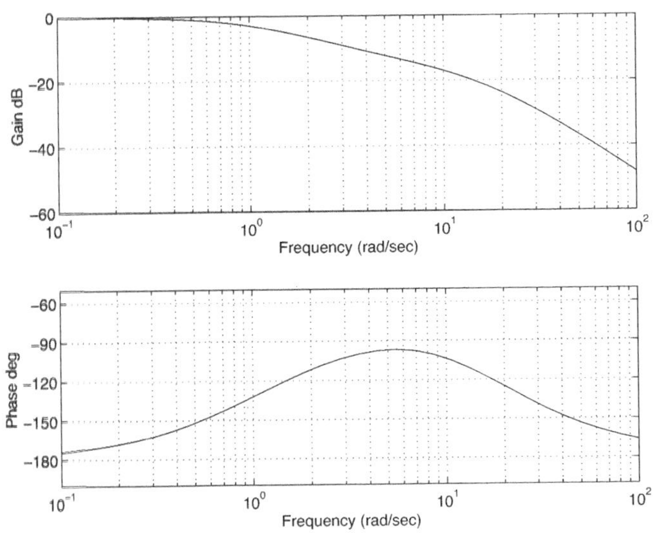


**Figure 7.17.** *Bode plot of* $\frac{H_4(p)}{K} = \frac{1+0.2p}{-(1+0.05p)(1+0.1p)(1-p)}$

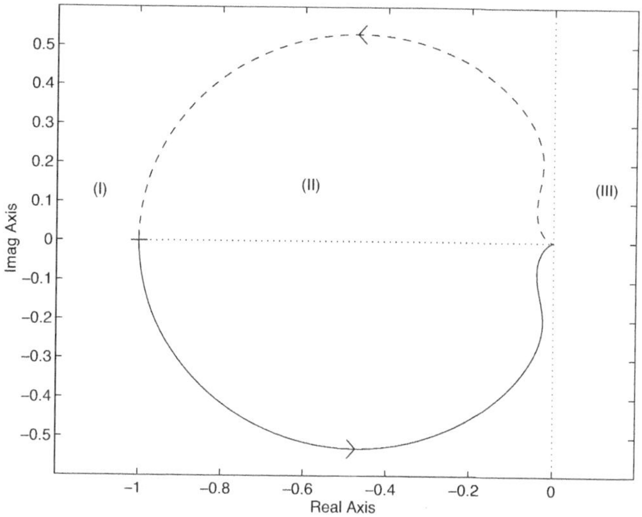


**Figure 7.18.** *Nyquist plot of* $\frac{H_4(p)}{K} = \frac{1+0.2p}{-(1+0.05p)(1+0.1p)(1-p)}$

Knowing that $P = 1$, then:

i) and iii) $\overset{\curvearrowright}{N} = 0 \Longrightarrow$ unstable system in CL, because $Z = 1$;

ii) $\overset{\curvearrowright}{N} = -1 \Longrightarrow$ stable system in CL, since $Z = 0$.

Knowing that $P = 0$, then:

i) and v) $\overset{\curvearrowright}{N} = 0 \Longrightarrow$ stable system in CL, since $Z = 0$;

ii) $\overset{\curvearrowright}{N} = 2 \Longrightarrow$ unstable system in CL, because $Z = 2$;

iii) $\overset{\curvearrowright}{N} = 3 \Longrightarrow$ unstable system in CL, because $Z = 3$;

iv) $\overset{\curvearrowright}{N} = 1 \Longrightarrow$ unstable system in CL, because $Z = 1$.

We are going to verify the stability of the CL system using the Routh criterion.

The transfer function of CL system no. 1:

$$F_1(p) = \frac{K(1+\tau p)}{(K\tau - T)p + K + 1} \qquad [7.48]$$

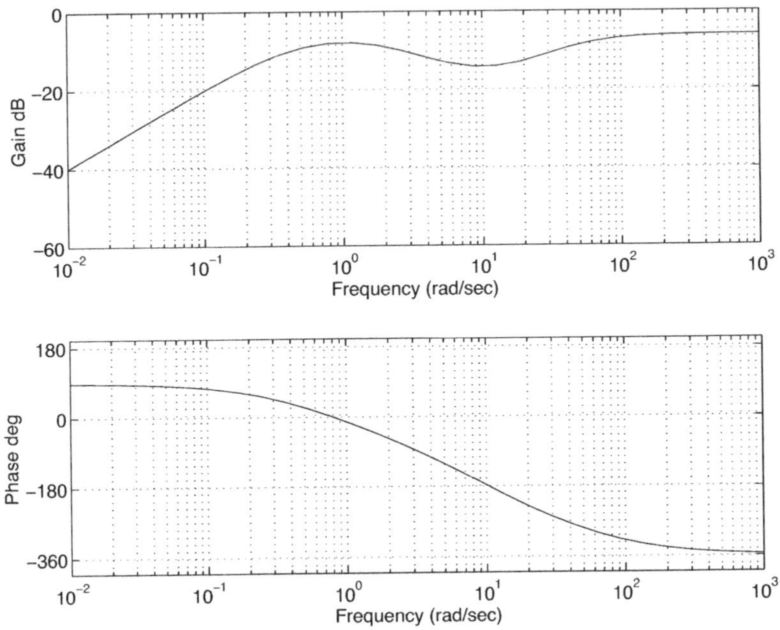


**Figure 7.19.** *Bode plot of* $\frac{H_5(p)}{K} = \frac{p(1-0.1p)^2}{(1+2p)(1+0.5p)(1+0.02p)}$

For CL system no. 1 to be stable, it is necessary that:

$$K\tau - T > 0 \Longrightarrow K > \frac{T}{\tau} \qquad [7.49]$$

The transfer function of CL system no. 2:

$$F_2(p) = \frac{K(1+\tau p)}{Tp^3 + p^2 + K\tau p + K} \qquad [7.50]$$

| | | |
|---|---|---|
| $p^3$ | $T$ | $K\tau$ |
| $p^2$ | $1$ | $K$ |
| $p^1$ | $K(\tau - T)$ | $0$ |
| $p^0$ | $K$ | $0$ |

**Table 7.9.** *Routh table, results system no. 2*

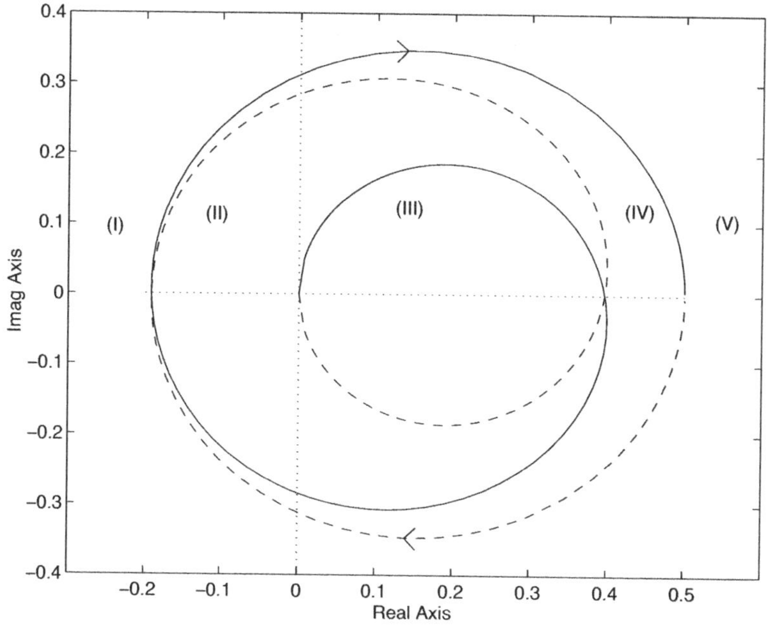


**Figure 7.20.** *Nyquist plot of* $\frac{H_5(p)}{K} = \frac{p(1-0.1p)^2}{(1+2p)(1+0.5p)(1+0.02p)}$

CL system no. 2 is still unstable because there are two sign changes, due to the fact that

$$(\tau - T) < 0 \text{ and } K > 0 \quad [7.51]$$

The transfer function of the third CL system:

$$F_3(p) = \frac{K}{p^2 - p + K} \quad [7.52]$$

CL system no. 3 is still unstable because all the coefficients do not have the same sign.

The transfer function of the CL system no. 4:

$$F_4(p) = \frac{K(1 + 0.2p)}{0.005p^3 + 0.145p^2 + (0.85 + 0.2K)p + K - 1} \quad [7.53]$$

| | | |
|---|---|---|
| $p^3$ | $0.005$ | $0.85+0.2K$ |
| $p^2$ | $0.145$ | $K-1$ |
| $p^1$ | $\frac{.1282-.024K}{0.145}$ | $0$ |
| $p^0$ | $K-1$ | $0$ |

**Table 7.10.** *Routh table, results system no. 4*

CL system no. 4 is stable if (first column of the same sign):

$$K \in ]1, \quad 5.34[ \qquad [7.54]$$

The transfer function of the CL system no. 5:

$$F_5(p) = \frac{Kp(1-0.1p)^2}{(0.02+0.01K)p^3+(1.05-0.2K)p^2+(2.52+K)p+1} \qquad [7.55]$$

| | | |
|---|---|---|
| $p^3$ | $0.02+0.01K$ | $2.52+K$ |
| $p^2$ | $1.05-0.2K$ | $1$ |
| $p^1$ | $\frac{2.626+.536K-.2K^2}{1.05-0.2K}$ | $0$ |
| $p^0$ | $1$ | $0$ |

**Table 7.11.** *Routh table, results system no. 5*

CL system no. 5 is stable if:

$$K \in ]-0.48, \quad 5.25[ \qquad [7.56]$$

EXERCISE 3. – The transfer function of an OL system is

$$H(p) = \frac{K(1+5p)}{p^2(1+0.8p+4p^2)} \qquad [7.57]$$

1) Plot the harmonic response of this system in Bode and Nyquist's planes.

2) A unity feedback control is achieved. Study the stability of the CL system. Determine the critical gain $K_c$ and the corresponding pulse $\omega_c$ using:

3) the Nyquist criterion;

4) the Routh criterion.

SOLUTION EXERCISE 3. – The OL transfer function of the system is given by:

$$H(p) = \frac{K(1+5p)}{p^2(1+0.8p+4p^2)} \quad [7.58]$$

The characteristic polynomial of the CL system is obtained by writing:

$$1 + H(p) = 0 \Longleftrightarrow p^2(1+0.8p+4p^2) + K(1+5p) = 0 \quad [7.59]$$

Hence, the characteristic polynomial

$$4p^4 + 0.8p^3 + p^2 + 5Kp + K = 0 \quad [7.60]$$

The Routh table is the following:

| | | | |
|---|---|---|---|
| $p^4$ | $4$ | $1$ | $K$ |
| $p^3$ | $0.8$ | $5K$ | $0$ |
| $p^2$ | $1-25K$ | $K$ | |
| $p^1$ | $K\frac{4.2-125K}{1-25K}$ | | |
| $p^0$ | $K$ | | |

**Table 7.12.** *Routh table results*

The CL system is stable if:

$$\left.\begin{array}{r} 1-25K>0 \quad \Rightarrow K<0.04 \\ 4.2-125K>0 \Rightarrow K<0.0336 \\ K>0 \end{array}\right\} \Longrightarrow 0<K<K_c=0.0336 \quad [7.61]$$

The OL transfer function $H(p)$ of the system is of the form:

$$H(p) = \frac{K(1+5p)}{p^2(1+2\frac{\xi}{\omega_0}p+\frac{1}{\omega_0^2}p^2)} \quad \text{with} \left\{\begin{array}{l} \omega_0^2 = \frac{1}{4} \Rightarrow \omega_0 = 0.5 \\ 2\frac{\xi}{\omega_0} = 0.8 \Rightarrow \xi = 0.2 \end{array}\right. \quad [7.62]$$

The following Bode plot is shown in Figure 7.21.

The Nyquist plot is thus derived based on the Bode plot.

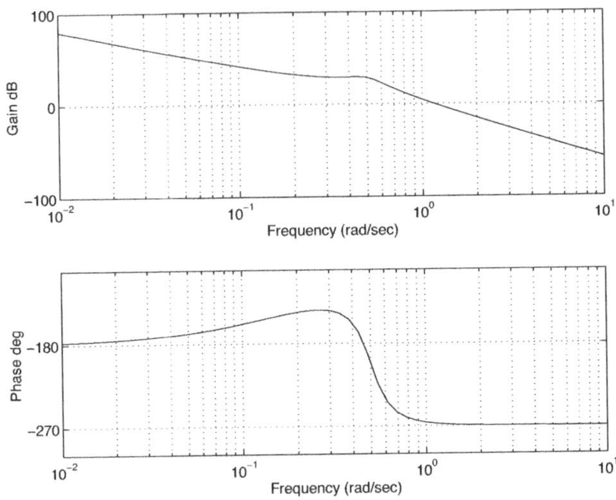


**Figure 7.21.** *Bode plot of* $\frac{H(p)}{K} = \frac{1+5p}{p^2(1+0.8p+4p^2)}$

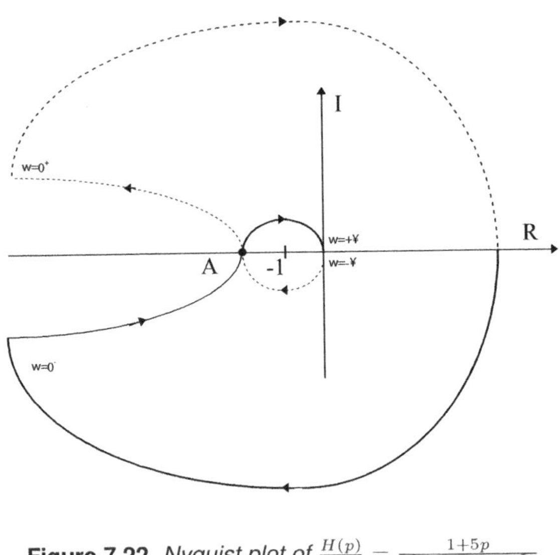


**Figure 7.22.** *Nyquist plot of* $\frac{H(p)}{K} = \frac{1+5p}{p^2(1+0.8p+4p^2)}$

We have $P = 0$, therefore for the system to be stable it is necessary that $N$ be equal to zero. For this purpose, it is necessary that

$$-\frac{1}{K} < A \qquad [7.63]$$

The critical point $A$ is such that $\phi(\omega = \omega_c) = -180°$. We are looking for the $\omega_c$ that corresponds to this:

$$\phi(\omega_c) = -180° \Longrightarrow -180° + arctg(5\omega_c) - arctg(\frac{0.8\omega_c}{1 - 4\omega_c^2}) = -180° \quad [7.64]$$

which yields:

$$arctg(5\omega_c) = arctg(\frac{0.8\omega_c}{1 - 4\omega_c^2}) \Longrightarrow 20\omega_c^2 = 4.2 \quad [7.65]$$

Given

$$\omega_c = 0.458 \text{ rad/s} \quad [7.66]$$

The system is stable if

$$\frac{1}{K} > \frac{1}{\omega_c^2}\sqrt{\frac{1 + 25\omega_c^2}{(1 - 4\omega_c^2)^2 + 0.64\omega_c^2}} = 29.76 \Longrightarrow K < K_c = 0.0336 \quad [7.67]$$

The result can be verified using the auxiliary equation

$$(1 - 25K_c)p^2 + K_c = 0 \Longleftrightarrow 0.16p^2 + 0.0336 = 0 \Rightarrow p = \pm j\omega_c = \pm j0.458 \quad [7.68]$$

EXERCISE 4.– The transfer function of a system is given by:

$$H(p) = \frac{0.64}{p(p^2 + p + 1)} \quad [7.69]$$

1) Plot the harmonic response of this system in Bode and Black–Nichols planes. What are the gain and phase margins?

2) Damping $\xi$ of a second-order system can be approximated by the following formula:

$$\xi \simeq \frac{MP}{100} \qquad (MP = \text{Phase margin in degrees}) \quad [7.70]$$

This approximation remains valid for a higher order system dominated by a pair of complex poles. Give an approximation of the damping of the CL system.

3) Determine the maximum modulus of the CL system based on the Black–Nichols nomograph and the corresponding pulse. Therefrom derive the equivalent damping coefficient $\xi$.

4) Show that the characteristic equation is written:

$$(p + 0.77)(p^2 + 0.225p + 0.826) = 0 \quad [7.71]$$

i) Deduce the equivalent value of $\xi$ from it.

ii) Compare damping $\xi$ obtained from the three methods.

SOLUTION EXERCISE 4.– The transfer function can be expressed in the form:

$$H(p) = \frac{0.64}{p(p^2 + 2\xi\omega_0 p + \omega_0^2)} \quad \text{with} \begin{cases} \omega_0^2 = 1 \Rightarrow \omega_0 = 1 \\ 2\xi\omega_0 = 1 \Rightarrow \xi = 0.5 \end{cases} \quad [7.72]$$

The second-order $\xi$ can be approximated by

$$\xi = \frac{\Delta\Phi}{100} \quad (\Delta\Phi \text{ Phase margin in degrees}) \quad [7.73]$$

This approximation remains valid for a higher order system *dominated* by a pair of complex poles. In the Black–Nichols chart, gain and phase margins can be read:

$$|H(j\omega)| = 0\ db \Longrightarrow \Phi(\omega) = -150° \Longrightarrow \Delta\Phi = 30° \quad [7.74]$$

The Bode and Black–Nichols plots are given as follows:

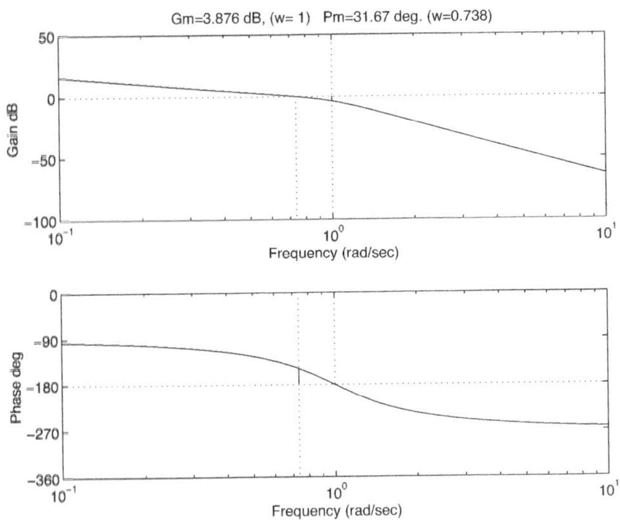


**Figure 7.23.** *Bode plot of* $H(p) = \frac{0.64}{p(p^2+p+1)}$

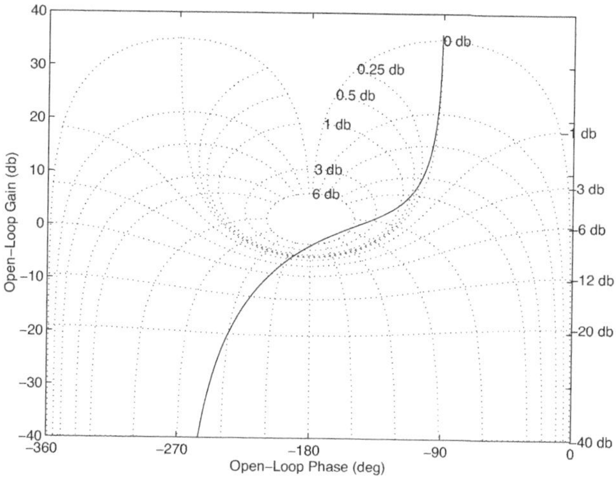


**Figure 7.24.** *Black–Nichols plot of* $H(p) = \frac{0.64}{p(p^2+p+1)}$

Therefore, the $\xi$ based on the phase margin is given by:

$$\xi = \frac{\Delta\Phi}{100} = 0.3 \qquad [7.75]$$

The maximum CL modulus $\left|\frac{H}{1+H}\right|_{max}$ occurs at $\omega_r = 0.88$ rad/s and is equal to $+9$ db, hence:

$$20\log_{10}(M_r) = 9 \text{ db} \Longrightarrow M_r = 2.8 \qquad [7.76]$$

With the help of the curves of a second-order $M_r = f(\xi)$, we can estimate the $\xi$ equivalent to

$$\xi = 0.175$$

The characteristic equation is given by:

$$(p+0.77)(p^2+0.225p+0.826) = 0 \Leftrightarrow (p+0.77)(p^2+2\xi\omega_0 p+\omega_0^2) = 0 \qquad [7.77]$$

This allows us to obtain

$$\xi = 0.124 \qquad [7.78]$$

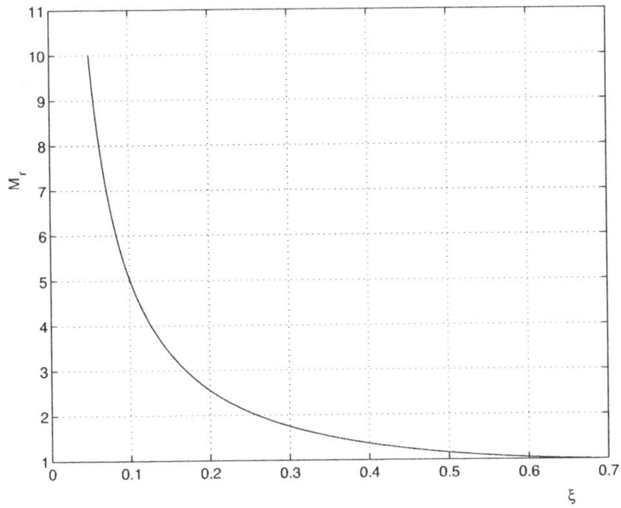


**Figure 7.25.** *Amplitude of the maximum of resonance $M_r$ with respect to $\xi$*

We can conclude that the complex roots of this CL system do not dominate and the real root adds a little damping. This is a system that does not have dominant poles (the real pole is close to the pair of complex conjugate poles). It is therefore advisable to choose the lowest value of $\xi$ obtained by $\Delta\Phi$ and $M_r$.

# Bibliography


[ARM 00] ARMSTRONG B., "Elastic-plastic friction model: contact compliance and stiction", *Proceedings of the 2000 American Control Conference*, Chicago, Illinois, June 28–30, 2000.

[ARQ 82] ARQUES, P.Y., *Décision en traitement du signal*, Masson, 1982.

[ASC 91] ASCH G. *et al.*, *Les capteurs en instrumentation industrielle*, Dunod, 1991.

[AST 97] ASTROM K., WINTTENMARK B., *Computer Controlled Systems*, Prentice-Hall, 1997.

[AST 89] ASTROM K.J., WITENMARK B., *Adaptive Control*, Addison-Wesley, 1989.

[AUB 84] AUBIN J.P. *et al.*, *Differential Inclusions, A Series of Comprehensive Studies in Mathematics*, Springer-Verlag, 1984.

[BAB 85] BABARY J.P., PELCZEWSKI W., *Commande optimale des systèmes continus déterministes*, Masson, 1985.

[BEL 12] BELANGER M., *Traitement numérique du Signal*, Dunod, 2012.

[BEN 87] BENVENIST A., METEVIER M., PRIOURET P., *Algorithmes adaptatifs et approximations stochastiques: Théorie et Applications*, Masson, 1987.

[BLA 81] BLANC A., PICINBONO B., *Fonctions aléatoires*, Masson, 1981.

[BLA 08] BLAU P.J., *Friction Science and Technologie: From Concepts to Applications*, 2nd ed., CRC Press T&F group, 2008.

[BOB 98] BOBROWS J.E. *et al.*, "Modeling identification and control of a pneumatically actuated, force controllable robot", *IEEE Trans. on Robotics and Automation*, no. 5, pp. 732–741, 1998.

[BOI 80] BOITE R., LEICH H., *Les Filtres Numériques*, Masson, 1980.

[BOU 92] BOUACHACHE B., *Time-Frequency Signal Analysis: Methods and Applications*, Longman-Cheshire, 1992.

[BOX 70] BOX G.E.P., JENKINS G.M., *Time Series Analysis: Forecasting and Control*, Holden-Day, 1970.

[CAN 86] CANDY J.V., *Signal Processing: The Model Based Approach*, McGraw Hill, 1986.

[CAN 98] CANUDAS DE WIT C., "Comments on a new model for control of systems with friction", *IEEE Transaction on Automatic Control*, vol. 43, no. 8, pp. 1189–1191, August 1998.

[CEL 96] CELLIER F.E. *et al.*, "Modeling from Physical Principles", in LEVINE W.S. (ed.), *The Control Handbook*, CRC Press, 1996.

[COS 97] COSTAS T., "Adaptive impedance control applied to a pneumatic legged robot", *Int. Journal of Intelligent and Robotic Systems*, vol. 20, pp. 105–129, 1997.

[CRA 89] CRAIG J.J., *Introduction to Robotics*, Addison-Wesley, 1989.

[DE 84] DE COULON F., *Théroie et Traitement des Signaux*, Dunod, 1984.

[DE 92] DE LARMINAT P., THOMAS Y., *Automatique des systèmes linéaires*, Flammarion, 1992.

[DI 94] DI STEFANO J.J., STUBERUD A.R., WILLIAMS I.J., *Systèmes Asservis*, McGraw Hill, 1994.

[DOR 95] DORF R., *Modern Control Systems*, Addison-Wesley, 1995.

[DUP 09] DUPONT P., "Torsional kinematic model for concentric tube robots", *Conf. Proc. IEEE International Conference on Robotics and Automation*, pp. 3851–3858, 2009.

[FAE 81] FAES Y., *Commande de processus industriels par calculateurs*, Masson, 1981.

[FAR 86] FARGEON C., *Commande Numérique des Systèmes*, Masson, 1986.

[FEM 97a] FEMMAM S., Méthodes et algorithmes de segmentation des signaux par détection de ruptures, Report, Laboratoire de robotique de Paris, 1997.

[FEM 97b] FEMMAM S., Méthodes et algorithmes en temps-fréquence et temps-échelle, Report, Laboratoire de robotique de Paris, 1997.

[FEM 11] FEMMAM S., *Analyse, traitement et diagnostic des signaux non-stationnaires par détection de ruptures et temps-Fréquence: application à la perception et à la caractérisation*, Editions Universitaires Européennes, 2011.

[FLA 93] FLANDRIN P., *Temps-Fréquence*, Hermes, Paris, 1993.

[FUK 99] FUKUDA T. *et al.*, "Recent topics on robotic actuation technologies", *Proc. of $2^{nd}$ Int. Conference on Climbing and Walking Robots*, Portsmouth, vol. 3-15, 1999.

[GIL 67] GILLES J.C., DECAULNE P., PELLEGRIN M., *Dynamique de la commande linéaire*, Dunod, 1967.

[GOO 84] GOODWIN G.C., SIN K.S., *Adaptive Filtering: Prediction and Control*, Prentice Hall, 1984.

[HAY 01] HAYKIN S., *Communication Systems*, 4th ed., John Wiley & Sons, 2001.

[KAR 06] KARNOPP C., *System Dynamics: Modeling and Simulation of Mechatronic Systems*, Wiley, 2006.

[KEL 87] KELLAL A., Contribution à l'étude des asservissements électropneumatiques, Application à la réalisation d'un asservissement numérique d'un axe de robot, Thesis, INSA, Lyon, 1987.

[KHA 86] KHALIL W., KLEINFINGER W., GAUTIER J.F., "Reducing the computational burden of the dynamical model of robot", *Proc IEEE Conf. on Robotics and Automation*, San Fransisco, pp. 525–531, 1986.

[KUN 06] KUNT M., *Traitement Numérique des Signaux*, 8th ed., Dunod, 2006.

[KUO 75] KUO B.C., *Automatic Control Systems*, Prentice Hall, 1975.

[LAC 94] LACROUX G., *Les Actionneurs Électriques*, Technique et Documentation, 1994.

[LAN 79] LANDAU I.D., *Adaptive Control: the Model Reference Approach*, Dekker, 1979.

[LAN 88] LANDAU I.D., *Identification et Commande des Systèmes*, Hermes, 1988.

[LIF 79] LIFERMANN J., *Les Méthodes Rapides de Transformation du signal*, Masson, 1979.

[LIN 92] LIN X.F., Contribution à la modélisation et à la commande d'un vérin pneumatique de grande longueur, Thesis, INSA, Lyon, 1992.

[LIU 88] LIU R. *et al.*, "An analysis of pneumatic servo system and its application to a computer-controlled robot", *Transactions of the ASME Journal Dynamic Systems, Measurements and Control*, vol. 110, pp. 228–235, 1988.

[LJU 87] LJUNG L., *System Identification: Theory for the User*, Prentice Hall, 1987.

[LJU 83] LJUNG L., SODERSTROM T., *Theory and Practice of Recursive Identification*, MIT Press, 1983.

[LOP 86] LOPEZ P., *Commande des Systèmes Robotique*, Test, 1986.

[MAX 81] MAX J., *Méthodes et Techniques du Traitement du Signal*, Masson, 1981.

[MIC 92] MICHAUT F., *Méthodes adaptatives pour le signal. Outils mathématiques et mise en oeuvre*, Hermès, 1992.

[MID 90] MIDDLETON R.H., GOODWIN G.C., *Digital Estimation and Control: A Unified Approach*, Prentice Hall, 1990.

[MOR 90] MOREAU N., *Modélisation et estimation spectrale*, ENST, 1990.

[NAJ 82] NAJIM K., *Commande adaptative de processus industriel*, Masson, 1982.

[NAJ 83] NAJIM K., MURATET G., *Pratique de la régulation de processus industriels*, Masson, 1983.

[OGA 03] OGATA K., *Modern Control Engineering*, Prentice Hall, 2003.

[ORT 08] ORTEGA-SANCHEZ M. *et al.*, "A global model of a tidal jet including the effects of friction and bottom slope", *Journal of Hydraulic Research*, vol. 46, no. 1, pp. 80–86, 2008.

[PAP 77] PAPOULIS A., *Signal Analysis*, McGraw Hill, 1977.

[PAY 61] PAYNTER H.M., *Analysis and Design of Engineering Systems*, MIT Press, 1961.

[PIC 86] PICINBONO B., *Eléments de Théorie du Signal*, Dunod, 1986.

[POW 80] POWELL F., *Digital Control of Dynamic Systems*, Addison-Wesley, 1980.

[PU 96] PU J. *et al.*, "Acceleration characteristics of servo controlled pneumatic cylinder", *ASME, Fluid Power Syst Technol.*, vol. 3, pp. 119–130, 1996.

[RAB 75] RABINER L., GOLD B., *Theory and Applications of Digital Signal Processing*, Prentice Hall, 1975.

[RAG 82] RAGOT J., *Exercices et problèmes d'automatique*, Masson, 1982.

[RAG 90] RAGOT J., *Validation de données et diagnostic*, Hermes, 1990.

[RIC 96] RICHARD E., SCAVADA S., "Comparaison between linear and nonlinear control of an electropeumatic servodrive", *Trans. of ASME Journal of Dynamic Systems, Measurements and Control*, vol. 1118, pp. 245–252, 1996.

[ROS 82] ROSENBERG R.C., "General friction considerations for engine design", *Society of Automative Engineers*, Warrendale, Paper no. 821576, 1982.

[ROS 83] ROSENBERG R.C., KARNOPP D.C., *Introduction to Physical System Dynamics*, McGraw Hill, 1983.

[ROU 79] ROUBINE E., *Introduction à la théorie de la communication*, Masson, 1979.

[SCA 89] SCAVARDA A S., *Les asservissements électropneumatiques de position*, Hermes, 1989.

[SEV 69] SEVELY Y., *Systèmes linéaires et asservissements échantillonnés*, Dunod, 1969.

[SHE 56] SHEARER J.L., "Study of pneumatic process in the continous control of motion with compressed air I, II", *Trans ASME Journal Dynamic Systems, Measurements and Control*, pp. 233–249, 1956.

[SOD 83] SODERSTROM T., STOICA P., *The Instrumental Varaible Approach to System Identification*, Springer, Berlin, 1983

[SOD 89] SODERTROM T., STOICA P., *System Identification*, Prentice Hall 1989.

[SPA 73] SPATARU A., *Théorie de la transmission de l'information*, Masson, 1973.

[SPA 87] SPATARU A., *Fondements de la théorie de transmission de l'information*, Presses Polytechnique et Universitaires Romandes, 1987.

[THO 96] THOMASSET D., *et al.* "Non linear controller with perturbation estimation for electropneumatic actuators", $9^{th}$ *Bath Fluidpower Workshop*, Bath, UK, pp. 72–86, 1996.

[TUS 47] TUSTIN W., Vibration and shock isolation trends and solutions, Equipment Reliability Institue, 1947.

[WAI 84] WALRATH C.D., "Adaptive bearing friction compensation based on recent knowledge of dynamic friction", *Automatica*, vol. 20, np. 6, pp. 717–727, 1984.

[WEL 79a] WELLSTEAD P.E., *Introduction to Physical System Modeling*, Academic Press, 1979.

[WEL 79b] WELLSTEAD P., ZARROP M., *Self-tuning Systems Control and Signal Processing*, Wiley-Blackwell, 1979.

[WID 75] WIDROW B., STEARNS S.D., *Adaptive Signal Processing*, Prentice Hall, 1975.

[WON 74] WONHAM, W.M., *Linear Multivariable Control: a Geometric Approach*, Springer Verlag, 1974.

http://www.aldebaran-robotics.com/en/

http://www.aldebaranrobotics.com/documentation/software/choregraphe/

http://robotdomotique.com/

http://fr.wikipedia.org/wiki/Robot_aspirateur,

http://www.maisonrobot.fr/comment-ca-marche/

http://www.mysti2d.net/polynesie//ETC/C041/32/capteursNAO/Documentstechniques6.html

https://community.aldebaran-robotics.com/doc/1-12/software/monitor/index.html

https://community.aldebaran-robotics.com/doc/1-12/software/naosim/index.html

http://www.aldebaran-robotics.com/en/Discover-NAO/Software/Webots-for-NAO.html

http://www.aldebaran-robotics.com/documentation/dev/matlab/index.html

http://www.cmake.org/cmake/resources/software.html

# Index

Other titles from

ISTE

in

Digital Signal and Image Processing

## 2015

BLANCHET Gérard, CHARBIT Maurice
*Digital Signal and Image Processing using MATLAB®*
*Volume 2 – Advances and Applications:The Deterministic Case – 2nd edition*
*Volume 3 – Advances and Applications: The Stochastic Case – 2nd edition*

CLARYSSE Patrick, FRIBOULET Denis
*Multi-modality Cardiac Imaging*

DUBUISSON Séverine
*Tracking with Particle Filter for High-dimensional Observation and State Spaces*

GIOVANNELLI Jean-François, IDIER Jérôme
*Regularization and Bayesian Methods for Inverse Problems in Signal and Image Processing*

MAÎTRE Henri
*From Photon to Pixel: The Digital Camera Handbook*

## 2014

AUGER François
*Signal Processing with Free Software: Practical Experiments*

BLANCHET Gérard, CHARBIT Maurice
*Digital Signal and Image Processing using MATLAB®*
*Volume 1 – Fundamentals – 2nd edition*

DUBUISSON Séverine
*Tracking with Particle Filter for High-dimensional observation and State Spaces*

ELL Todd A., LE BIHAN Nicolas, SANGWINE Stephen J.
*Quaternion Fourier Transforms for Signal and Image Processing*

FANET Hervé
*Medical Imaging Based on Magnetic Fields and Ultrasounds*

MOUKADEM Ali, OULD Abdeslam Djaffar, DIETERLEN Alain
*Time-Frequency Domain for Segmentation and Classification of Non-stationary Signals: The Stockwell Transform Applied on Bio-signals and Electric Signals*

NDAGIJIMANA Fabien
*Signal Integrity: From High Speed to Radiofrequency Applications*

PINOLI Jean-Charles
*Mathematical Foundations of Image Processing and Analysis*
*Volumes 1 and 2*

TUPIN Florence, INGLADA Jordi, NICOLAS Jean-Marie
*Remote Sensing Imagery*

VLADEANU Calin, EL ASSAD Safwan
*Nonlinear Digital Encoders for Data Communications*

## 2013

GOVAERT Gérard, NADIF Mohamed
*Co-Clustering*

DAROLLES Serge, DUVAUT Patrick, JAY Emmanuelle
*Multi-factor Models and Signal Processing Techniques: Application to Quantitative Finance*

LUCAS Laurent, LOSCOS Céline, REMION Yannick
*3D Video: From Capture to Diffusion*

MOREAU Eric, ADALI Tulay
*Blind Identification and Separation of Complex-valued Signals*

PERRIN Vincent
*MRI Techniques*

WAGNER Kevin, DOROSLOVACKI Milos
*Proportionate-type Normalized Least Mean Square Algorithms*

FERNANDEZ Christine, MACAIRE Ludovic, ROBERT-INACIO Frédérique
*Digital Color Imaging*

FERNANDEZ Christine, MACAIRE Ludovic, ROBERT-INACIO Frédérique
*Digital Color: Acquisition, Perception, Coding and Rendering*

NAIT-ALI Amine, FOURNIER Régis
*Signal and Image Processing for Biometrics*

OUAHABI Abdeljalil
*Signal and Image Multiresolution Analysis*

## 2011

CASTANIÉ Francis
*Digital Spectral Analysis: Parametric, Non-parametric and Advanced Methods*

DESCOMBES Xavier
*Stochastic Geometry for Image Analysis*

FANET Hervé
*Photon-based Medical Imagery*

MOREAU Nicolas
*Tools for Signal Compression*

## 2010

NAJMAN Laurent, TALBOT Hugues
*Mathematical Morphology*

## 2009

BERTEIN Jean-Claude, CESCHI Roger
*Discrete Stochastic Processes and Optimal Filtering / 2nd edition*

CHANUSSOT Jocelyn *et al.*
*Multivariate Image Processing*

DHOME Michel
*Visual Perception through Video Imagery*

GOVAERT Gérard
*Data Analysis*

GRANGEAT Pierre
*Tomography*

MOHAMAD-DJAFARI Ali
*Inverse Problems in Vision and 3D Tomography*

SIARRY Patrick
*Optimization in Signal and Image Processing*

## 2008

ABRY Patrice *et al.*
*Scaling, Fractals and Wavelets*

GARELLO René
*Two-dimensional Signal Analysis*

HLAWATSCH Franz *et al.*
*Time-Frequency Analysis*

IDIER Jérôme
*Bayesian Approach to Inverse Problems*

MAITRE Henri
*Processing of Synthetic Aperture Radar (SAR) Images*

MAITRE Henri
*Image Processing*

NAIT-ALI Amine, CAVARO-MENARD Christine
*Compression of Biomedical Images and Signals*

NAJIM Mohamed
*Modeling, Estimation and Optimal Filtration in Signal Processing*

QUINQUIS André
*Digital Signal Processing Using Matlab*

## 2007

BLOCH Isabelle
*Information Fusion in Signal and Image Processing*

GLAVIEUX Alain
*Channel Coding in Communication Networks*

OPPENHEIM Georges *et al.*
*Wavelets and their Applications*

## 2006

CASTANIÉ Francis
*Spectral Analysis*

NAJIM Mohamed
*Digital Filters Design for Signal and Image Processing*